HOW TO TROUBLESHOOT & REPAIR AMATEUR RADIO EQUIPMENT

BY JOSEPH J. CARR, K4IPV

TAB BOOKS

BLUE RIDGE SUMMIT, PA. 17214

FIRST EDITION

FIRST PRINTING—MARCH 1980

Printed in the United States of America

Library of Congress Cataloging in Publication Data

Carr, Joseph J.
How to troubleshoot & repair amateur radio equipment.

Includes index.
1. Amateur radio stations—Maintenance and repair.
2. Radio—Repairing—Amateurs' manuals. I.Title.
TK 9956.C353 621.3841'66'028 79.25806
ISBN 0-8306-9721-7
ISBN 0-8306-1194-0 pbk.

Cover photo courtesy of Dentron Radio Co., Inc. Twinsburg, Ohio

Contents

Introduction

There was a time when a Radio Amateur did not hesitate in the slightest to delve into his station equipment with the purpose in mind of fixing whatever was wrong with it. This was only natural, because, after all, he built it in the first place. Receivers, transmitters, antennas—even transmission lines—were put together from the bare essentials, and there were no mysteries lurking in the open breadboard type of construction then in use.

Even as numbers grew, and factory-built gear started to show up in a few hamshacks, the Amateur was still prone to find and fix any fault that showed up—sometimes as a matter of pride in being able to do so, but as often as not because it came natural to him. Most Amateurs had a technical background, and all had gone quite thoroughly into the theory of how things worked in order to pass the license exam. You had to draw and explain schematic diagrams in those days—sometimes answer questions orally, as well.

All this is not to detract from the value of theory learned by today's Amateur, however. The exam today encompasses more things than ever dreamed of in yesteryear—television (and TVI), facsimile, FM, SSB, crystal filters, transistors, remote control, and dozens of other topics. The Amateur is certainly exposed to enough electronics to know what is going on.

Perhaps it is just this very large range of theory that prevents many Amateurs from going into their receiver, transmitter, or transceiver to find out what needs fixing; they fear getting in over their head. This, coupled with admonitions from the manufacturer, usually in fear-inspiring bold type, to have service performed only by qualified personnel, could be the reason many hams would never dream of taking screwdriver and voltmeter to their pet rig.

This book is designed to show you that these fears are groundless, the rigs not as mysterious as you may think. Armed with a few basic instruments (and the knowledge of how these instruments can be used), with an understanding of what any given circuit in your equipment should be doing, and with a logical approach to the problem, you are capable of troubleshooting your own gear.

So, read, enjoy, and never let your rig mystify you again.

Joseph J. Carr, K4IPV

Chapter 1
Troubleshooting Electronic Equipment—Who, Me?

Yes you! It is the basic assumption of this book that troubleshooting is not a hidden, arcane art, known only to a few professional technologists...the high priests of a modern electronic age. It is a skill, and skills can be developed! Sure, the professional has the edge over you in both knowledge and "savvy," but that does not mean that you cannot be successful in troubleshooting...it means only that the professional will probably beat the socks off you in a troubleshooting-speed race. But if your main purpose is getting back on the air with minimum expense (and no two-week shipping time during which you nervously wonder if the shipper will demolish the thing!), then this book is for you.

Electronic equipment, which is capable of giving various types of pleasure to millions of people, jobs to millions more, and performing seeming miracles, sometimes breaks. Amateur Radio equipment is no exception. But, most consumers have almost no electronic knowledge, so they must refer *all* servicing to the professionals.

Amateur Radio people, however, are required to possess at least a familiarity with electronic principles (of course, I know you may have just memorized the license manual...but don't sell yourself short, something *had* to rub off!) in order to pass the FCC Amateur Radio license examinations.

For some reason, however, many (perhaps most) amateurs are frightened of troubleshooting their equipment. This attitude is, for the most part, unjustified. Troubleshooting is a rational, mostly common-sense, procedure that is imminently knowable. Some basic knowledge of electronics *is* required, make no mistake about that! But the level required to perform elementary repairs is actually quite a bit lower than the majority of hams believes. This does not mean that I am recommending that you jump in and randomly perform some circuit magic! It doesn't work that way at all. But if you are willing to learn a few basic rules, and proceed in a reasonable manner, then success will most likely be yours.

Hint: When you are stuck on a problem, call the manufacturer's service department and bug a service technician. Some companies will try to shunt you off to sales, "customer service" (sales in disguise), or even engineering. The engineer may be able to help you, but it is likely that the service technician, squirreled away in a shop in the back of the plant, will have already seen your type of problem! Some of the best pros in the field make judicious use of long distance telephone calls to the factory, thereby making themselves look very big in the eyes of their unsuspecting customers!

TROUBLESHOOTING—OVERSIMPLIFIED

All electrical circuit troubleshooting involves one of two activities:

1. Finding a *lost,* but required, path for current
2. Finding a new, unwanted, path for current.

Simple? Well, almost.

In other words, all, or at least most, electronic troubleshooting involves around finding either, a) short circuits, or, b) open circuits. That's not so hard, is it? The techniques used in troubleshooting are, for the most part, merely ways of finding a) and b) above.

The exact procedures used may differ in different types of equipment, or in different situations, but in general we *divide to conquer.*

1. Determine what the problem is *not!* This involves thinking about the symptoms, and what could and could not cause them. By studying the service manual, or even just the schematic diagram, you can often pinpoint some areas where the problem could not possibly be located. If a receiver is dead, but the dial lamps come on, then it is very unlikely that the ac-line fuse or power-on switch is at fault. You may extend this principle even further when dealing with transceivers. If a problem affects only one section,

Fig. 1-1. Well-equipped amateur radio station (courtesy of The Heath Company).

i.e., the receiver *or* the transmitter, then it is improbable (note: I said *improbable,* not *impossible!*) that the problem is in any stage common to both sections.

2. Locate the defective stage. This is the principle area in which troubleshooting varies from equipment to equipment, and between situations. You can use a variety of techniques—signal injection, signal tracing, oscilloscopes, VTVMs, etc. This subject will be developed more fully in chapters to follow.
3. Locate the bad component. There are two main current paths in any circuit: the AC path and the DC path. The DC path is for the operating potentials (biases, plate voltage, etc.) derived from the DC power supply. If these are normal, and the stage is still defective, then it is likely that the problem is in the AC path; capacitors, transformers, etc.
4. If analysis techniques fail to locate a defective component, then you can always check each component in the stage individually, or "shotgun" the stage (replace all of the

components in the stage). Neither of these are terribly elegant, but they are very practical. (I can already hear some of my fellow pros telling each other "Carr *actually* told them to shotgun!") But the truth is that everybody does it! Why waste $40 per hour of commercial service time to find out which of a half-dozen 50-cent transistors is the culprit? I believe that, whenever the elegant and the practical courses are at odds, then it is wise to go the practical way—let the high priests of technology be elegant, while you go back on the air. My techniques have worked for me for twenty years—and even the high priests use them!

Chapter 2
Basic Troubleshooting

Troubleshooting is both a skill and an art. The skill is developed mostly by doing it, while the art comes with time and expertise. Troubleshooting is *not* an arcane art, known only to a few of the elite. In order to introduce you to troubleshooting in a gradual way, let us limit our discussion to the Amateur-Radio and general-coverage shortwave communications receiver.

One thing that most easily distinguishes the novice troubleshooter from the experienced professional is the beginner's tendency to jump right in and start "doing something." The true pro will first size things up, and take note of any problems or symptoms.

I know that this sounds pretty silly, but the first job is to determine whether or not you really *have* a problem! I once drove a service truck more than 35 miles on a television service call...only to plug in the AC power cord! It had become unplugged when the lady of the house was vacuuming, and never noticed by anyone. The only thing that they saw was the lack of operation of the TV! That mistake cost them twenty bucks. Similarly, many people take a car radio out of the car at the first sign of trouble, and carry it to a service technician. The idea, I think, is to save the "RnR" (removal and replacement) charge. Once the technician checks out the radio (charging the mandatory minimum labor fee) and finds out that there is nothing wrong, then, perhaps, the consumer will realize that a blown fuse, open or "stuck" speaker, or bad antenna can also cause the same symptoms!

On an amateur communications receiver, check all of the items connected to the receiver before unbuttoning the cabinet. Check the speaker or earphones, antenna connection, T/R relay (if any), and so forth.

The next step in the troubleshooting process is to make some preliminary observations, assuming that there is nothing wrong with the loudspeaker, the AC cord is plugged in, etc. These checks required virtually no test equipment, only your own natural senses of sight and sound. Answer these questions: does it light up? Is there *any* sound coming out of the speaker/earphones? If so, does it sound like a "hiss"? Is it 60 or 120 Hz AC hum? Is there an oscillation present (whistles, birdies, "motorboating," etc)? Taking proper note of these kinds of symptoms can give you a time-saving jump-off point from which to begin troubleshooting. For example, if the dial lamps and vacuum tube filaments fail to light up, then don't waste your time troubleshooting the i-f amplifier with an oscilloscope...look on the primary side of the power supply.

In cases where the lamps and filaments *do* light up, then you can still look for more subtle clues: For instance, operate the *audio gain* or *volume* control through its entire range from minimum to maximum. It is rare that any receiver more than a few months old will have a control in perfect shape, and free of dust, so therefore it will make a subtle "scratching" sound. If you hear this sound, then it is a safe bet that the trouble is ahead of the control, thereby *probably* exonerating the entire audio section of any blame.

The next step is to switch on the *beat frequency oscillator* (BFO) or turn the *mode* switch to the SSB/CW position (this turns on a BFO). If the background noise, or apparent sensitivity, increases, or there are static crashes as you turn on the switch, then it is probable that your problem is ahead of the detector.

Now turn the bandswitch through its entire range. Listen for signals, oscillations, and static crashes as you operate the switch. If the receiver is not dead on all bands, then look for components that are used only on the dead bands. These components include inductors in the front-end, switches and switch contacts, converter crystals (double- and triple-conversion models), loose connections, etc.

Also note that the existence of static crashes as you operate the bandswitch temporarily eliminates the i-f amplifier section. Be aware, however, that static crashes are broad-spectrum rf signals, so will pass through converter and mixer stages even if the conversion oscillator is *not running.*

Next, remove the antenna connection, and then "scratch" the antenna terminal on the receiver chassis with a screwdriver, the end of the antenna wire, or some other tool. The idea is to create a few static crashes in the receiver. If they are heard loud and clear (very distinctly), then the rf amplifier is probably in good order. This means that a dead receiver will probably mean a nonworking local oscillator. An "old hand" at this troubleshooting game can often gain information from the level of hiss coming from the receiver output. This hiss is generated inside of the receiver, and is thermal noise. The more amplification, then naturally the more hiss. A loud hiss means that many stages of amplification are present past the fault point. Remove the antenna sometime, when your receiver is working properly, and note the hiss level. If the same (or nearly the same) hiss level is noted when the receiver is not working, then look to the rf-amplifier and antenna-input circuits.

Now, let us take stock of what we are doing. Without any more test equipment than we were born with, we have actually gained a lot of information about the problem. That wasn't so hard was it? I guarantee that you will find *some* problems that are difficult to solve—almost like trying to skin an amoeba! But for the most part, the problems that we shall encounter are simple.

DEAD: NO DIAL LIGHTS OR TUBE FILAMENTS

Let's first consider how to troubleshoot the vacuum tube receiver that does not even light up. To many people, this is a sign of major catastrophe! But, wait a minute, before you moan and groan, heed some advice: there is nothing quite so easy to troubleshoot as a piece of electronic equipment that fails to light up. It can only be in the power supply, and is usually on the primary side of the power supply! After you have determined that the set was plugged in, and the power switch to the outlet was on, and the fuse in the wall box was not blown, then proceed to troubleshoot the power supply primary circuit.

CAUTION: The 110-volt AC power mains can be *lethal* (it will kill you). Be very careful! Never handle live AC power mains wires, or touch points inside of the equipment when power is applied. Do *not* trust switches, even the wall switch (it could break the neutral instead of the hot wire...if you doubt that this is possible, then prepare for some slow marching and sad music). Always remove the equipment's AC power plug from the receptacle before touching the circuit.

Now that I have (hopefully) given you some caution, let's proceed with checking the primary side of the power supply.

1. Remove the fuse and check it with a continuity tester or ohmmeter. Don't rely on your vision to check the fuse, unless you can plainly see that it is blown (smoked), use the meter. Many open fuses develop hairline cracks in the element, that are invisible to the naked eye.
2. Replace the fuse if it is bad, but remember a tried and true bit of wisdom about fuses:
   "A fuse does not *cause* trouble—it *indicates* trouble!"
3. Use an ohmmeter or continuity tester to test the AC powerline continuity. Place the probes of the tester or meter across the prongs of the line cord, and turn the power switch to the *on* position. The resistance will be low, on the order of only a few ohms. Do not be concerned; that low resistance is backed up by loads of reactance so the power company isn't blown apart by your 2 ohms!
4. If there is no continuity, then check the two halves of the line cord individually, the AC power switch, fuse and its holder, all of the wiring in this circuit, THE AC power plug (yes, I did say *plug*!), and the primary winding of the power transformer—they do sometimes go open.
5. If all of this fails (rarely), then look to the transformer secondary windings.
6. If all of this fails, then it is likely that, a) you goofed, or, b) there is something else in the circuit that I overlooked in preparing this discussion... (I goofed).

We will discuss the power supply in greater detail in Chapter 10. At that time you will learn what to do about a power supply that feels it must pop fuses in order to find true happiness.

APPROACHES TO TROUBLESHOOTING

If you have read this far, then it is assumed that there is something wrong with the old receiver that isn't in the primary side of the AC power supply. The technique now is to "divide and conquer."

There are two basic philosophies in troubleshooting electronic circuits by using signals: One is called *signal tracing,* and the other *signal injection.* Both of these methods have their own dyed-in-the-wool supporters, to the exclusion of the other. Both are, however, useful, and no pro will rely totally on one of them. It is true that one will work better than the other in some situations, but then, you will also find the reverse to be true.

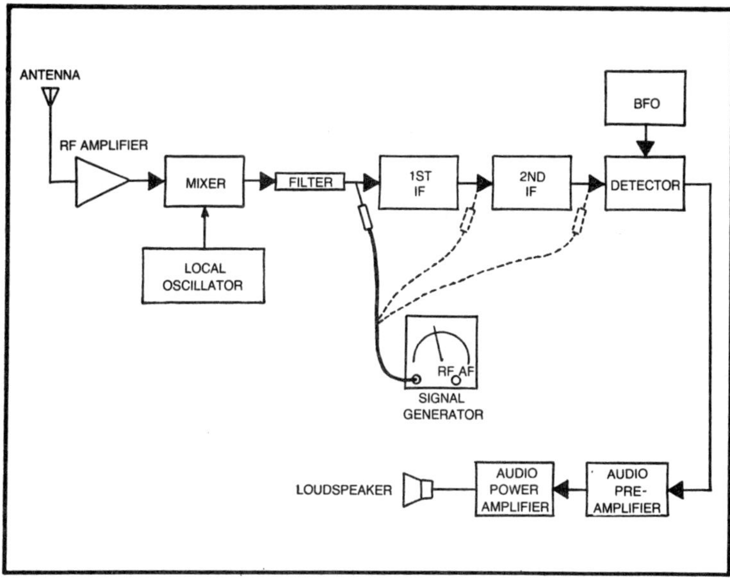

Fig. 2-1. The signal-injection technique.

Signal injection is shown in Fig. 2-1. An output monitor is used to listen or look for the output signal. This could be an oscilloscope, AC voltmeter, or even a loudspeaker or earphones. The idea is to aware of when an output signal is present.

In signal injection we use a generator or other signal source to inject a signal into each stage, in succession, beginning with the output stage. Check each stage backwards from the audio-output stage, in succession, until you find a stage that either passes no signal, or only a weak signal. Note that the audio power amplifier normally requires a fairly hefty input signal, so it may sound attenuated if a weak signal generator is used. If you can pass a loud audio signal through one or more preamplifiers, or from the volume control (audio gain), then it is a relatively safe bet that the problem is *not* in the audio stages.

Generally speaking, the fault will be found between the last point where a signal passed, and the first point that failed to pass the signal. Consider, for example, the vacuum-tube audio preamplifier circuit shown in Fig. 2-2. Assume that you are able to hear a loud output signal when the audio signal is applied to point "A," but not when the signal is applied to point "B." This means that the fault is most likely between points A and B, i.e., V1, R1, R3, R4, C1, C2, C3 or C4.

You should now test the tube, preferably by substitution with a new tube that is known to be good. Do not trust any of those old "junk box" and "hamfest" specials of uncertain origin. They may be in worse shape than the one in the set. I am a firm believer in tube testers for screening all of the tubes in the set. In fact, I recommend testing all of the tubes as a first step in the troubleshooting process. Even if you have to use a drug-store emission type of tester. It is better than nothing at all! But, keep in mind another old-timer's axiom: If the tube tester says the tube is bad, then believe it. But, if it says it the tube is *good,* believe it only with reservations. Go on with the troubleshooting procedure, but if all else fails, then test the tube again by substitution. There are many faults that will not show up on a tube tester—and many circuits will fail to function even when the tube is technically within the limits of a "good" tube. This is especially true of rf oscillator circuits that are poorly designed; when the tube gain drops off a little, the circuit will fail to oscillate.

Another thing that I firmly believe in is that all amateurs should keep brand new, bought-from-a-distributor, tubes to fit every socket in their equipment. This does not mean sufficient stocks to completely retube the beast, only enough to troubleshoot. If, for example, your rig has sixteen 6AU6s, then only a couple will be needed.

Also, do not fall for that old rumor about preventive maintenance, i.e., completely retubing the receiver every year or two. You should *check* all of the tubes on yearly intervals, but replace only those with low gain. This can perk up the receiver's performance quite a bit. But shotgunning all of the tubes is asking for more trouble than it is worth! It seems that most of the premature failures in any large batch of tubes will occur in the first 90 or so days of operation! Frequent total retubing only sets you up for a large percentage of these "green tube" failures.

Unless you will be exposed to dangerous electrical potentials, it is wise to keep the power on while removing the tube from its socket. If you hear a "click" or static crash, then you know the tube was drawing current. This means that you can temporarily forget the plate load and cathode circuit; resistors, coils, transformer primaries, etc.

If replacing the tube does not restore operation, then it will be necessary to make some voltage measurements. Note that all circuits in a receiver may have two paths for current; AC and DC. The DC path is taken by currents derived from the DC power supply, and fed to the tube as biases, and operating voltages. In Fig.

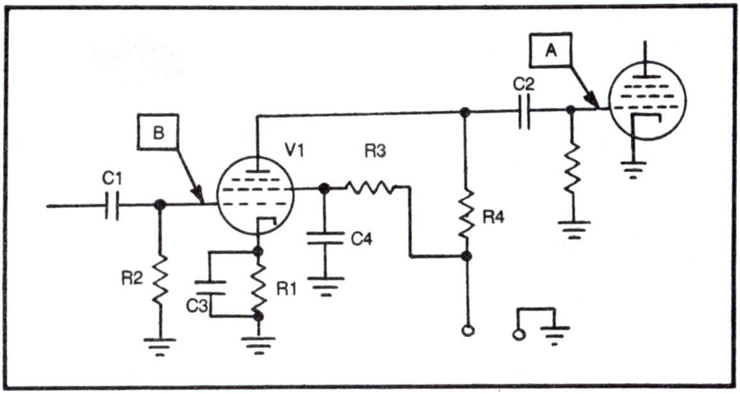

Fig. 2-2. A vacuum-tube audio preamplifier.

2-2 the DC path includes R1, V1 and R4 in the plate-cathode circuit, and R3 in the screen circuit. The AC, or signal, paths include capacitors C1, C2, C3, and C4 and the tube, V1.

Use a VTVM or high-sensitivity VOM to make the DC measurements. Most receiver service manuals will either give the voltages to expect at various points on the schematic, or will provide a chart of voltages at some key points. Although you will have to use the manual, or your knowledge of vacuum tubes, to determine that the voltages are "correct," there are some generalizations that can be made: the plate is (+), the cathode is slightly (+) with respect to ground, and the grid will be zero or slightly (−).

Always try to work from a service manual, or at least a schematic diagram of the receiver. In fact, should you not have one, obtain one before servicing becomes necessary. If you cannot locate a manual from the same model, then obtain one from a similar model. You would be surprised how little creativity there is in some companies; they frequently use the same (or similar) circuits in several different models. This may be lack of imagination, or a convervative desire to "go with what's worked before," but it also works to your advantage in obtaining a service manual.

If a service manual or schematic is not available, then at least obtain a receiving-tube manual such as RCA's *Receiving Tube Manual,* GE's *Essential Characteristics,* or any pre-1979 *ARRL Radio Amateurs Handbook* (the ones with the tube charts in the back!).

Voltages, resistances, and currents may normally vary 10 to 20 percent from the nominal value listed in the manual. Fortunately, most faults cause variations greater than this range. In cases of doubt, then measure the voltages in other stages and note if they tend to be off in the same direction. This could indicate that the variation is normal. But, if the voltages are nearer the mark in other stages, then the fault may be subtle enough to only cause small changes.

Most problems tend to involve a bad tube, the DC power supply (a lot of problems!), and the voltages on the tube elements. A shorted or leaky screen bypass capacitor, for example, will cause the screen voltage to drop almost to zero (and may well overheat the screen resistor), while an open cathode resistor will cause the plate and screen voltages to rise almost to the full B+ voltage. An open plate resistor will cause the plate and cathode voltages to drop to zero, or nearly so, and the screen voltage to rise slightly. All of these become readily apparent from measuring the voltages at these key points.

If no problems are found in the DC path, then look to the AC path for problems. In the example of Fig. 2-2, the main signal path includes capacitors C1 and C2, the tube path includes capacitor C3, while C4 acts as a screen bypass.

The loss of either C1 or C2 (by being open) will cause a complete loss of signal. The loss of C3 may cause complete loss of signal, but may also cause only a severe loss of gain. If C4 is open, then the stage will probably oscillate. Simple substitution with known-good equivalent capacitors is the best troubleshooting technique. Use alligator clip leads, or "solder-tack," to temporarily connect the new capacitor into the circuit. *Do this with the power turned off!*

SIGNAL TRACING

Signal tracing is shown in Fig. 2-3. In this technique, a strong station, or a signal generator, is used as the signal source. The signal is injected into the circuit at the input, at the antenna terminals in a receiver. A *signal tracer* is used to locate the defective stage.

A signal tracer may be an oscilloscope equipped with appropriate input probes, or may be a special device (either homebrew or bought). I peronally like the Heathkit and Eico signal tracers, especially the older types. These can often be picked up at hamfests and ham auctions for a couple of dollars. In the most simple form, a signal tracer is merely an audio amplifier and a loudspeaker. Almost any audio amplifier can be used, such as a public-address amplifier, hi-fi amplifier, a homebrew IC or transistor audio amplifier, or the audio stages from an old radio or TV receiver. The only real

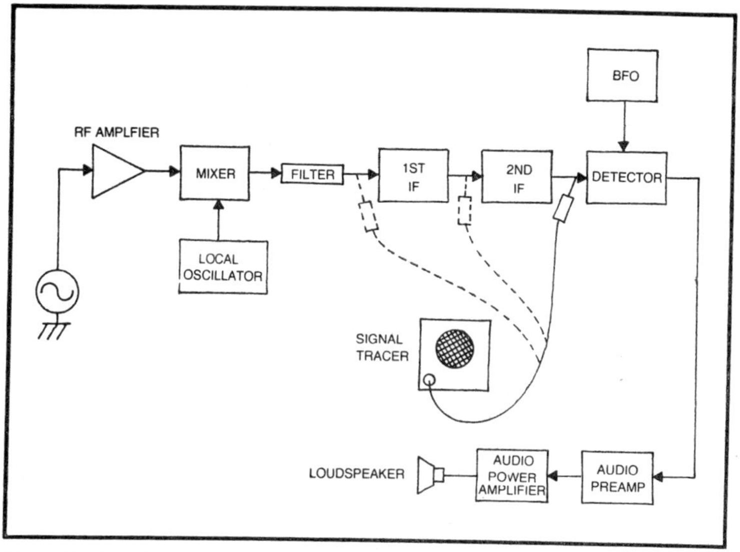

Fig. 2-3. Signal tracing.

requirement is for enough power (100 - 500 mW) to drive the loudspeaker, and enough gain to produce a decent output with an input signal of only a few millivolts.

Superheterodyne receivers use converter or mixer/oscillator circuits to change the rf frequency to an intermediate, or I-F frequency. In some cases, should these types of stages become defective, the fault can be located by using techniques already discussed. But, in other cases, the problem may be a little more difficult to solve. One of these types of problems is when the oscillator is running, but is running way off the correct frequency. If the local oscillator (LO) is a variable frequency oscillator (VFO), then you can often detect operation of the oscillator by using a voltmeter. Measure the small negative DC voltage on the grid of the oscillator tube as you tune the VFO from one end of the band to the other. If the LO is oscillating, then the voltage will change as the dial is tuned. In the case of a crystal oscillator, there will be an abrupt change in the voltage as the crystal is removed from its socket, or disconnected. In both types of oscillator, the output waveform can be viewed on an oscilloscope that has adequate frequency response.

One of the most difficult problems to spot with simple test equipment is the off-frequency oscillator. If the oscillator is so far off frequency as to push signals out of the I-F passband, then the result will be a dead receiver. This topic will be developed more fully in chapter 14 (Troubleshooting the Dead Oscillator), but I can give you a few pointers here.

The absolute best way to find this problem is to use a digital frequency counter to measure the operating frequency of the oscillator. The oscillator frequency should be the receiver's dial frequency plus or minus the intermediate frequency (I-F). There will always be some small error, but if a great error is noted, then suspect troubles. It is, incidentally, crucial to observe good procedure when connecting the counter to the oscillator. Most receivers do not use a buffer amplifier between the LO and the mixer. If there is a buffer, then measure the frequency at its output. But in unbuffered cases, the capacitance of the probe and the input of the counter, will shift the oscillator's running frequency, thereby spoiling your measurement. Use a very small value capacitor in series with the probe, or a low capacitance probe from an oscilloscope or rf voltmeter.

When I first started giving this advice in magazine articles aimed at professional servicers, the readers would often complain that counters were prohibitively expensive—over a kilobuck for a "cheap" model. However, today, counters can be purchased at very low cost, and many Amateur-Radio operators own digital frequency counters.

As an alternative to the frequency counter, you can often find a heterodyne frequency meter such as the Army's BC-221, the Navy LM-series, and commercial models such as those by Gertsch, etc. These are no longer useful to commercial servicers, so are seen at hamfests at very low prices.

You can also use a substitute signal in testing for the off frequency LO. Almost any signal source can be used, even the old fashioned grid (or gate) dip oscillator can be used, although a good quality signal generator will prove easier to work with. The frequency of the substitute should be the receiver's dial frequency plus or minus the i-f. Tune the signal source back and forth across the correct frequency. If the LO is running on the correct frequency, then you should hear whistles and "birdies" in the receiver output. If the LO is not running on the correct frequency, then you will hear stations in the output, just as if you were tuning across the band!

TEST EQUIPMENT

A bench full of multi-kilobuck electronic test equipment is a joy to behold, and gives one a sense of great satisfaction to use. But the question here is not prideful satisfaction of owning such equipment, but rather is "can we get back on the air in reasonable time with less?" The answer is, of course you can.

Commercial radio servicers use the expensive test equipment for only two valid reasons: 1) to meet FCC specifications in their measurements, and, 2) to make their operation more efficient (i.e., profitable). To be sure, some operators indulge in purchases based on emotion, but that is rare in successful businesses.

The most basic piece of test equipment is the multimeter, either a VOM or an electronic voltmeter (VTVM, TVM, FETVM, DVM, DMM, etc). If a simple VOM is selected, it should be a model with a sensitivity of at least 20,000 ohms volt, and higher if possible. Models with sensitivity figures in the 50,000 ohms/volt range are not overly expensive.

There is no longer any reason why every amateur radio operator shouldn't have a VOM. There are many commercial ready-built models, kits, imports, and other models on the market at very low cost. Do not scorn the "lowly" VOM in favor of more sophisticated digital multimeters. The digital jobs have their place, of course, but have you ever seen what many of the lower cost models do when a transmitter is keyed anywhere near to them? They go bananas! I personally favor owning *both* types. The VOM can be used around transmitters, and where portable operation is needed. Even battery operated electronic types give problems in the field—have you ever tried to buy exotic batteries on Field Day?

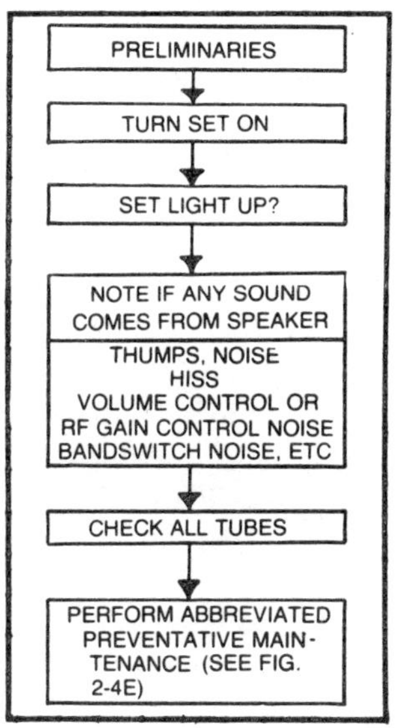

Fig. 2-4. Preliminary troubleshooting chart.

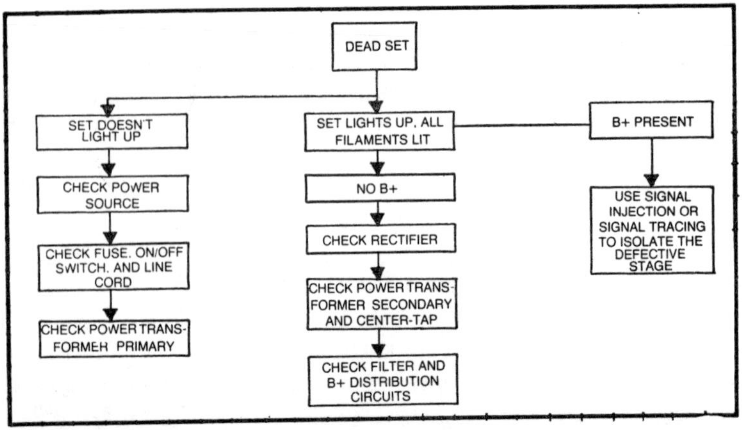

Fig. 2-5. Dead receiver troubleshooting chart

The digital types, on the other hand, are ever-so-easy to read; little interpretation is required. The digital, and other electronic types, also have a very high input impedance, and this is much better when working on transistor equipment.

In chapters to follow I will point out some of the other test equipment that proves useful, or even necessary, when trying to troubleshoot amateur radio equipment.

Some Advice

1. Troubleshooting is a rational procedure. Follow procedures such as charted in Figs. 2-4 through 2-8. Go in for the "try anything" approach only when all else has failed. To be sure, there are exotic repairs that become possible only through "gut feeling" diagnosis, but 99.99 per cent of all troubleshooting success results from a disciplined procedure, reasoning out the problem in light of information obtained, and well-organized work.
2. You have one advantage when troubleshooting a receiver that has suddenly gone dead: *it once worked properly!* This is a great advantage, so do not confuse troubleshooting with the types of problems that you may have encountered in "debugging" new construction projects. Those problems may have been due to some false premise in the design, an error in the magazine article it is built from, components problems, layout, etc. Do not try to "philosophize," do not hunt for exotic problems (until the obvious solutions have failed). Deep, subtle, and occult-caused problems are in the tiniest minority!

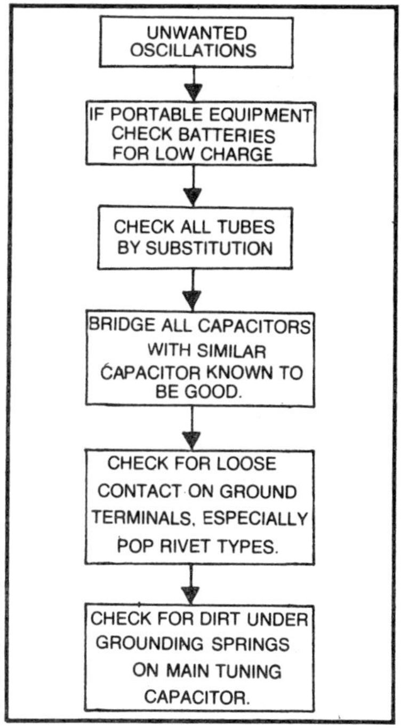

Fig. 2-6. Unwanted oscillation troubleshooting chart.

3. Avoid bizarre repairs! A popular pseudo-fix seen from time to time is the bending of variable capacitor plates to make the oscillator track properly with the dial. Since it is a good bet that the oscillator alignment was once very nearly correct, this is a fool's method. If the receiver

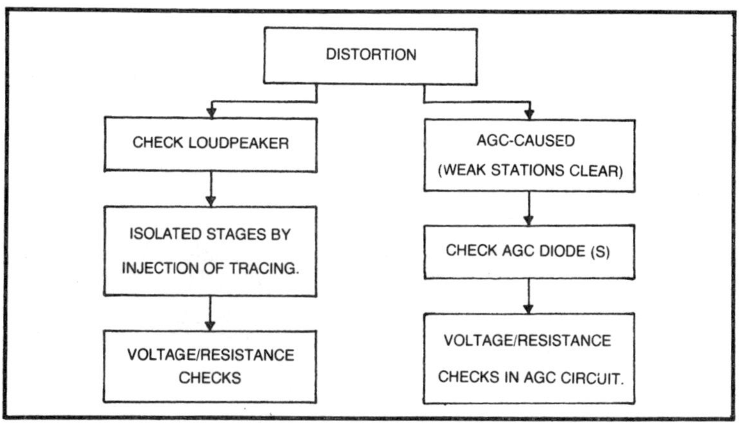

Fig. 2-7. Distortion troubleshooting chart.

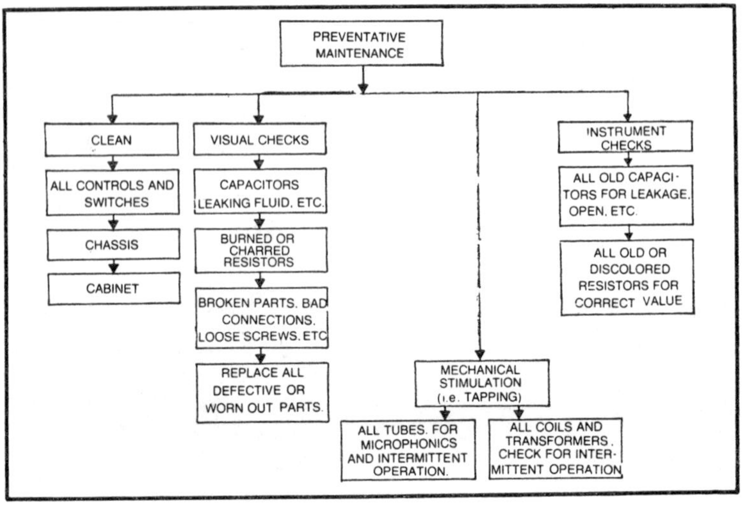

Fig. 2-8. Preventative maintenance chart.

mistracking is noticed only long after manufacture, then it is probable that the problem is a bad component, misalignment, or some other malignancy! The brand of LO tube, for instance, is a common problem. The same tube, from different makers, may have different interelectrode capacitances. These differences have been known to cause mistracking in communications receivers.

4. Don't *EVER* use alignment as a troubleshooting technique. Misalignment never causes sudden failures, only gradual deterioration. Contrary to certain "semi-official" advice, I do not recommend automatic realignment every year or two. True, some minor improvement is affected, but the cost in loosened tuning components and outright damage is not worth it!
5. Use safe working procedures. You are dealing with potentially lethal voltages!
6. Use only high grade replacement components. Ask many servicers why they will not work on Amateur-Radio equipment, and they will answer "because the fellows will not want to pay the price!" Be prepared to spend some money to repair your equipment. I have seen a fellow use 20¢ rectifier diodes, from a "bargain" house, in an SSB transceiver that cost him nearly $1000! The correct diodes cost only a buck or so, so the fellow saved 80 cents—and the supply burned out again within the month!

Chapter 3

Troubleshooting Transistor Circuits

Almost every fault that can occur in a transistor will show up as radically changed DC voltages. In this chapter we will discuss the elements of troubleshooting transistorized circuits. But first, let's take a quick review of transistor fundamentals.

SOME BASICS

Figure 3-1A shows a stylized schematic representation of an NPN transistor. Note that the power supplies connected to the transistor cause it to be reverse biased. In this case, the charge carriers are drawn away from the junctions, creating large depletion zones. The transistor cannot pass current when it is reverse biased. Only a relatively tiny "leakage" current will flow across the junctions.

In Fig. 3-1B we see the opposite situation, the NPN transistor is forward biased. When the transistor is forward biased, charge carriers are driven away from the battery, towards the base-emitter junction. Here they can combine with oppositely charged carriers from across the junction.

Figures 3-2A and 3-2B show the proper supply polarities to forward bias both NPN and PNP transistors, respectively. In the NPN transistor shown in Fig. 3-2A, the base terminal is only slightly more positive than the emitter; approximately 0.2 volts in germanium types and 0.7 volts in silicon transistors. Note that the "normal" voltages may actually be 0.2 to 0.3 volts and 0.6 to 0.7 volts, respectively.

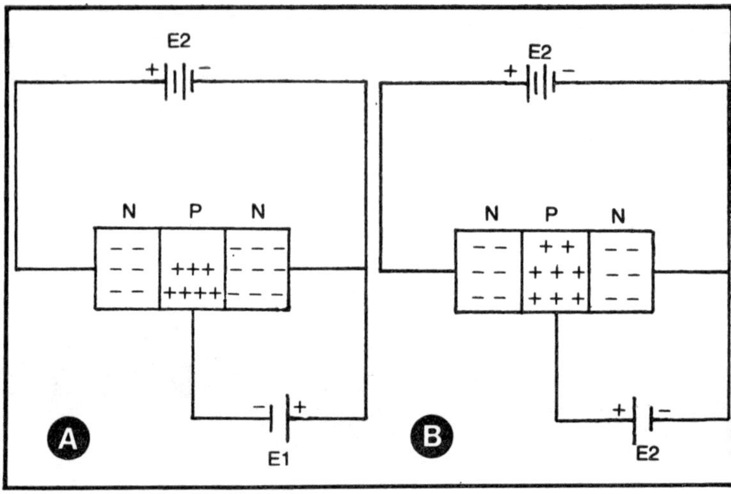

Fig. 3-1. (A) NPN transistor, (B) PNP transistor.

PNP transistors have the same voltage levels as NPN (the b-e voltages are set by the type of material used in the transistor); but the polarity is *opposite*. In the PNP transistor, the base will be slightly more *negative* than the emitter.

Be aware that polarities are relative, i.e., "more negative" could also mean "less positive." In some cases, for example, a PNP transistor may be found in a circuit where the voltages measured on the electrodes relative to ground are positive. But, on closer inspection, we may find that the emitter is at +14.4 volts, and the base is at +14.2 volts. This is the situation found in early transistor car radios. Most audio power transistors in those days (early 1960s) were PNP germanium types ((2N176, etc). The automobiles, however, used negative-ground electrical systems, so the power supply was positive with respect to ground. The solution was to make the base and emitter positive with respect to ground, but the emitter was more positive than the base! Figure 3-3 shows such a circuit. This circuit is representative of the circuits used in those car radios. The PNP transistor is still correctly biased, even though the voltages measured on the terminals are positive with respect to ground.

The correct relationships of measured voltages between the three elements of the transistor are tabulated in Table 3-1. Note, however, that these relationships apply to linear circuits (such as amplifiers). They may or may not apply in nonlinear circuits, such as control, squelch, pulse circuits, or oscillators.

Table 3-1. Relationships of Measured Voltages Among Transistor Elements.

Junction	NPN Si	NPN Ge	PNP Si	PNP Ge
base-emitter	0.6 to 0.7	0.2 to 0.3	-0.6 to 0.7	-0.2 to -0.3
Collector-base	++++	++++	--	--
Collector-emitter*	+++++	++++++	---	---

*In most cases the C-E voltage will be approximately equal to C-B + B-E

DC VOLTAGE CHECKS

All transistor stages have several voltage levels that are of interest when troubleshooting electronic equipment. Most electronic service technicians recognize that it is important to measure the voltage drops across the collector and emitter load resistors, and the base to emitter voltage. For most problems, you will be able to spot the defective stage (and probably component) with just these measurements. Of course, a proper service manual or schematic will allow you to know exactly what voltages to expect. But even without the aid of such information, you can infer proper performance from ballpark values.

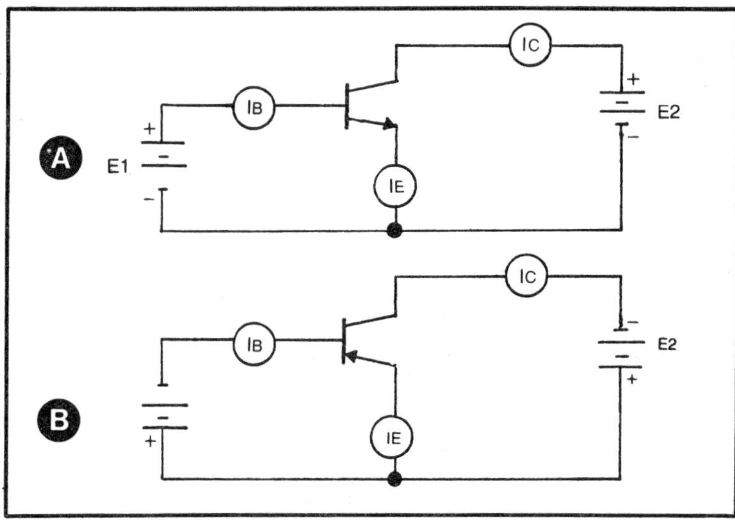

Fig. 3-2. (A) Bias for an NPN transistor, (B) bias for a PNP transistor.

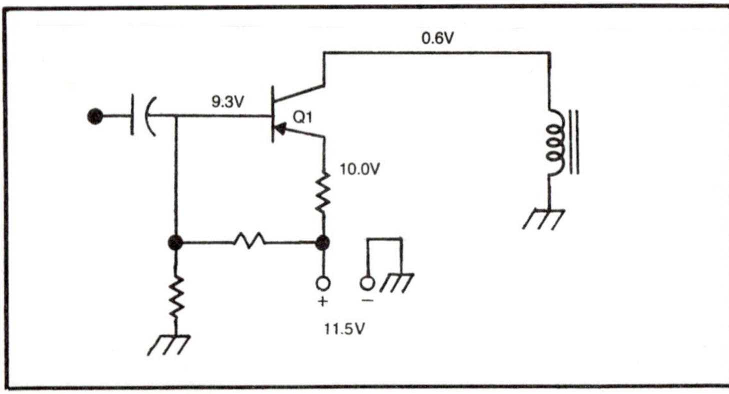

Fig. 3-3. PNP bias network in a negative-ground circuit.

In chapter 2, I covered the use of signal tracing and signal injection to locate a defective stage. These methods also work in transistor circuits, but a DC check is performed first. It seems that high amplitude signals can sometimes "shock excite" certain types of intermittent transistor problems into normal operation. You are then left with a hidden problem that will not show up until later...but it *will* show up again.

DC path tracing requires only an electronic voltmeter or high-impedance VOM. Figure 3-4 shows a typical chain of NPN transistors in cascade, such as might be found in a mobile rig or 2-meter transceiver. Since the rig uses negative ground, and the transistors are NPN, the negative probe of the voltmeter will be grounded to make measurements. Loss of voltage drop across the emitter resistor, in this example the emitter voltage, proves that the transistor is not drawing any current. An excessive emitter-resistor voltage drop, on the other hand, would indicate that the transistor is drawing too much current. The emitter-resistor voltage drop thus serves to point the way, to enable us to tell which stage is defective, and how it is defective.

In general, we can make the statement that zero voltage-drop across the emitter resistor means that the transistor is open, or is reverse biased. A low voltage-drop indicates, usually, that a low-bias condition exists. An excessive voltage-drop, on the other hand, indicates either that the transistor is leaky or shorted, or that the transistor is heavily forward biased i.e., it is turned on too hard.

In rf amplifiers, or other gain-controlled stages, such as ALC in transmitters, this test may yield false results because of the action of the agc circuit. On receivers, if you cannot easily disable the agc, then try tuning the receiver dial back and forth (in transceivers that are channelized, switch to different channels) while noting the meter readings. The agc-controlled stage will show variations in meter readings as the dial is tuned across the stations. AGC can foul you up badly in troubleshooting, but it can also be used to your advantage. An alternative would be to tune the receiver to an unused frequency, where the rf-amplifier gain is maximum, so the voltage drop across the emitter resistor will be maximum.

Incidentally, it is usually true that variations in the agc voltage, the rf-amplifier emitter-resistor voltage drop, or the S-meter indicates that the fault is *not* between the antenna terminals and the agc- or S-meter take-off point.

Equipment using PNP transistors may have either positive- or negative-ground power supplies. If the power supply is negative ground, then use the above technique, but expect the voltages to have a minus polarity. In positive-ground circuits that use the PNP transistor, however, the technique must be modified a little. In this type of equipment, the voltmeter positive terminal should be connected to the B+ (or Vcc+) line. Then use the minus probe to measure the voltage drop across the PNP emitter resistors. In many circuits, the power supply will use a large value electrolytic capacitor as a filter/decoupler, and this capacitor can often be used to identify the B+ line, in the absence of a service manual. Figure 3-5 shows the use of a DC voltmeter in the PNP circuits.

The emitter-resistor voltage drop is useful in solid-state oscillator circuits to determine whether or not the stage is oscillating. The emitter-resistor voltage drop will vary as the receiver is tuned

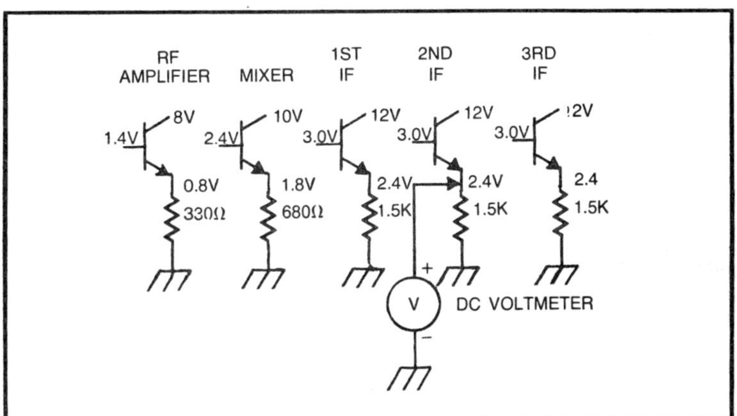

Fig. 3-4. Using a voltmeter to troubleshoot a cascade chain of NPN transistors.

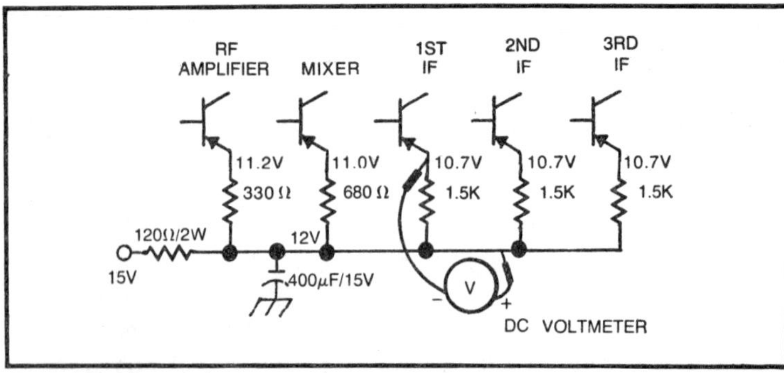

Fig. 3-5. Using a voltmeter to troubleshoot a cascade chain of PNP transistors.

through its range. This change is greater in general coverage receivers than on Amateur-band-only models, because of the wider frequency range covered. In crystal oscillators, a sudden change in this voltage will be noted when the crystal is removed from the circuit. Note that the voltage change only indicates oscillation, not that the stage is oscillating on the correct frequency.

Caution

If you have ever worked on both transistor and vacuum tube circuits, then it should be apparent to you that tubes are a lot more forgiving of errors than are transistors. The use of ungrounded, AC-powered test equipment often results in transients (voltage spikes) that will destroy transistors.

SOME CASE HISTORIES

People involved in training electronic service technicians have long recognized the validity of using case histories, examples, and so forth to illustrate the types of problems one gets into, and their solution. Therefore, we will now consider some of the common faults seen in transistor amplifier circuits. Keep in mind, however, that these are linear circuits, not pulse, switching, or oscillator circuits.

Case No. 1

The receiver is dead. Using the technique of measuring emitter voltage drops we locate the defective stage. Figure 3-6 shows the voltages that we encounter. The 0-volts potential measured on emitter tells us that the transistor is not conducting. We can also make the claim that the collector current is zero, because the collector voltage is very nearly the same as the source voltage,

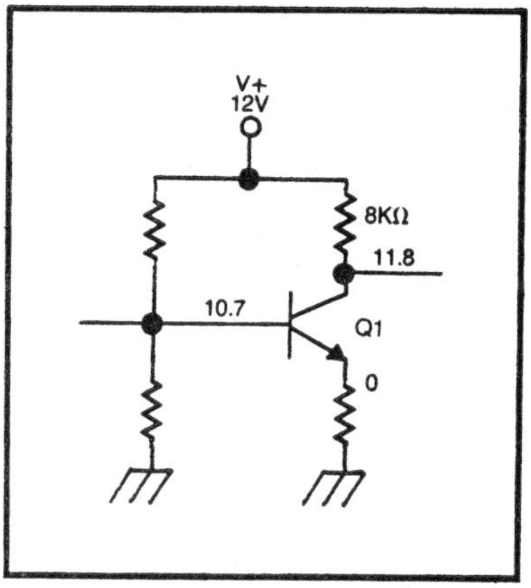

Fig. 3-6. Typical voltages for an open base-emitter junction.

Vcc+. Normally, the collector voltage would be in the neighborhood of Vcc/2. Measuring the base to emitter voltage, we find a potential of 10.7 volts DC, instead of the 0.6 to 0.7 volts that we would expect.

These symptoms usually point to an open base-emitter junction; that is, 1) no collector-emitter current, 2) base-emitter voltage almost as high as the collector voltage. Replacement of the transistor further proved this to be the case.

Case No. 2

Again, we are presented with a dead receiver. In this case, however, we find the emitter-resistor voltage drop to be very low, indicating that the stage is passing very little current, but is not totally cut-off. Measurement of the DC voltages on the base and collector (Fig. 3-7) showed that the collector voltage was very near

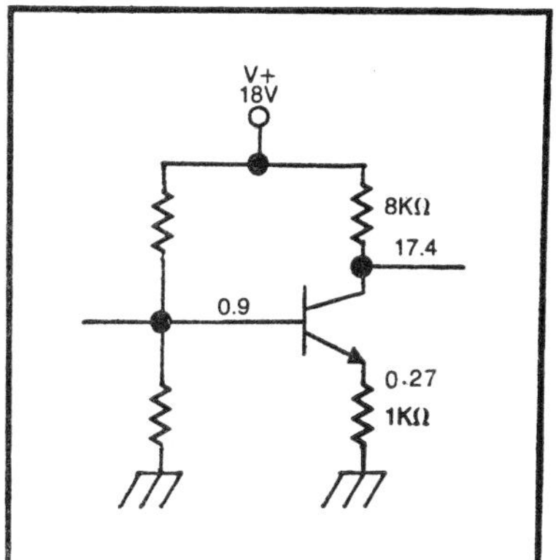

Fig. 3-7. Typical voltages for an open collector-base junction.

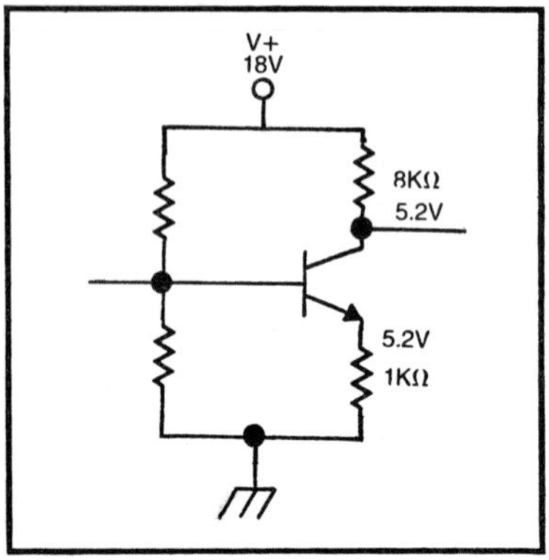

Fig. 3-8. Voltages that could result from an open ground on the emitter resistor, or a collector-to-emitter short.

the Vcc+, very low emitter voltage, and a near-normal base voltage.

Case No. 3

In this case the dead radio did not have a low, or zero, emitter-resistor voltage drop, but had a high voltage drop (see Fig. 3-8). In fact, the collector voltage and the emitter voltage was nearly the same. There are two main faults that could cause this symptom: 1) collector-emitter short circuit, and, 2) an open ground connection on the emitter resistor. If the resistor appears to be very hot, then it is most likely that the transistor is shorted between the collector and emitter. Note well, however, that burning of the resistor is not necessary, especially if a high value and/or high wattage resistor is used. The transistor could be shorted even if there is no evidence of burning or overheating. One quick way to find out if the ground is open would be to measure the voltage on the ground end of the resistor. But, this would not tell us if the resistor was open! An ohmmeter between the collector and emitter terminals of the transistor, however, would tell us if the transistor is shorted!

Chapter 4
Voltmeters, Ammeters, Ohmmeters, etc.

No piece of equipment is more basic to any attempt at servicing electronic equipment than a means for measuring voltage and current. Although this can be done on an oscilloscope, it is frequently more convenient to use one of the more standard forms of voltmeter or ammeter on the market.

The purchaser today has a wide variety of instruments to select from, and the choice is often confusing. The scale runs from volt-ohm-milliammeters (VOMs) designed before WW-II (but still sold today), to modern digital multimeters. Unfortunately, no one meter will solve all of your problems, and it is often wise to include two or more different models in your collection of test equipment.

In this chapter, you will learn the basic types of meter movement and the instruments in which they are used. Digital voltmeters are covered in another section, so here we will cover the analog forms of VOM and electronic multimeter.

METER TYPES

There are a number of different meter movments in use, but the two best examples are both of the permanent-magnet moving-coil (PMMC) design shown in Fig. 4-1. The two types are called the D'Arsonval and Taut-band movements. Both use a fine coil of wire on a bobbin in the field of a large permanent magnet. A pointer is attached to the bobbin, and it indicates points on a scale calibrated in current units. The pointer must be free to move, therefore the

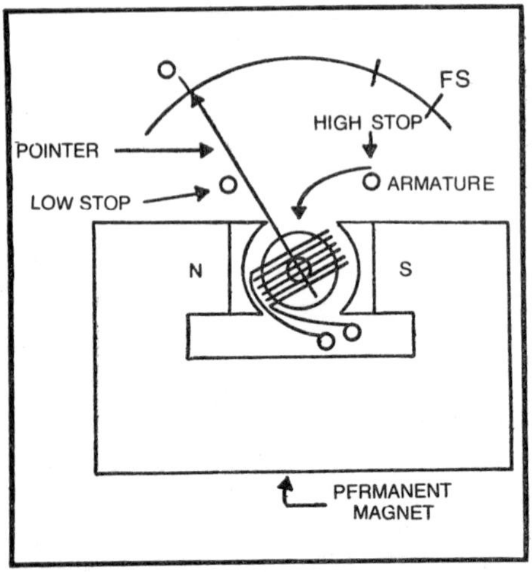

Fig. 4-1. Permanent magnet moving coil (PMMC) meter movement.

bobbin must be mounted so that it can rotate on its axis. Therein lies the difference between the two types of PMMC meter movement. In the D'Arsonval movement, the bobbin is mounted on jewel bearings, while in the taut-band it is suspended between supports on a thin metallic band that is kept under tension.

When a current flows through the coil on the bobbin, it creates a magnetic field that interacts with the magnetic field of the permanent magnet. Since magnetic fields will either repel or attract each other, the bobbin will be given a force that tends to deflect it, proportional to the current flowing in the coil. This, then, will deflect the pointer across the scale, indicating the strength of the current.

The PMMC meter movements are for direct current (DC) only. One terminal will be marked positive with either a "+" symbol, a dot, or a splash of paint. The polarity must be observed, or it is possible to damage the meter.

Basic Rules For Using DC Current Meters

1. *Always* connect the *current* meter in *series* with the load, or the circuit branch, in which you want to measure current.
2. Use a meter, or switch setting in a multimeter, that has a greater full-scale reading than the maximum expected current. If the meter is a multimeter, then start at the highest current range and work down, until a usable deflection is obtained.
3. Use a current meter with an internal resistance that is *low* compared to the circuit resistance.

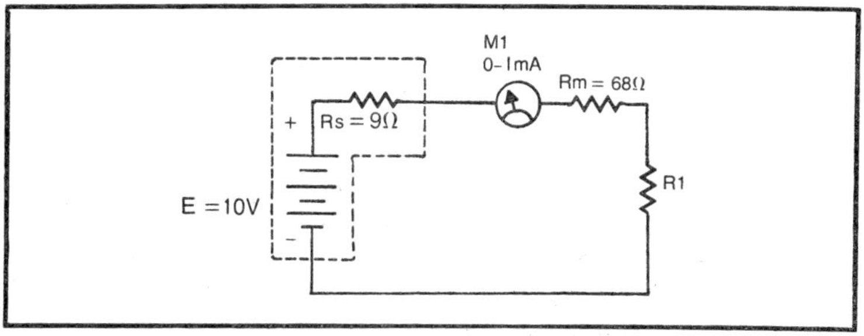

Fig. 4-2. Meter and power-supply resistances create error.

It is important, even critical, to remember that the current meter is always connected in series with the load. If the meter is accidentally connected across (i.e. in parallel with) the load, then it shorts out the supply, and allows a very high current to flow. This current might destroy the meter movement so fast that you will not be able to make any correction!

Figure 4-2 shows the correct connection for the current meter. M1 is connected in series with load R1. You can also see the reason for the third rule in this circuit. Note that there are three resistances in the circuit, although most will only "see" the load R1. Resistance R_s is the internal resistance of the power supply, while resistance R_m is the internal resistance of the meter (the coil resistance). As long as the meter and power supply internal resistances are much less than the load resistance, then the meter will read the current in the load with reasonable accuracy. But, if the sum of the other two resistances is more than approximately 10 percent of the load resistance, then large errors result.

Increasing Current Scale. The basic meter movement will be designed to pass a certain maximum current level. It is difficult, maybe impossible to make the meter movement deflect full scale with any other current level. You can, however, increase the apparent current scale by using an external shunt resistor, as shown in Fig. 4-3. In this circuit, you are bypassing most of the current around the meter movement. You know, from basic knowledge of parallel circuits, that current I_{fs} (full scale) is equal to the sum of the meter current (I_{M1}) and the shunt resistor current (I_s):

$$I_{fs} = I_{M1} + I_s$$

Either Ohm's law, or the current-divider equation, may be used to calculate the value of the shunt resistor.

If you know the full-scale meter-movement current, and the meter coil resistance, then the voltage drop across the meter at *full scale current* is their product:

$$E_{m1} = I_{M1\,(max)} \times R_m$$

You can see from the circuit in Fig. 4-3 that the shunt resistor is connected directly across the meter, so this same voltage will appear across the shunt resistor. Also, the full-scale current value of R_s will be the desired full-scale current (I_{fs}) less the meter full-scale current. The value of R_s, then, is:

$$R_s = E_{M1} / (I_{fs} - I_{M1\,(max)})$$

Step-by-step:

1. Determine I_{fs}, I_{M1}, and R_m
2. Calculate I_s from $I_{fs} - I_{M1}$
3. Calculate E_{M1} using Ohm's law
4. Calculate R_s from E_{M1} and I_s using Ohm's law

This procedure is reduced to an equation for those who are interested in such things:

$$R_s = \frac{I_{M1}\,R_m}{I_{fs} - I_{M1}}$$

The *current divider equation* tells us that, by Kirchoff's current law, the current flowing in resistance R_s (the shunt resistor) is a fraction of the total current I_{fs}:

$$I_s = \frac{I_{fs}R_m}{R_m + R_s}$$

The value of R_s may be calculated by algebraically rearranging the current divider equation above to solve for R_s.

Both methods have their fans, and both will yield the same result. It is, therefore, up to you which is more suitable for you.

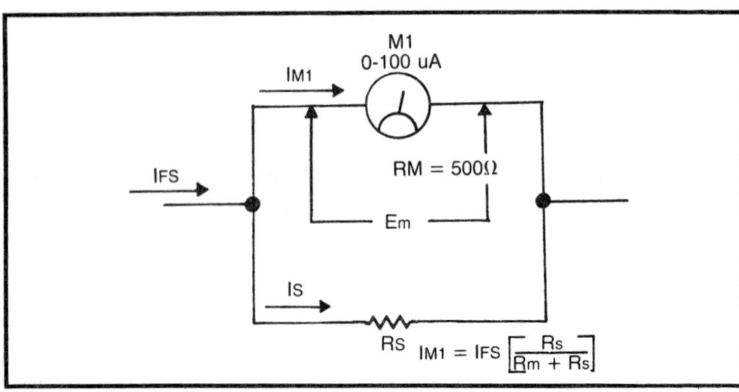

Fig. 4-3. Using a shunt resistor to increase full-scale reeding.

VOLTMETERS

There is not any actual voltmeter movement, but voltmeters are created from current meters by using a series multiplier resistor as in Fig. 4-4A. The equivalent circuit, that takes into account the internal resistance of the meter movement, is shown in Fig. 4-4B.

If a voltage, E, is applied across the series combination of R_{mx} and the meter, then a current will flow in the meter equal to

$$I1 = E\ /\ (R_{mx} + R_m)$$

Where:

I1 is the current in amperes
E is the applied voltage
R_{mx} is the resistance of the multiplier resistor.
R_m is the internal resistance of the meter movement

Calculate the multiplier resistor needed for any given full-scale voltage E by setting I1 in the equation above equal to the full-scale rating of meter movement M1 and then solving for R_{mx}. This, of course, assumes that we know R_m.

We may also use the voltage-divider equation to calculate the value of the multiplier resistor, R_{mx}. The full-scale voltage drop across meter movement M1 will be I1 X R_m.

A good indicator of the quality of a voltmeter is the sensitivity, given in ohms-per-volt. The sensitivity of the voltmeter can be found from the reciprocal of the full-scale current rating of the meter movement:

$$\text{Sensitivity} = 1/I_{fs}$$

If, for example, the meter has a 0 - 1 mA movement, then the full scale current is 1 mA, or 0.001 A. The sensitivity is

$$\text{Sens.} = 1\ /\ 0.001$$
$$\text{Sens.} = 1000 \text{ ohms/volt}$$

In general, the higher the sensitivity, the better the instrument. A 1000 ohms per volt instrument is not too good for any but the roughest measurements, such as the voltage of your car battery. In electronic circuits, the sensitivity must be higher. For most vacuum-tube circuits, the sensitivity should be at least 20,000 ohms/volt, and for transistor circuits, 30,000 ohms/volt. 100,000 ohms/volt instruments are available, and should be used if you can afford the cost.

Voltmeter impedance (Z) can be calculated from the sensitivity and the full-scale voltage:

$$Z = E_{fs} \times \text{sensitivity}$$

Example:

Find the impedance of a voltmeter with a sensitivity of 20,000 ohms/volt if the full-scale range is 100 volts.

$$Z = E_{fs} \times \text{sensitivity}$$
$$Z = (100\ \text{V})\ (20{,}000\ \text{ohms/volt})$$
$$Z = 2{,}000{,}000\ \text{ohms}$$

Rules For Using DC Voltmeters

1. Always connect the ***voltmeter*** in ***parallel*** with the load.
2. Select a voltmeter with a full-scale range greater than the maximum expected voltage.
3. Make sure that the voltmeter impedance is ***very high*** compared with the circuit resistance (more than 10 times, for instance).

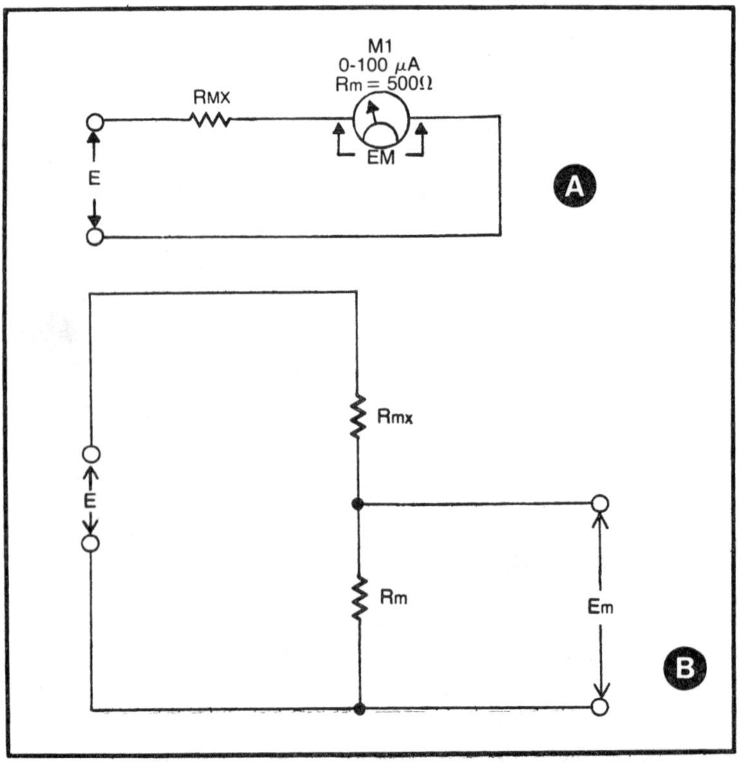

Fig. 4-4A. Using a multiplier resistor to measure voltage with a current meter movement. B. Equivalent circuit.

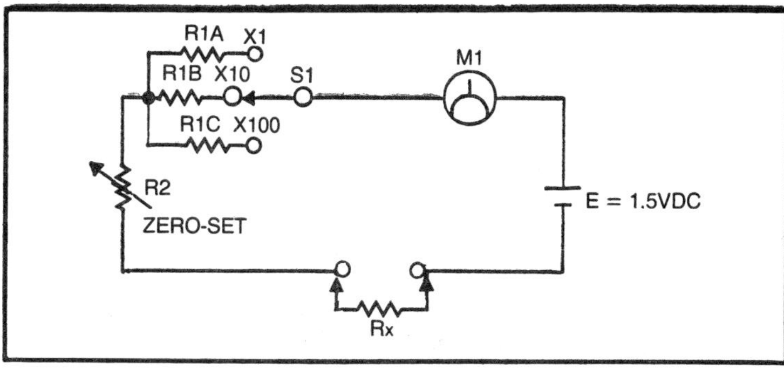

Fig. 4-5. Simple ohmmeter circuit.

Ohmmeters. Resistance can be measured using a dc current meter, a calibrated voltage source, and some precision resistors. A simple ohmmeter circuit is shown in Fig. 4-5.

Resistors R1A through R1C set the basic range of the instrument, while resistor R2 sets the zero reference point. R2 is a potentiometer, and is adjusted for a full-scale current flow when the resistance across points A and B is zero (the two points are shorted together). If an unknown resistance is then connected across A-B, then the current through the meter movement will reduce by an amount proportional to the resistance of the unknown.

Multimeters. It is both impractical and prohibitively expensive to keep meters for each and every current, voltage, or resistance range that might be needed. A multimeter will use a single meter movement and a switch to select a large variety of current, voltage, and resistance ranges.

AC Meters. Neither of the two basic types of PMMC meter movement will measure alternating current. But, if we rectify the ac, to produce a pulsating dc, then it will be possible to measure ac—especially 60-Hz sinewave ac as obtained from the power mains. Although some instruments may use a half-wave rectifier scheme, most use the full-wave bridge circuit of Fig. 4-6. In fact, special preassembled bridges, called instrument rectifiers, are often used for this application. Most of these are copper-oxide rectifiers to take advantage of their low forward-voltage drop.

Copper oxide instrument-rectifiers cannot operate at high frequencies, ac voltmeters intended to operate in the rf region must use small-signal diodes, such as the 1N60, 1N34, 1N914, to gain frequency coverage at the expense of low-voltage operation. Germanium diodes have a 200 to 300 mV junction potential, while silicon diodes have a 600 to 700 mV junction potential.

ELECTRONIC MULTIMETERS

Both VOMs and electronic voltmeters have their own set of fans who swear by their particular favorite. But the truth is that *both* have their own areas in which they are best suited.

The old VOM, which was designed prior to World War II, is a rugged and very portable instrument. It will take a lot of abuse, yet it suffers from certain problems that have only become intensified by the advent of solid-state circuitry. Most of the problems revolve around sensitivity and impedance. The impedance of a voltmeter is directly related to the sensitivity, as discussed earlier. But, it is also proportional to the full-scale voltage range. This means that the impedance of a voltmeter in a VOM configuration is not only *low*, but *varies* from range to range.

The electronic voltmeter (VTVM, FETVM, DVM, DMM, etc.), on the other hand, is capable of very high imput-impedance levels that remain constant over most voltage ranges. Most vacuum-tube voltmeters have input impedances of 10 or 11 megohms on all ranges, while certain models with MOSFET front-ends have input impedances as high as 100 megohms!

On the negative side, however, the electronic voltmeter is less portable. Even battery-operated types can get you into trouble on a mountain top, because they tend to use non-standard batteries that are not available at every country general store. Also, the electronic voltmeter is not capable of operation in a strong rf field, such as might be present around your transmitter or near a flyback transformer in a TV set. The rf field will tend to bias the input amplifier of the electronic voltmeter, and cause it to read erroneously. One popular portable FETVM will read approximately one-third to one-half scale all the time when a nearby transmitter is turned on.

An electronic voltmeter, or electronic multimeter, might use any of the technologies; transistors, ICs, digital electronics, or even

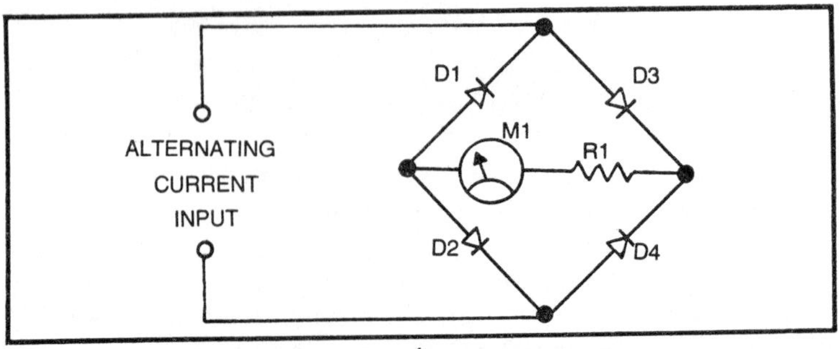

Fig. 4-6. Bridge rectifier to measure AC voltage on a DC meter.

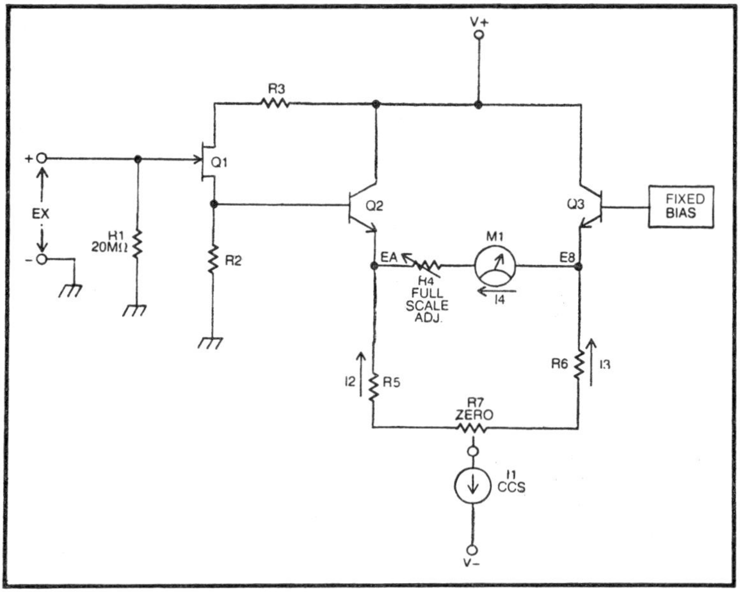

Fig. 4-7. Balanced circuit used in electronic voltmeters.

the old vacuum tube. Some instruments use an amplifier to allow the reading of voltages smaller than the basic full-scale circuit, and almost all have an attenuator that allows reading voltages at full-scale ranges much higher than the basic reading. The one common point among all of the types, however, is that they use active devices, so require a power supply. Some may use batteries for portable operations, while others may be operated only from the ac-power mains. Others will use both battery and ac power so that both portable and bench use is allowed.

The different types of electronic voltmeter go by names that reflect the type of technology used inside. The vacuum tube voltmeter, for example, is called a VTVM, while a similar instrument using a junction field-effect transistor is called an FETVM.

The basic principle behind all non-digital electronic voltmeter circuits is the balance network shown in Fig. 4-7. In this circuit, a differential amplifier consisting of transistors Q2 and Q3 form a balanced bridge circuit. Field effect transistor Q1 is used as a source-follower to increase the input impedance of the circuit. The output of Q1 drives the base of bridge transistor Q2.

Current I1 is a constant current, so we know that the following relations are true:

$$I1 = K = I2 + I3$$

The bias on transistors Q2 and Q3 is such that I2 = I3 when E_x (the unknown input voltage) is zero. In this condition, voltages E_a and E_b are equal, so the current through meter M1 is zero (I4 = 0).

The bias on transistor Q3 is fixed by a stable reference supply, but the bias on Q2 is a function of the voltage drop across R2, which in turn is a function of E_x. When an unknown voltage E_x is applied, then the bias on Q2 increases and that causes voltage E_a to increase. Now, E_a is greater than E_b, so current I4 is no longer zero. The magnitude of current I4 flowing in the meter movement is proportional to E_x.

The value of E_x that causes maximum deflection of M1 is the basic range of the instrument, and is usually the lowest voltage range listed on the voltage-selector control. Higher input-voltage ranges are obtained through the use of an *input attenuator*. Figure 4-8 shows a typical input-attenuator network that uses resistors to provide voltage division, and capacitors for ac compensation at higher frequencies. Compensation is required in part because of the use of wirewound precision resistors that have a certain amount of inductance.

AC EVMs. Alternating currents may be accommodated in much the same way as in passive voltmeters. Although a load (R1 in Fig. 4-9) is required for the rectifiers, and, sometimes, a capacitor is needed to smooth out the peaks.

Ac multimeters may also use an averaging circuit, such as Fig. 4-10. In this circuit, an ideal rectifier is used to derive a full-wave-rectified, pulsating-dc waveform, and a low pass filter is used to

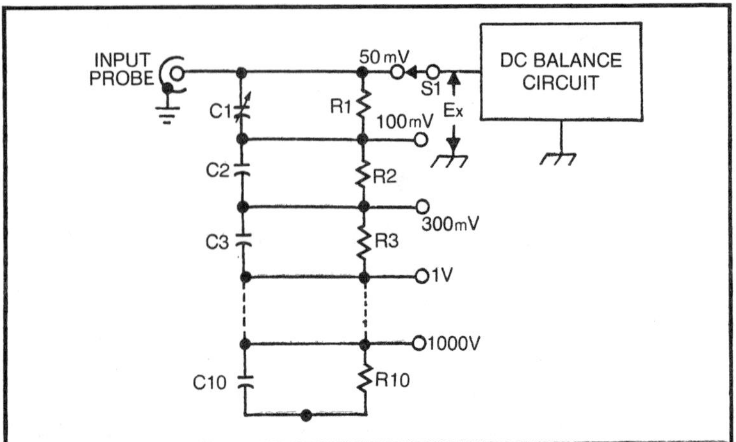

Fig. 4-8. Input attenuator circuit.

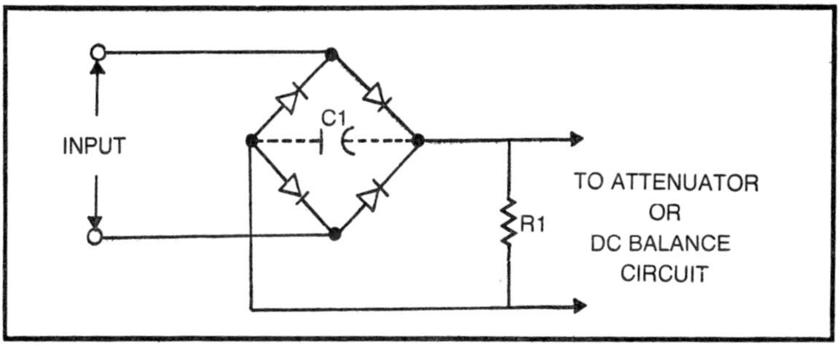

Fig. 4-9. Instrument-rectifier input-circuit to allow AC voltage measurements on electronic voltmeter.

average the peaks to produce an output voltage closer to the rms value of the waveform. An "ideal rectifier" is an operational amplifier circuit that compensates for the nonlinearity of the diode at low voltages. These circuits are covered in my book, *OP-AMP Circuit Design & Applications* (TAB book No. 787).

Electronic meters may measure resistance the same way as other meters, but, in most cases, they use a slightly modified technique. A constant current is passed through the unknown resistance, and this will generate a voltage drop across the unknown resistance proportional to the resistance. If, for example, 1 mA

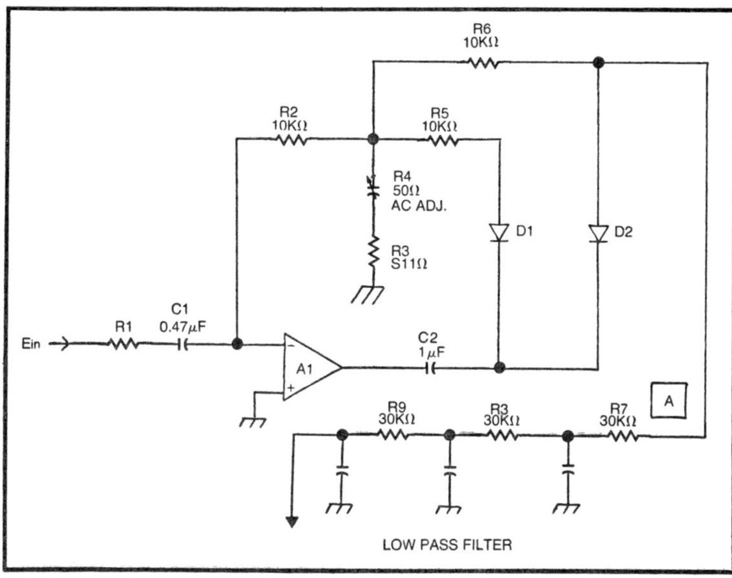

Fig. 4-10. AC-averaging-type rms measuring circuit.

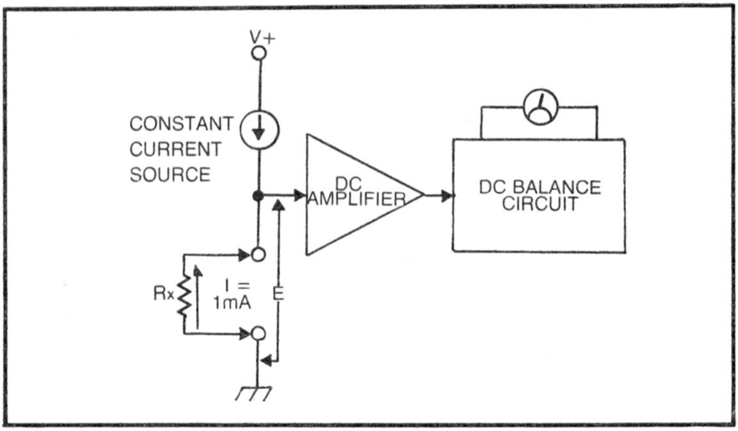

Fig. 4-11. Electronic ohmmeter.

were passed through a 1000-ohm resistor, we would see a voltage of (100 ohms) (0.001 amperes), or 1 volt. The 1-mA constant current would, therefore, produce a voltage drop proportional to kohms, in which E = 1V/kohm.

Figure 4-11 shows a block diagram of an electronic ohmmeter. An amplifier is used because the constant current source produces a very small current (not usually the 1 mA of our simplified example above). The output of the amplifier will then be applied to the voltage-input of a balance circuit, such as Fig. 4-7, to deflect the meter-movement proportional to the resistance across the input terminals.

Note that when the resistance terminals are open, the voltage seen by the input of the amplifier is full-scale. In this case, the circuit is overranged, and will indicate maximum voltage. If the voltmeter section is a digital voltmeter, then the reading will be +1999, or whatever is full scale in the particular instrument. Many will cause a series of bars (-----) or dots (. . . . .) to appear when the resistance is overranged.

DIGITAL VOLTMETERS

Digital voltmeters use analog-to-digital converters (ADC) to produce a digital output proportional to the applied voltage. In most currently sold instruments, the ADC is a dual-slope integrator that produces a pulse train with a number of pulses gated into a digital counter. The operation of the dual slope integrator, as well as certain other forms of ADC circuit, are covered in my book, *How to Design & Build Electronic Instrumentation* (TAB book No. 1012).

Chapter 5
Signal Generators

A signal generator is an instrument, containing an oscillator, that provides a controlled output signal for testing, aligning, measuring, or troubleshooting electronic circuits. Like most electronic instruments, signal generators produced in the past decade or so tend to be more sophisticated, complex, and capable of more applications, than instruments costing many times as much only a few years before.

Some modern signal generators are little more than transistor and IC versions of circuits that were used prior to 1940. Others are the products of modern innovations in the engineer's art.

Regardless of the newness, oldness, of a particular design, the use of solid-state devices has improved performance, reduced cost, reduced power consumption, and made the instruments smaller than ever.

Where a laboratory, or communications-service-shop grade rf-signal generator formerly required several cubic feet of cabinetry, the modern rf-signal generator may get lost on your workbench!

It used to be simple to discern the different types of signal generators. Audio generators were audio generators, and rf generators were rf generators. We now find "audio" generators operating into the rf range (11 MHz), and some rf generators operate down to 10 kHz, clearly an "audio" frequency.

Keeping in mind that distinctions are sometimes blurred, let's try to define the various types of signal generator that you might encounter in troubleshooting Amateur-Radio equipment, and the kinds of circuits Amateurs are likely to build.

AUDIO GENERATORS

This group of instruments has traditionally covered the frequency range 20 Hz to 20 kHz, many operate to 100 kHz and a few models exist that operate up to 11 MHz. Some are found that operate at frequencies as low as 1 Hz.

Audio signal generators usually produce sinewave outputs, although some might also produce squarewave outputs. An audio signal generator typically uses either precision output metering, or a precision output attenuator, to produce controllable output voltages down to approximately −40 or −50 VU.

FUNCTION GENERATORS

Function generators are often difficult to distinguish from audio generators, because they typically cover the same frequency range. Most function generators, however, cover to 100 kHz or more, and operate to sub-hertzian levels. One known function generator operates up to 80 MHz, and others operate down to frequencies as low as 0.001 Hz.

The principle difference between an audio signal generator and a function generator is in the number of output waveforms produced. The audio generator provides sinewaves, and possibly squarewaves, while the typical function generator provides these two, plus triangle waveforms. Some function generators also produce sawtooth and pulse outputs.

Another difference between these two generators is in the output attenuator. The function generator usually has a roughly calibrated output attenuator. The output range is typically 0-10 volts RMS, but the control over the output voltage is coarse. The audio generator might produce output potentials over the same range, but is precisely controllable, allowing you to set very accurate output levels. Two different approaches are used. In some instruments, the output meter has a fixed point scribed on it, usually in the middle to upper third of the scale. A coarse output control is adjusted until the meter reads at that point. The output level is then set by a precision attenuator. The meter allows the input to the attenuator to be set to the same point every time. Since the attenuation ratios are the same, we can then predict the output voltage to very good precision. In the alternative system, the attenuator is still a precision one, but the level is set by metering on a precision ac voltmeter. If we want, say, 1 mV rms, then we would adjust the *range* control (coarse attenuator) to the appropriate setting, and then the precision attenautor (or vernier) to the exact output level desired. The coarse selector not only sets the approximate output range, but also adjusts the sensitivity of the meter as needed.

Both audio and function generators usually provide a 600-ohm resistive output impedance. Function generators are most often equipped with a BNC type of output connector, while audio generators use the standard five-way banana binding posts mounted on ¾-inch centers. These distinctions, however, are becoming blurred, because more and more audio generators use BNC output connectors.

RF GENERATORS

An rf signal generator produces output frequencies that are in the rf range. But, this means little, today, because there are Navy radio transmitters operating in the 13 kHz region—which is in the "audio" spectrum. Of course, the wave produced is electromagnetic instead of acoustical, so *that* qualifies it as "rf."

One aspect of a signal generator that does place an instrument into the "rf signal generator" category is the matter of the output impedance. In rf generators, the output impedance will be 50 ohms resistive, unless the device is intended for the television receiver or television antenna service market, in which case 75 ohms is used.

Also, all serious rf signal generators have precision output attenuators. Those low cost "service grade" instruments that use primitive attenuators, or level controls, are not actually suited for the service shop—so much signal leaks around the cabinet flanges, attenuator, and output connector that this, alone, can drive a 2-meter FM receiver into 20-dB of quieting!

The attenuator used in rf signal generators might be calibrated in microvolts (0.1 to 100,000 μV) or dBm (decibels power, referred to 1 mW in a 50-ohm load taken as 0 dB). Many signal generators use both the dBm and μV calibration on different scales of the same dial.

Although there are some new PLL-controlled rf generators that seem to cover the spectrum from DC to daylight, most surplus service instruments cover from 1 or 2 MHz to 400 MHz, or will cover a VHF or UHF range independently.

Audio Generator Circuits

The audio oscillator is typically used in tests of high-fidelity equipment, public-address systems, and communications equipment. In the latter context, we find low-grade instruments used for troubleshooting, and high-grade instruments used to test tight-tolerance radio transmitters. The FCC, for example, has something to say about the total harmonic distortion (THD) produced by radio-broadcast stations, so the audio generator used in THD measurements must be low-distortion, otherwise oscillator distortion will contaminate the results.

In general, keeping in mind the exceptions discussed earlier, audio signal generators are oscillators that produce output frequencies in the range 20 to 20,000 Hz, at levels of −50 dBm to +4 dBm.

There are two methods of frequency selection typically used in these instruments: continuous tuning and step tuning. The continuous type of readout dial is much like the frequency-selector dial on a communications receiver (non-digital, of course,). You must turn the knob to the desired frequency.

Many continuously tuned signal generators have dial markings such as "20 - 200," or "2 - 20." A range selector switch determines whether the output frequencies will be 20 - 200 Hz, 200 - 2000 Hz, or 2000 - 20,000 Hz.

In a step-tuned generator, the dial is replaced with a series of decade (10 position) switches. These will be marked with integer values 0 - 9 (or sometimes 0 - 10, in which case the maximum value of any given switch is the same as the minimum value of the next most significant switch). In some cases, all of the switches will be marked with the 0 - 9 digits, but very often we see them marked according to their decimal weight; i.e., 0,100, 200, ... 900; 0, 10, 20, ... 90, etc. The least-significant-digit may be a continuously variable "trimmer," or a step switch calibrated in positions of 0.1, 0.2 ... 0.9. The range selector switch becomes a multiplier switch in step-selector models.

Consider, as an example, a step-selected audio signal generator in which the multiplier is set to X1000, the hundreds switch is set to 5, the tens to 3, and the units to 2. A fractionally calibrated, continuously variable control is set to 0.4. Find the output frequency.

$$(1000) \times (500 + 30 + 2 + 0.4) = (1000)\ (532.4) = 532{,}400 \text{ Hz}$$

Although there are a wide variety of oscillator circuits that will operate into the audio range, only a few different types find practical application. An *LC* oscillator (Hartley, Colpitts, etc), for example, can be *made* to operate at audio frequencies quite easily, but the *L* and *C* values become too large. Making such a beast variable (except in the heterodyne oscillator, of which more later), is almost out of the question (where do you find a variable capacitor of 0 - 2.2 μF?). Most audio oscillators used for service and laboratory work tend to be *RC* designs.

An *RC* phase-shift oscillator is shown in Fig. 5-1. In any oscillator, the criteria for oscillation are, a) greater than unity gain around the feedback loop (gains overcome losses), and, b) an in-phase (360 degree) feedback signal between output and input.

An active element, such as a bipolar or field-effect transistor, or an IC operational amplifier, provides the needed gain. In Fig. 5-1, a JFET is used as the active element, but this could have just as easily been an operational amplifier, linear IC of another description, NPN or PNP transistor, or a vacuum tube.

Since Q1 in Fig. 5-1 is operated in the common-source mode, it will provide 180 degrees of the needed 360 degrees of phase shift for the feedback signal. The remaining 180 degrees is provided by a three-section *RC* network, R1C1, R2C2, and R3C3. Each of these sections provides a 60-degree phase shift at some particular frequency determined from

$$F_o = \frac{1}{2\pi RC}$$

Where:

F_o is the frequency of oscillation

R is the resistance of R - R3 in ohms, each the same value

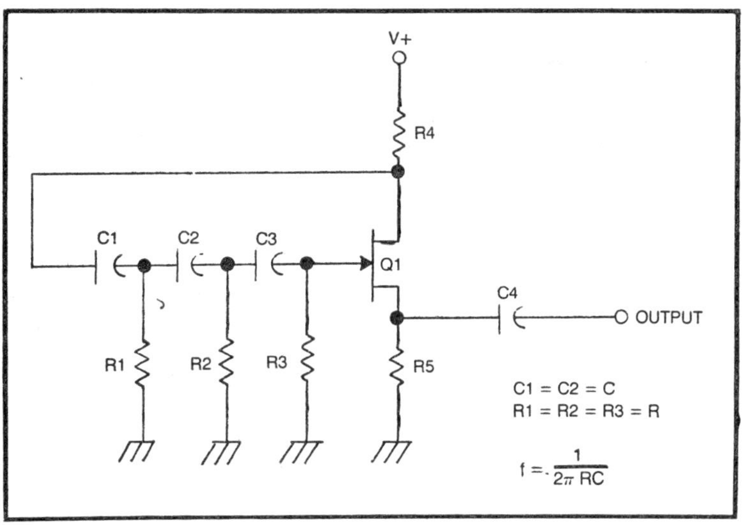

Fig. 5-1. RC phase-shift oscillator.

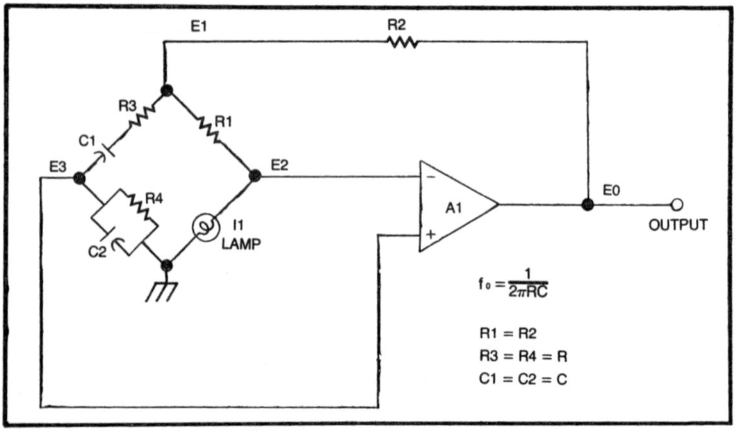

Fig. 5-2. Wein-bridge oscillator.

C is the capacitance of C1 - C3 in farads, each capacitor having the same value.

In most cases, the resistors will be part of the decade selectors, while the capacitors will be selected by the multiplier.

An example of the Wien-bridge oscillator is shown in Fig. 5-2. A differential amplifier, A1, is connected across the output nodes of an *RC* Wien bridge consisting of four legs: R1, I1, R3C1, and R4C2. At all times E_o and E_1 are in-phase with each other, and 180 degrees out of phase with E_2. The feedback in this loop is degenerative, therefore the circuit is stable at all frequencies except F_o (which is also described by Eq. 5-1.)

Voltage E_3 is out of phase with E_o, but at frequency F_o, voltages E_3 and E_o are in-phase. Since E_3 is applied to the noninverting input of the amplifier, this circuit will oscillate.

Lamp I1 is used to stabilize the output amplitude, and prevent saturation of the amplifier device. This job is accomplished by the changing resistance of the lamp as E_o increases, causing more current to flow in the lamp filament. In most cases, I1 is operated at current levels below incandescence, although I have seen a few signal generators in which the lamp glows (but dimly).

The Wien-bridge oscillator is used in many audio signal generators because of its relatively pure output waveform. Although R3/R4 may be made variable by using either a potentiometer or decade switch and resistor bank, it is the latter that is most popular.

A twin-tee oscillator is shown in Fig. 5-3A. Transistor Q1 provides 180 degrees of phase shift at all frequencies. The remain-

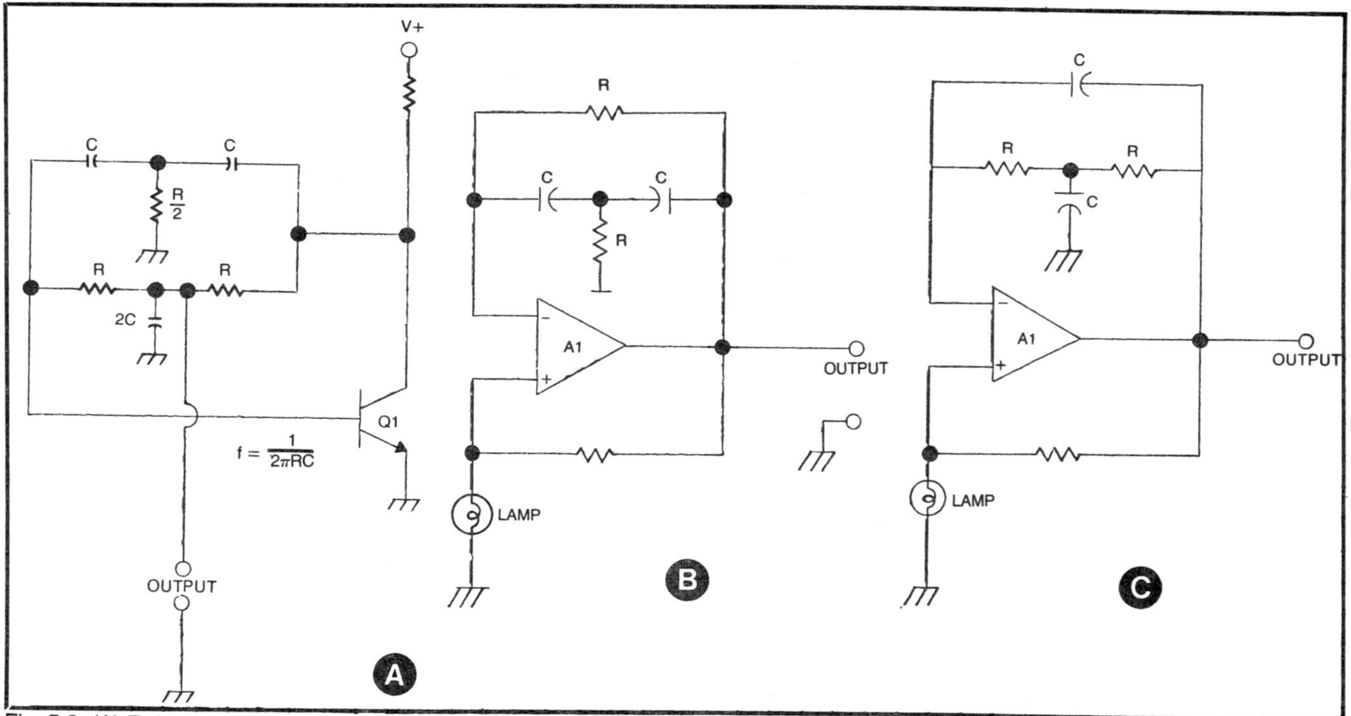

Fig. 5-3. (A) Twin-tee oscillator, (B) Resistor bridged-tee oscillator, (C) Capacitor bridged-tee oscillator.

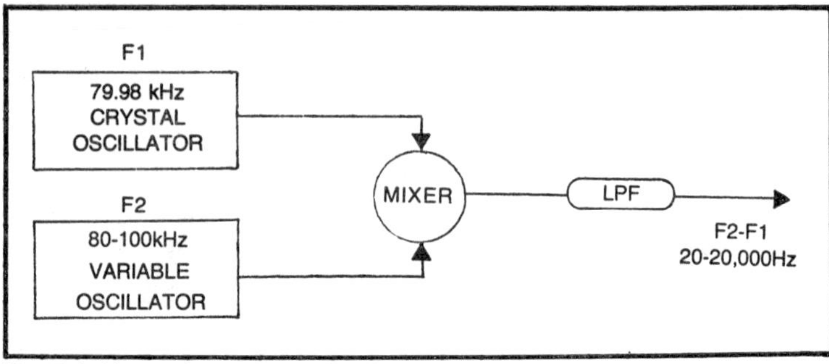

Fig. 5-4. Heterodyne-type audio-signal source.

ing 180 degrees of phase shift is provided by the twin-tee *RC* network.

There are two simple ways to describe this circuit and its operation. In one sense, we point out that there is a 180-degree phase shift through the twin-tee network at some frequency, F_o. Therefore, the input of the amplifier sees an in-phase feedback signal (180 degrees from the amplifiers, +180 degrees from the network).

Some other textbooks view the circuit in a slightly different way. They will point out that the twin-tee network is, in essence, a notch filter, and that it passes all frequencies except F_o, and those very near to F_o. Because of this, all frequencies other than F_o receive a tremendous dose of degenerative, or negative, feedback. The circuit gain at all frequencies other than F_o is very low. But F_o receives very little negative feedback, because only a small amount of signal at F_o will pass through the network. For some reason, I prefer the first view.

There are two variations of the twin-tee design that tend to make the circuit very stable. These circuits are called the *bridged-tee* oscillators, and are shown in Figs. 5-3B and 5-3C. The circuit in Fig. 5-3B is the resistor-bridge version, while that of Fig. 5-3C is a capacitor-bridge circuit.

The last oscillator which we will consider is the ***heterodyne*** design of Fig. 5-4. This signal generator has not been popular in recent years, but was once very common, especially in vacuum-tube designs.

In the heterodyne system, a variable-frequency oscillator (F1) is heterodyned (hence the name) against a fixed-frequency oscillator (F2) to produce a difference frequency (F2 - F1).

Most heterodyne audio-signal generators used *LC* oscillators operating in the 50 to 200-kHz range, although a few were known

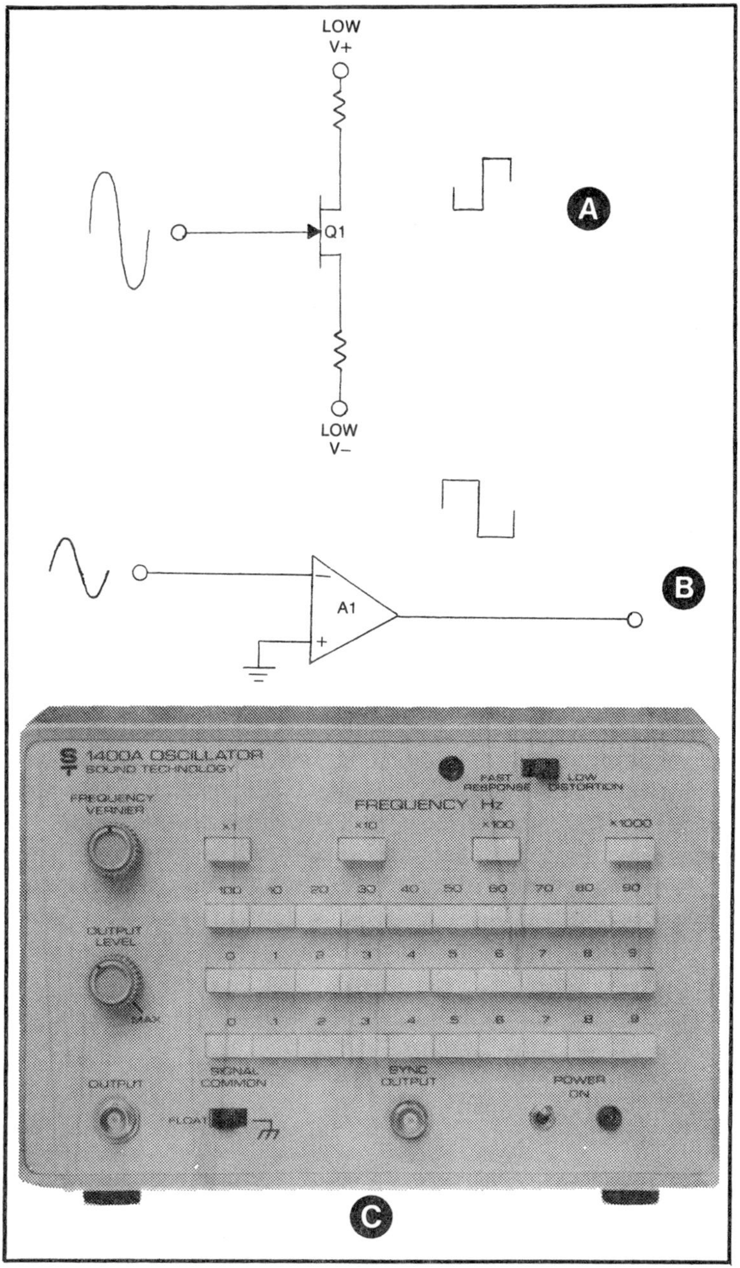

Fig. 5-5. (A) Overdriven amplifier to produce squarewaves, (B) Comparator to produce squarewaves. (C) Commercial audio signal generator (courtesy of Sound Technology Inc.).

that operated at much higher frequencies. In general, though, lower frequency *LC* oscillators produce more stable signals than do higher-frequency versions. The fixed oscillator might be a VFO, but in most cases it was a crystal oscillator for greater stability.

GENERATING SQUAREWAVES FROM SINEWAVES

Squarewaves produced by an audio signal generator must be symmetrical (not always a requirement of function generators, however), not only in the left-right sense but across the zero baseline as well.

The squarewave can be produced by an astable multivibrator that is frequency-locked to the sinewave oscillator, but this is difficult to achieve in practice. It is, then, considered the best practice to derive the squarewave directly from the sinewave.

Figure 5-5 shows two methods for deriving a squarewave from a sinewave. In Fig. 5-5A we see the use of an overdriven amplifier stage, a technique that is popular in low cost models. Transistor Q1 is operated with low V− and V+ voltages, so is easily over-driven by the input signal. The output signal is severely clipped because of these low supply voltages. Unfortunately, good rise times are a little difficult (but not impossible) to achieve.

A somewhat superior approach is the circuit of Fig. 5-5B, in which an operational amplifier is used as a voltage comparator. The lack of negative feedback around the amplifier causes the amplifier gain to be very high (25,000 - 1,000,000 in common commercial-grade IC devices). Since the noninverting input is grounded, or at a potential of zero volts, the output saturates anytime the input is more than a few microvolts. On negative excursions of the input sinewave, the amplifier output snaps positive, while on positive excursions of the input sinewave, the output snaps negative. This inversion is due to the use of the (−) input of the operational amplifier. The use of a bipolar power supply accounts for the symmetry about the zero baseline.

A high-frequency comparator will produce output squarewaves with very good risetimes. But even this circuit can be improved by the addition of small amounts of positive feedback to speed up the output-waveform transition time, thereby improving risetime.

An example of a commercial audio signal generator featuring very low distortion is shown in Fig. 5-5C. This instrument is the model 1400A by Sound Technology, Inc., and is used to test modern, very low distortion, high-fidelity equipment.

AUDIO GENERATOR OUTPUT SECTIONS

The audio signal generator must be able to deliver as much as 10 or 15 volts RMS into a 600-ohm load. It must also be capable of delivering controllable potentials of only a few millivolts, all without changing the generator's output impedance. Additionally, the output signal must be symmetrical about zero, so the actual output range will be ± 10 volts rms, or ± 14.14 volts peak. The output stage of the signal generator, then, must be operated from bipolar power supplies (as in Fig. 5-6). This amplifier will be required to deliver up to several hundred milliwatts at a very low total-harmonic-distortion figure. Additionally, it should have a certain amount of short circuit protection, because users *will* accidentally short the device out.

In any event, regardless of whether or not an output amplifier is needed to provide the output parameters demanded, a buffer amplifier between the audio oscillator and the output terminals is always highly desirable. This will prevent both output-level and output-frequency changes with changes in the output-load impedance.

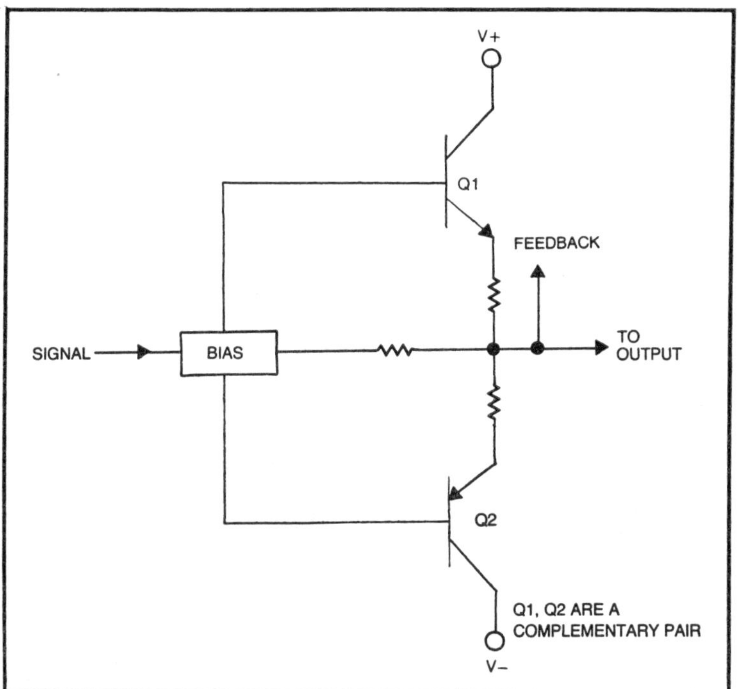

Fig. 5-6. Power output stage.

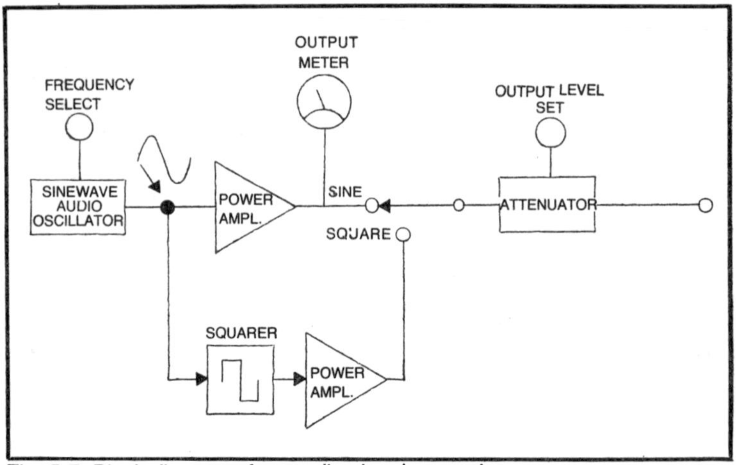

Fig. 5-7. Block diagram of an audio signal generator.

Figure 5-7 shows the block diagram of a typical audio signal generator. The oscillator section will use one of the circuits given earlier. A power-amplifier stage provides buffering between the load and the oscillator, and develops the output-signal amplitude.

The ac voltmeter at the output is optional, but in some models it is used with a vernier level control to precisely set the signal to the precision attenuator. Not all quality audio signal sources use this technique, however, so the existence or nonexistence of an output meter is not, in itself, adequate reason to judge the quality of the instrument.

A pair of typical attenuator circuits are shown in Fig. 5-8. The circuit in Fig. 5-8A is used in some instruments, and may take the form of a potentiometer, or switched resistor bank (as shown). This circuit is not often used in high-quality audio signal generators, but is quite common in lower-grade instruments, and in homebrew projects published in the Amateur Radio press.

One of the major faults of this type of attenuator is that the output impedance of the signal generator changes drastically as the output level is changed. A proper attenuator would produce a constant output impedance at all settings. In addition, whatever load is connected to the output severely affects the amplitude of the output signal (voltage divider action). This could lead to severe output voltage errors if nonstandard loads were used.

An unbalanced ladder network is shown in Fig. 5-8B. This network, and several popular variations (balanced and unbalanced), are used in most commercial audio signal generators with any claim to quality. It provides a constant output impedance, so will eliminate the output-voltage errors inherent in the other design.

A squarewave is created from the sinewave in a squaring circuit. In some models, the squarewave section will have its own output amplifier and output terminals. In others, the output *function selector* switch is used to select squarewaves or sinewaves. The output calibration, however, is for sinewaves and may be totally meaningless, except as a relative indicator, for squarewaves.

FUNCTION GENERATOR CIRCUITS

We have already told you that the function generator and the audio signal generator are very similar in many respects. In this section, therefore, we will cover only those points of difference that might exist.

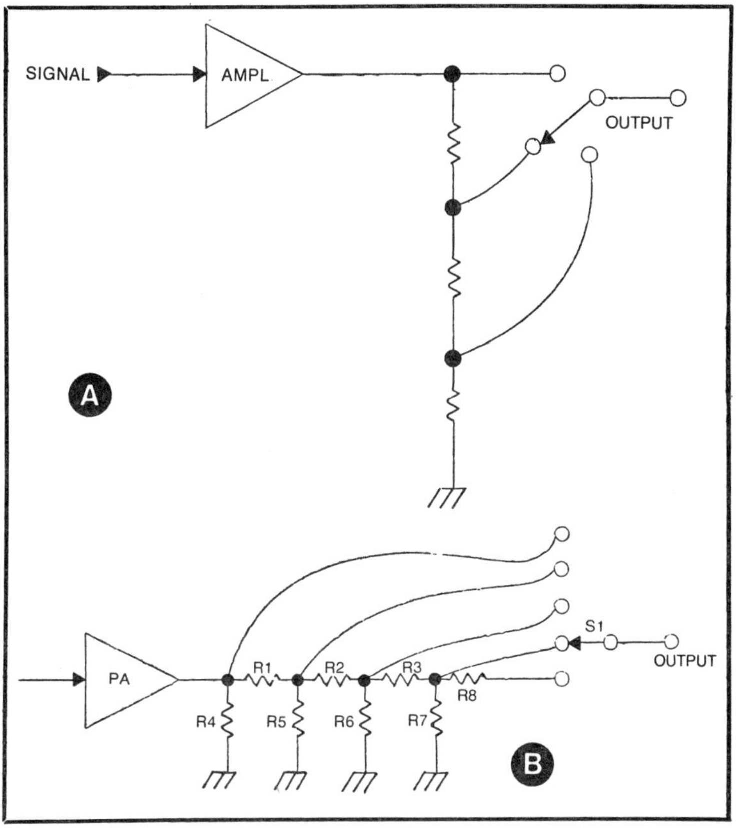

Fig. 5-8. (A) Simple output attenuator, (B) constant-impedance output attenuator.

Figure 5-9A shows a simplified schematic of a typical function generator that produces sinewaves, squarewaves and triangular waves. In most of these instruments, a multivibrator generates the basic signal, and all other signals are derived from it. Most of the sinewave oscillators are a little difficult to design with a constant output amplitude over a wide frequency range, or with good frequency stability at all covered frequencies. But we also find that certain squarewave oscillators, on a cost-for-cost basis, are able to produce better performance in these areas.

In Fig. 5-9A, amplifier A1 is used as a voltage comparator (and in practice may be a voltage comparator IC instead of an operational amplifier), while amplifier A2 is used as an integrator. When the circuit is initially turned on, the output of amplifier A1 will snap HIGH, creating current I_2. The voltage at point "A" will also cause the feedback capacitor in the integrator to begin charging, thereby causing voltage at point "B" to begin rising. Voltage E_B rises to the point where current I_1 is equal to I_2, then voltage E_A snaps back to zero, making I_2 also zero. But E_B is still at the same level, so I_1 is greater than I_2. This makes the amplifier output pass through zero to become high in the negative direction. With E_A now negative, the integrator capacitors will begin to discharge, so E_B reduces towards zero. The waveforms at points A and B are shown in Fig. 5-9B. The upper trace shows the waveform at point A (a squarewave), while the lower trace shows the waveform at point B (a triangle).

A sinewave can also be generated from a squarewave, because the squarewave is a composite of a fundamental-frequency sinewave and an extremely large collection of odd harmonics. A low pass filter, as in Fig. 5-9A, will remove all of the harmonics, leaving only the fundamental. If filtering is good enough, then the sinewave will have very low distortion.

Figure 5-9C shows the effects of using a low-pass filter on the squarewave signal. Although it is not apparent in the drawing, there is a very great loss of amplitude. An amplifier following the low pass filter is therefore needed to build the signal amplitude back up to reasonable levels.

SAWTOOTH GENERATORS

A sawtooth waveform (Fig. 5-10) consists of a ramp leading edge, followed by a fast drop to the zero baseline. The sawtooth is used to drive the horizontal sweep of an oscilloscope, or the VCO in a sweep generator such as might be used to align and troubleshoot FM receivers.

Figure 5-11A shows one traditional approach to generating a sawtooth waveform: the unijunction transistor (UJT) relaxation oscillator.

The UJT remains turned off until voltage *E* exceeds a threshold voltage, at which point the UJT emitter breaks over and the emitter to base 1 path becomes nearly a short circuit.

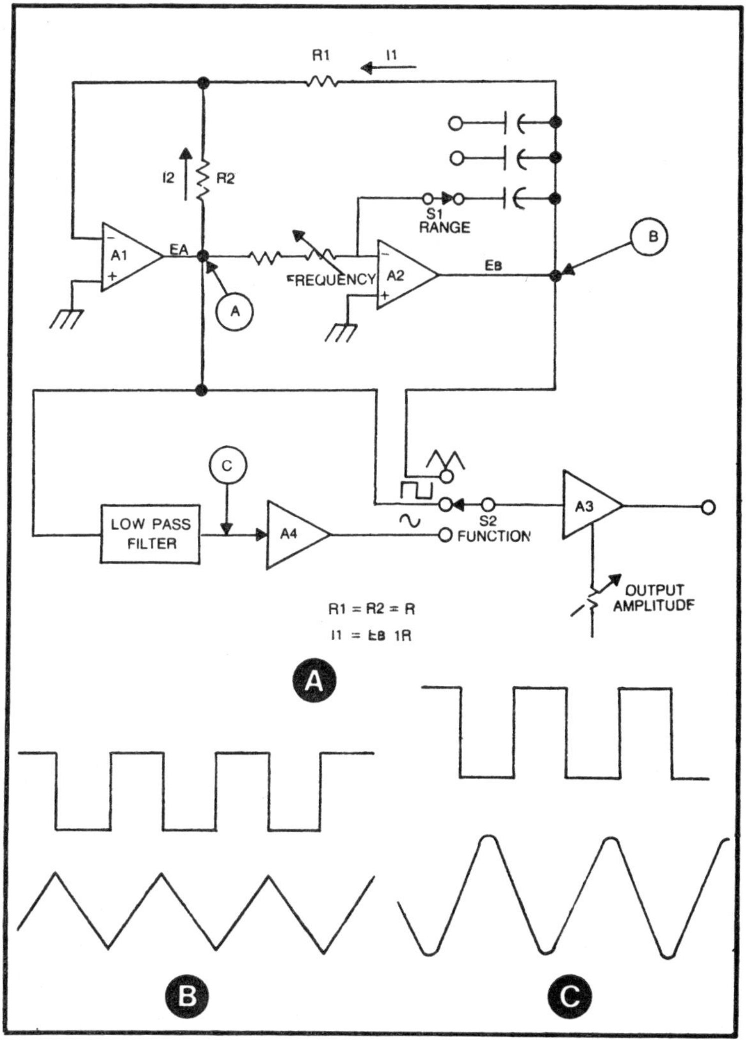

Fig. 5-9. (A) Block diagram of a function generator, (B) Triangle waves are created by integrating squarewaves, (C) Sinewaves are created by filtering squarewaves.

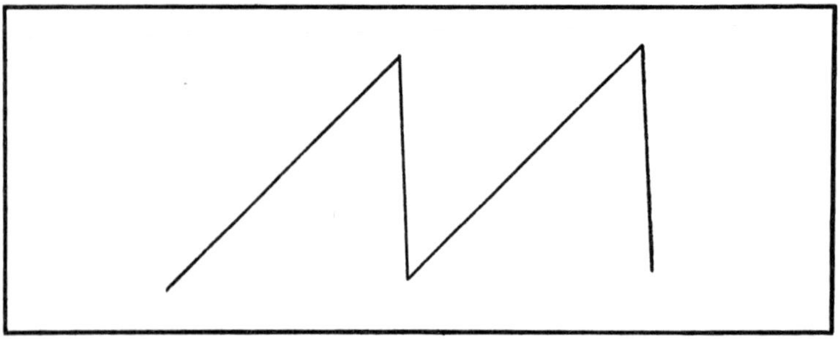

Fig. 5-10. Sawtooth waveform.

Current *I* charges capacitor *C* until the voltage across *C* exceeds the UJT breakdown voltage. The low resistance across the base 1-emitter junction rapidly discharges the capacitor.

Unfortunately, the simple circuit of Fig. 5-11A produces a poor example of a sawtooth. The leading edge of a good sawtooth must be linear, i.e., straight. In the circuit shown, the leading edge would be an exponentially rising capacitor-charge waveform (as in Fig. 9-11B).

We could make this waveform linear by using only the first 10 or 15 per cent of the charging waveform, a tactic that is common in the design of RC integrator circuits. But this is not usually the most practical approach. A somewhat better approach is the use of a constant current source (CCS) (Fig. 5-12A) to charge the capacitor. A CCS can be a special diode-connected JFET, or it may be a simple two-transistor bipolar circuit. The CCS, regardless of form, however, must be able to maintain *I* constant, so that the output waveform will have a linear leading edge.

A circuit in which we commonly employ the first 10 to 20 per cent of the *RC* time-constant is the Miller integrator sawtooth generator of Fig. 5-13. This is the circuit used in many modern function generators to create a sawtooth (as well as triangles). Note that it differs from the circuit used to generate the triangle in that a CMOS electronic switch (S1) is used to short out the feedback capacitor *C*. A voltage comparator is used to turn the switch on and off. When the output voltage E_0 rises to a point equal to E_1, then the comparator output snaps HIGH, turning on the CMOS switch to rapidly discharge capacitor *C*. When E_0 drops back to zero, the comparator output drops LOW, opening the switch. This will allow the procedure to repeat itself. The input to the integrator is a constant reference voltage, and this keeps both I_1 and I_2 constant also.

PULSE GENERATORS

A pulse differs from a squarewave mostly in the matter of symmetry. A pulse needs neither baseline, nor right-left symmetry (which a squarewave requires), and need not be periodic. In fact, a perfectly symmetrical squarewave that occurs just once, instead of in a wave train, will behave more like a pulse than a squarewave. The pulse *may* be repetitive, even periodically so, but need not be.

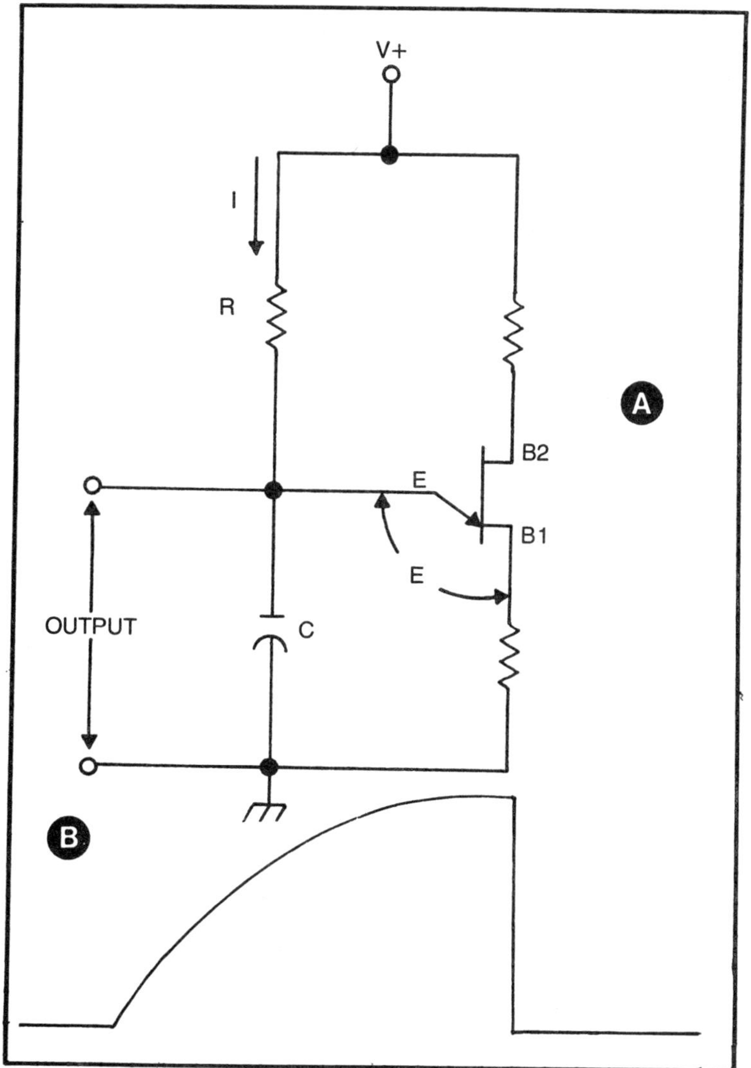

Fig. 5-11. (A) Simple UJT sawtooth generator, (B) Capacitor-charge waveform.

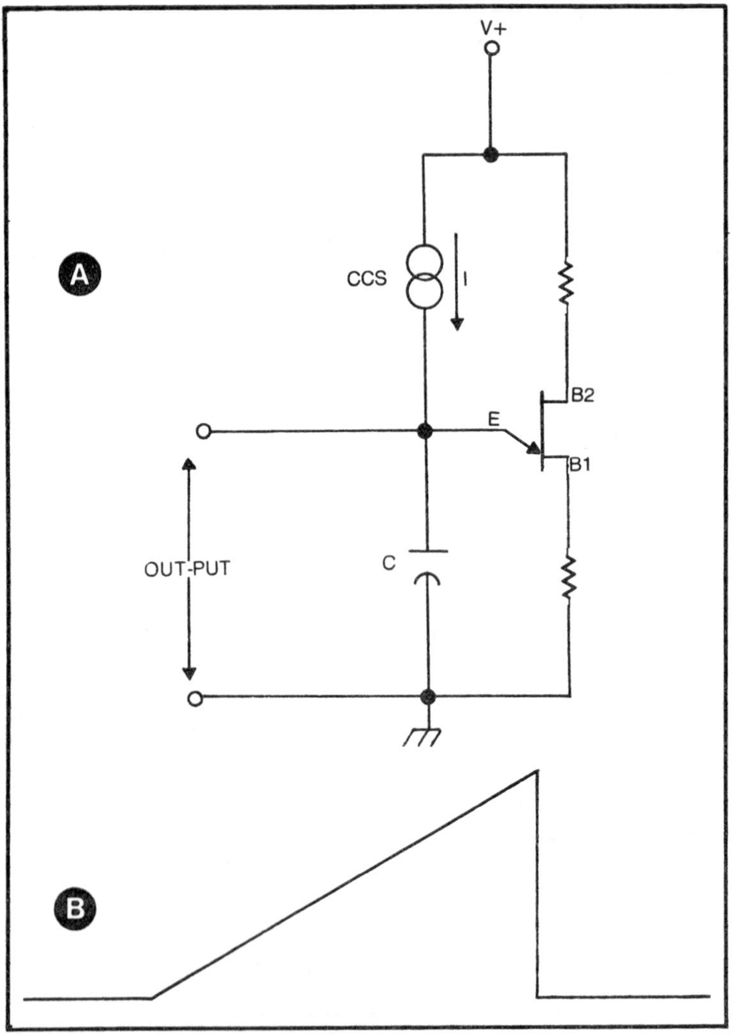

Fig. 5-12. (A) UJT sawtooth generator in which a constant-current source charges the capacitor, (B) capacitor charging waveform.

Figure 5-14A shows one popular method for generating a pulse output from a function generator. A monostable multivibrator (i.e. one-shot circuit) follows a squarewave oscillator, or an astable multivibrator of some sort. The pulse repetition rate is set by the squarewave frequency. The one-shot triggers on the leading edge of the squarewave, and produces exactly one output pulse for each squarewave.

The duration of each output pulse is set by the on-time of the one-shot circuit, and may be either very short (even down to micro- or nanoseconds!), or may approach the period of the squarewave. The relationship between these waveforms is shown in Fig. 5-14B.

RF SIGNAL GENERATOR CIRCUITS

An rf signal generator uses an oscillator to produce ac signals in the region above approximately 13 kHz. Most rf signal generators also produce A-M or FM (or both) modulation of the output signal. This will allow radio receivers to be aligned or serviced.

Figure 5-15A shows the most basic form of rf signal generator. Here we see a bandswitched, tunable oscillator and buffer amplifier driving the attenuator directly (in cheaper models the buffer is deleted). The output meter determines the level of signal applied to the input of the attenuator. In most cases, the meter is not calib-

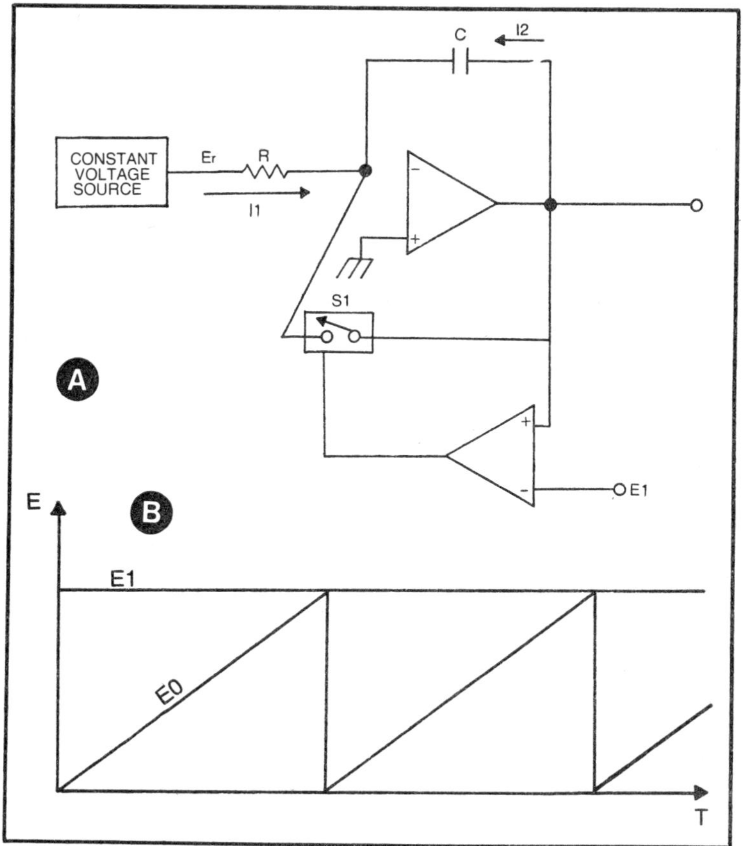

Fig. 5-13. (A) Miller integrator, (B) capacitor charging waveform.

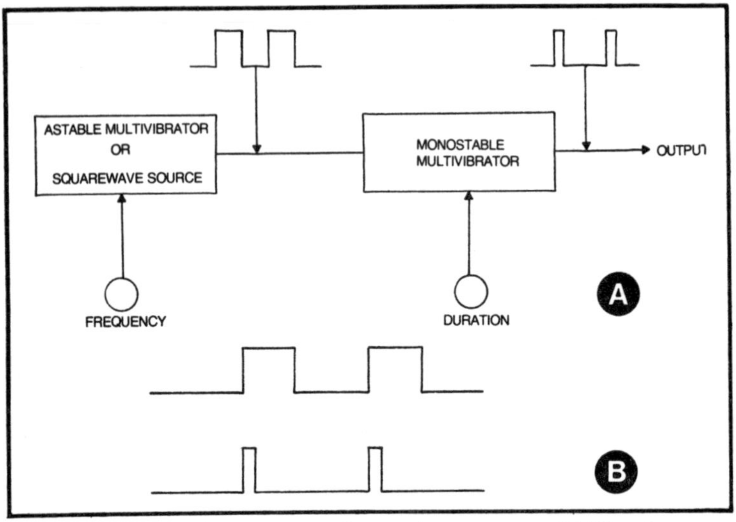

Fig. 5-14. (A) Basic block diagram for a pulse generator, (B) Waveforms.

rated, but will have a calibration point marked on the scale (usually in the form of a red line). When the coarse output-level control is set so that the meter lies on that line, then the output voltage, or dBm markings on the precision attenuator will be accurate. By making the *input* to the attenuator repeatable and precise, we are able to set the output level.

In some older models, the output frequency would change slightly when the attenuator is adjusted. But the calibrations will be accurate if the output level is set to the red line.

In Fig. 5-15B we see an attempt to provide automatic control of the output level. In this circuit an automatic gain control (AGC) circuit is used to keep the level constant. An amplitude modulator stage is placed between the output of the oscillator and the input of the attenuator. The oscillator signal is the "carrier" and the output of an AGC rectifier is the modulating signal. The rectifier output is proportional to the rf signal level at the modulator output, and by feedback action it keeps the level constant. Since the stage is an amplitude modulator, we may also produce an A-M function by applying the 400- or 1000-Hz audio sinewave to one input of the modulator.

FREQUENCY SYNTHESIZERS

In the past ten or fifteen years a new type of rf signal generator has come into its own: the rf frequency synthesizer. Various attempts at frequency synthesis have been tried in the past, with varying degrees of success. Most of these suffered from too many faults, or were too costly to be seriously considered in commercial productions. In one type, for example, a wideband frequency spectrum resembling white noise was generated and amplified. A series of costly LC filters were then used to pick off components to be mixed together to construct any given frequency. Most recently, "crystalplexers" have been used. In these devices, also used in some early "synthesized" 2-meter FM and 23-channel CB sets, a bank of crystal oscillators (each containing several crystals) were heterodyned together to produce the desired frequency. One common problem with this design, and the broad spectrum models, was too many spurious output products. It seems that is was difficult to adequately suppress all of the possible mixer products

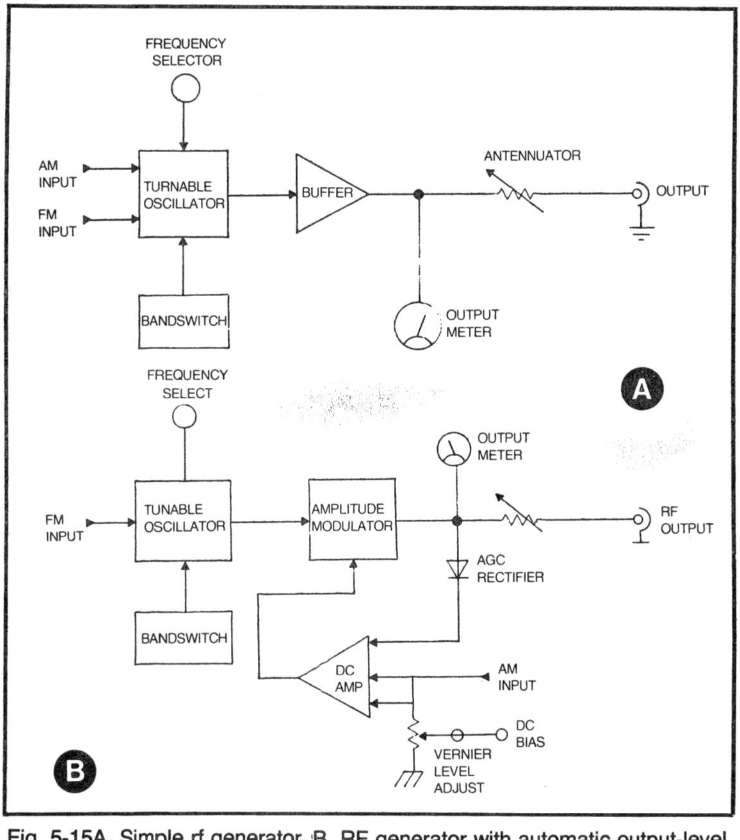

Fig. 5-15A. Simple rf generator. B. RF generator with automatic output-level control.

that might be produced. In one notorious crystalplexed 2-meter FM rig, for example, you had the possibility of keying up several repeaters (on different frequencies) every time you went on the air. In a busy repeater area, this was not calculated to make you popular with other users. Also, the crystalplexer design becomes prohibitively expensive if more than a very few frequencies must be covered. A 23-channel CB or 2-meter rig may prove economical, but a wide-range rf signal generator could not use this method at a price anyone could afford.

Most modern synthesized rf signal generators use the phase-locked loop to stabilize the output signal. A block diagram of a typical PLL rf generator is shown in Fig. 5-16. The principle parts of this circuit are: voltage controlled oscillator (VCO), divide-by-N counter, phase detector, reference frequency oscillator, a low pass filter and a dc amplifier.

The VCO oscillates at rf frequency, F_0, that is a function of the dc control voltage, E. Voltage E is created by comparing the output frequency from the divide-by-N counter (a subharmonic of F_0) to a low-frequency reference-oscillator signal. The reference signal is crystal controlled, and very precise. A phase-comparator circuit is used to compare the F_0/N signal with the reference signal.

The divide-by-N counter is a digital circuit that divides the input frequency (F_0 in this case) by an integer N. The integer division ratio is set by the binary word applied to the program-N

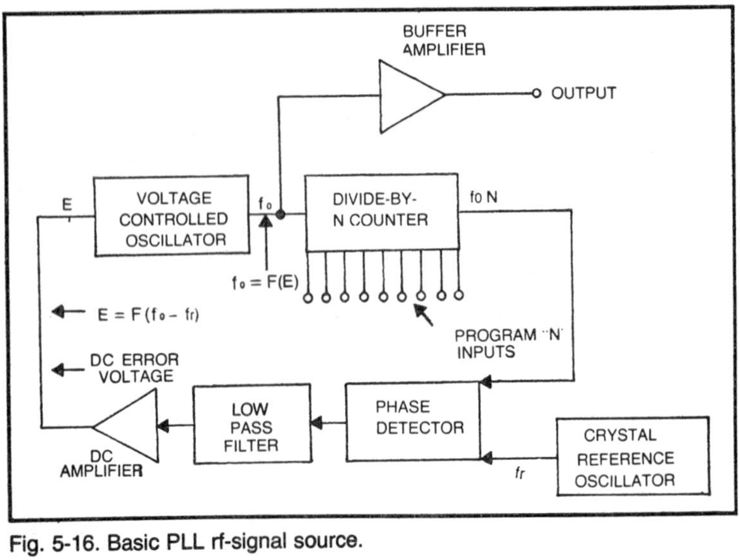

Fig. 5-16. Basic PLL rf-signal source.

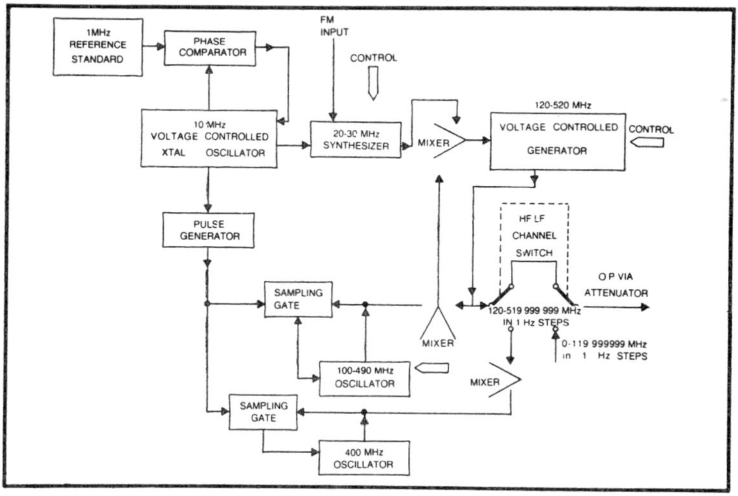

Fig. 5-17. Commercial PLL synthesized rf-signal generator.

inputs of the counter. The output of the counter will be a frequency that is equal to F_0/N.

The outputs of the N-counter and the reference oscillator (usually 1, 2, 5, or 10 kHz) are applied to opposite inputs of a digital phase-detector circuit. As long as the reference frequency F_r is equal to F_0/N, then the VCO will oscillate at the frequency determined by the dc-amplifier offset-control and the N-code applied to the inputs of the divide-by-N counter.

But if the VCO should attempt to drift off frequency, then a dc error voltage will appear at the output of the phase detector, and this changes in a direction and amplitude proportional to the direction and magnitude of the drift. The dc error voltage is filtered, amplified by the DCA, and then used at the input of the VCO to pull the frequency back to the correct point. The PLL, then, is constantly correcting for changes in VCO frequency. The stability and precision of this type of signal generator is essentially equal to the precision and stability of the crystal reference oscillator.

A deliberate frequency change can be made by reprogramming the divide-by-N counter to a different division ratio. The signal generator may use programmable switches, or a numerical keyboard (a modern approach) to change the N-code.

In most wide-range, PLL-based signal generators, several different PLL subsections are used, and their respective output signals are heterodyned together to produce the actual output signal. An example of this approach is seen in Fig. 5-17.

Chapter 6
Oscilloscopes

The cathode ray oscilloscope has been called the most useful of all electronic measurement instruments. Many professional service technicians would rather do without almost any other instrument and keep their oscilloscope, if they were allowed only *one* piece of equipment. The "scope" can measure the voltage amplitude, frequency, and period of a signal directly, and, if a little cleverness is used, it will also measure current and resistance.

The cathode ray oscilloscope (CRO) is a lot like a television set. It produces a "picture" of the signal on the viewing screen of a cathode ray tube (CRT)

CATHODE RAY TUBES

The operation of an oscilloscope depends entirely on the CRT. Figure 6-1 shows the basic elements of a typical CRT. These are: the electron gun, focus electrode, accelerating electrode, vertical and horizontal deflection plates, and a phosphorescent viewing screen. Some models will also have a second accelerator electrode after the deflection system, and this is called a post-deflection accelerator.

The electron gun consists of a heated cathode inside a metal cylinder. The end of the cylinder facing the viewing screen is closed, except for a tiny hole through which electrons emitted from the heated cathode can escape.

Electrons come out of the gun in a stream, and are accelerated toward the viewing screen by the positively charged accelerator grid. The electron beam is also focused and controlled by special electrodes in this section.

The power supply of the oscilloscopes must provide a high voltage to the CRT. The electrons from the gun and focus assemblies must be accelerated to a speed that will cause the phosphor atoms on the viewing screen to emit light when they are struck by the electrons. There will be a high negative power supply voltage applied to the elements of the electron gun and focus assembly. A high positive potential is applied to the post-deflection accelerator. This potential will be from about 1000 volts in low cost oscilloscopes to around 7000 volts in the more expensive ones. Color television receivers might use as much as 30,000 volts, although because of potential X-ray production at those voltages, most are restricted to the 25,000 volt range. In most scopes used in amateur work, the post-deflection accelerator will be in the +1000 to +2500-volt range.

DEFLECTION SYSTEMS

There are two ways to deflect the electron beam: magnetic and electrostatic. The magnetic system uses a pair of electromagnetic coils, inside a common *yoke* assembly, to provide a magnetic field to deflect the electon beam. The yoke is positioned on the neck of the CRT, and is an external device. The signals are applied to the yoke in the form of a current that creates the magnetic field.

The inductance of the yoke coils limits the upper frequency response of magnetic deflection systems. All television receivers use magnetic deflection, but they use 60 Hz and 15,734 Hz as the vertical and horizontal deflection frequencies, respectively.

For service, engineering, and most other applications, however, the frequency-response limitations of the magnetic deflection system render it useless. All of the oscilloscopes used in these applications use electrostatic deflection systems.

The CRT diagrammed in Fig. 6-1 uses electrostatic deflection. The structures inside of the CRT called the *deflection plates* provide this function. There are two sets of plates, one pair each for vertical and horizontal planes. By varying the potentials to these plates, we create *electrostatic* fields that will deflect the electron beam left and right or up and down.

The most basic type of oscilloscope is the X-Y oscilloscope, in which both vertical and horizontal signals are derived from sources outside of the oscilloscope. The vertical and horizontal amplifiers receive signals from outside, and build them up to amplitudes capable of deflecting the electron beam.

The most common oscilloscope is the Y versus time, or Y-T oscilloscope, in which the horizontal, or X-axis, signal is an internally generated time-base signal. Many Y-T oscilloscopes are also capable of X-Y (X axis versus Y axis) operation because they have an external horizontal input that may be selected by a switch.

LISSAJOUS PATTERNS

One of the principle applications of the X-Y oscilloscope is the production of patterns called Lissajous figures. Lissajous patterns result when harmonically related signals are applied simultaneously to the vertical and horizontal inputs. Figure 6-2A shows examples of Lissajous patterns produced when the vertical and horizontal signals have the same frequencies, but different phase angles with respect to each other.

Note the first Lissajous pattern, a straight line. This pattern occurs when the signals are in-phase (i.e., 0 or 360 degrees) with each other. The line tilts from lower left upwards to the right. The exact angle of the tilt will be 45 degrees from the horizontal when the signals applied to the two sets of plates have equal amplitudes. The angle will vary from 0 degrees when the vertical component is zero, to 90 degrees when the horizontal component is zero.

The phase angle is indicated by the "fatness" of the ellipse. The second pattern shown is for signals in 0 to 90-degree, or 270 to 360-degree, range.

When the phase angle is exactly 90 degrees, the ellipse fattens into a perfect circle, as shown in the third Lissajous pattern.

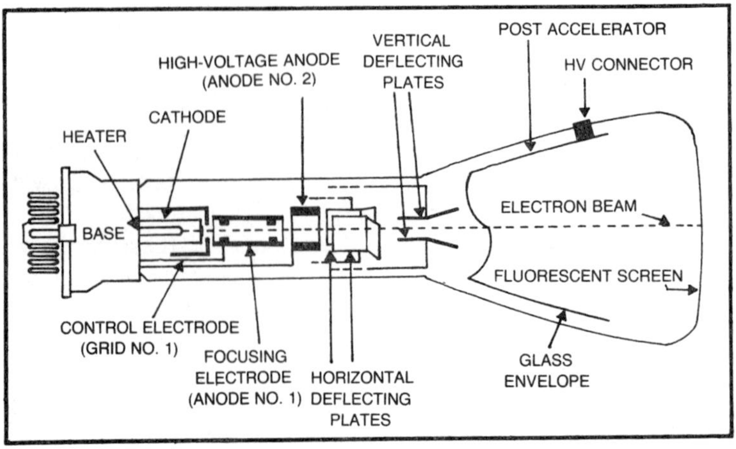

Fig. 6-1. Cathode-ray tube.

As the phase angle passes 90 degrees, the ellispse returns to the former shape, but its tilt angle is reversed. The pattern will continue to become thinner at increasing phase angles, until it is once again a straight line at phase angle of 180 degrees. Note that this is the same as in the 0-degrees case, but the tilt angle is reversed.

It is a little difficult to discern the phase angle from a quick observation of the Lissajous pattern. But if the screen of the oscilloscope is marked with a graticule, then we may measure it as in Fig. 6-2B. The phase angle is

$$\text{Phase Angle} = \text{Arcsin } (a/b)$$

(using the notation of Fig. 6-2B).

We can also calculate the tilt angle (especially in the 0 and 180 degrees cases) from the applied voltages, assuming that the gains of the vertical and horizontal amplifiers are the same. Similarly, if we can measure the tilt angle and one of the two applied voltages, we can calculate the second voltage. The formulas for both measurements is

$$\text{Tilt Angle} = \arctan (E_v/E_h)$$

Where:

E_v is the voltage applied to the vertical input

E_h is the voltage applied to the horizontal input

The tilt angle is measured in degrees, and the two voltages must be measured in the *same units* (i.e., volts, millivolts, etc).

The Lissajous pattern in Fig. 6-3 shows what to expect when we apply harmonically related signals to the vertical and horizontal inputs.

We know that the actual *shape* of the Lissajous pattern is determined by phase differences between the two signals, while the number of loops in each plane is determined by the two frequencies. The relationship between the two frequencies is given by

$$\frac{F_v}{F_h} = \frac{N_h}{N_v}$$

Where:

F_v is the frequency to the vertical input

F_h is the frequency applied to the horizontal input

N_v is the number of loops along the vertical edge of the pattern

N_h is the number of loops along the horizontal edge of the pattern.

The use of Lissajous patterns is covered in the chapter on frequency measurement. It is limited in practical applications to those cases where the unknown frequency and the known fre-

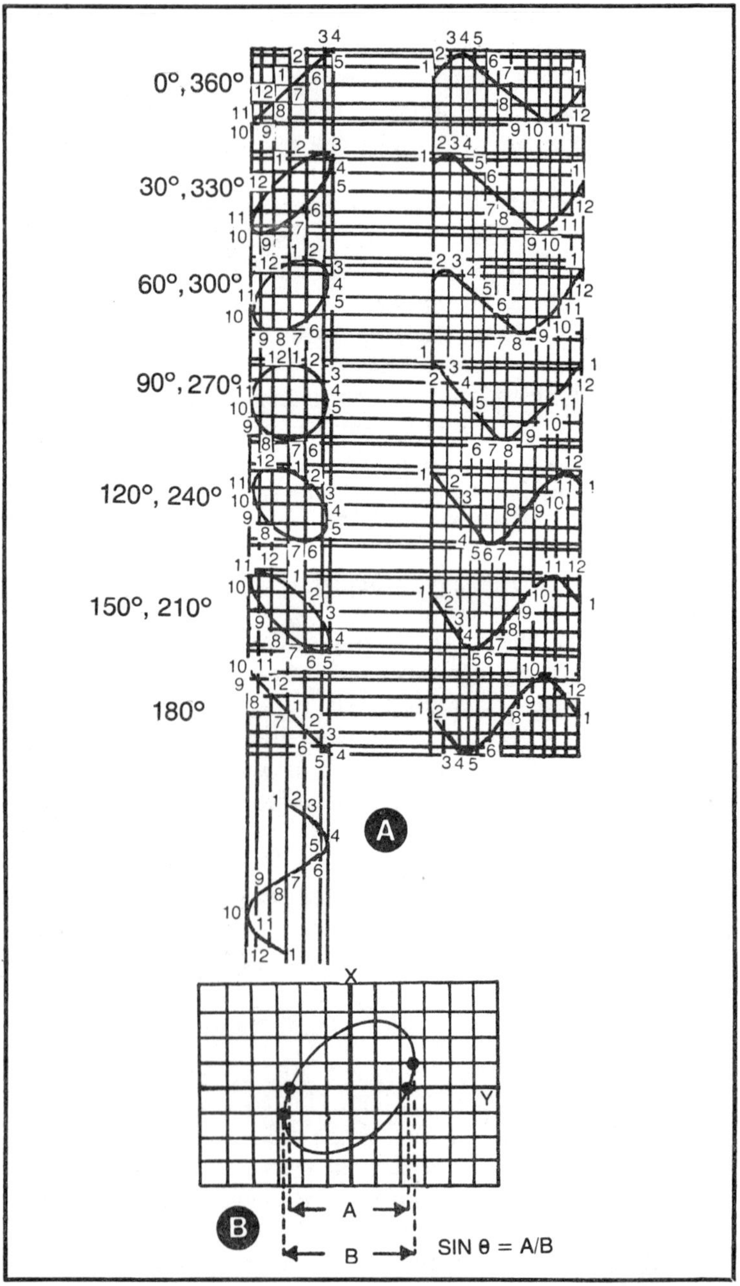

Fig. 6-2. Lissajous figures, (A) for various phase angles (1:1 frequency ratio), (B) Measuring phase angle.

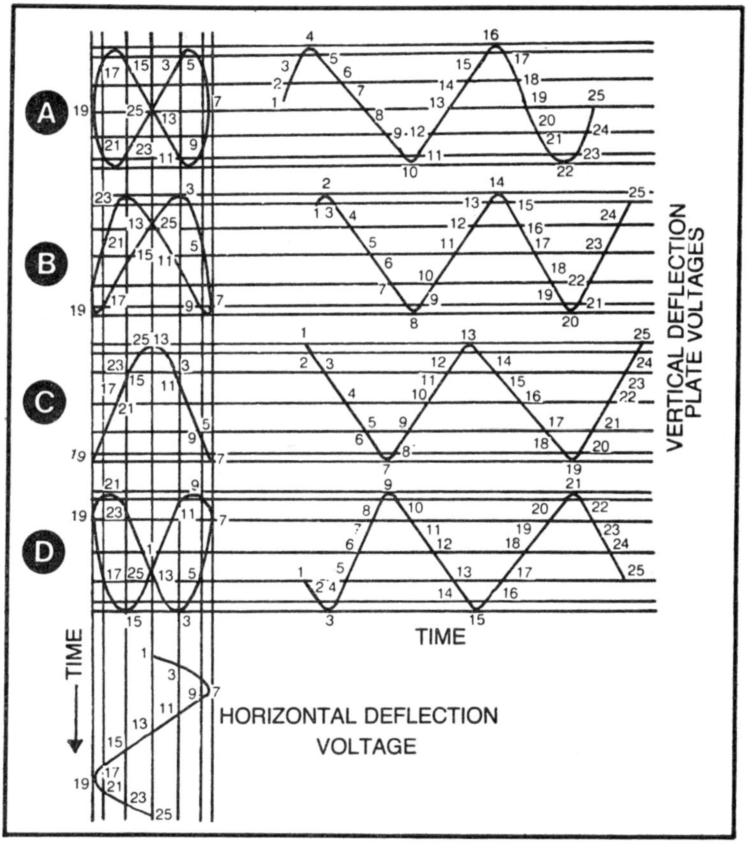

Fig. 6-3. 2:1 Lissajous figures.

quency have harmonic relationship, and the pattern can be locked in to become stable. In short, it is limited to rough determinations of frequency in the low rf- and audio-frequency ranges.

Y-T OSCILLOSCOPES

The Y-T oscilloscope is the type of scope most people identify in their minds at the word "oscilloscope." It is, of course, only one of several types, but is the most common (and to servicers the most useful).

Figure 6-4 shows the basic block diagram of a simple Y-T oscilloscope. The vertical amplifier channel is exactly like those in X-Y models. The horizontal channel, however, is a little different. The input signal to the horizontal amplifier is an internally generated sawtooth.

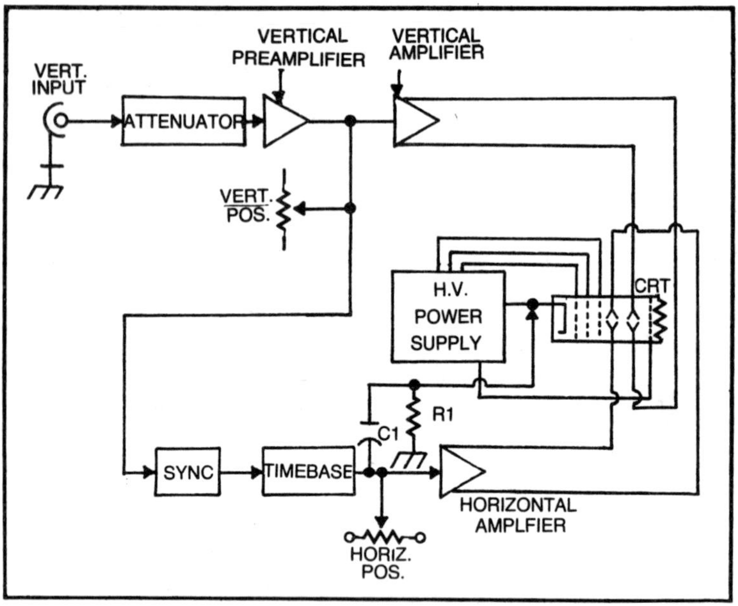

Fig. 6-4. Block diagram of a simple free-running-sweep scope.

Figure 6-5 shows the way the sawtooth causes the waveform of the vertical signal to be traced onto the screen of the CRT. The static potentials on the CRT horizontal deflection plates are ar-

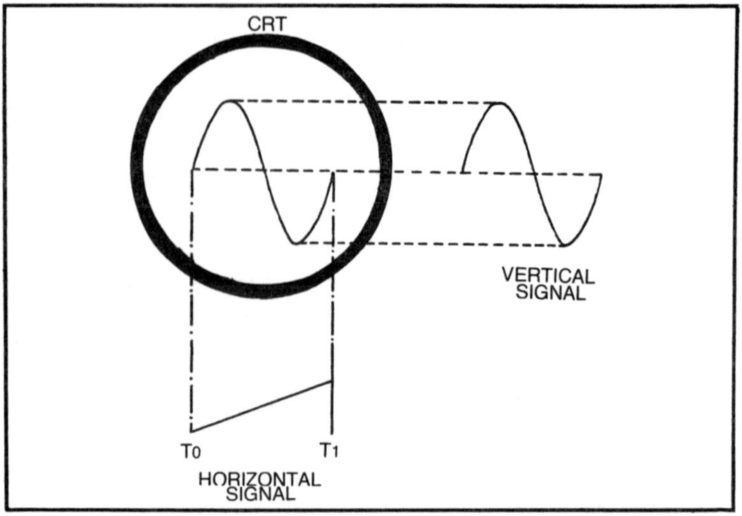

Fig. 6-5. Relationship between sawtooth and vertical signal to the CRT trace.

ranged such that the electron beam will be on the left-hand side of the screen when the horizontal signal (a sawtooth waveform) is zero. At time T_0, however, the sawtooth begins to rise, so the beam will be pushed to the right across the CRT screen. At time T_1 the beam has reached the extreme right-hand edge of the CRT, so the sawtooth waveform abruptly drops back to zero. This will yank the beam back to the left-hand side of the CRT screen. A blanking capacitor (C1 in Fig. 6-4) will create a pulse during this retrace period. This pulse blanks the CRT so that the viewer will not see the retrace line. Some advanced model scopes will use a one-shot pulse generator, that is triggered on the falling edge of the sawtooth, for the same purpose.

There are two basic types of Y-T oscilloscopes, *free-running sweep* and *triggered sweep*.

The free-running sweep is the type used in low-cost oscilloscopes. The horizontal controls will be marked on the front panel in terms of *frequency* (i.e., hertz and kilohertz). The sawtooth oscillator in these models is essentially free-running, although it usually can be synchronized to the vertical signal.

In the triggered-sweep models, the horizontal sweep is controlled by a trigger circuit that can be adjusted through front panel controls to permit the sweep to begin only when the vertical amplitude has reached a certain level. The trigger control will usually be labeled "+," "–," and "0." Some will also have a delay feature that holds off the start of the horizontal sweep for a set number of microseconds. This control will be labeled in terms of

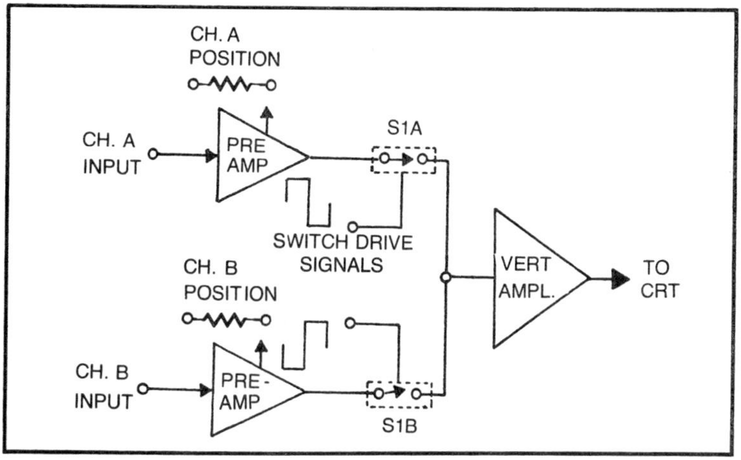

Fig. 6-6. Two-channel electronic chopper.

time. Note that the horizontal-sweep selector of triggered-sweep models will be calibrated in terms of time per division, such as 0.5 ms/cm. If the fine control is set to a certain position, usually indicated by a detent in the knob, then the main selector switch will be calibrated. Many models have a lamp that comes on when the horizontal control is *not* calibrated.

DUAL BEAM

Multiple-trace oscilloscopes usually do not have more than one electron beam because it would be difficult, if not impossible, to use separate deflection systems for each, without causing interaction! The usual procedure is to use an electronic switch to select between the beams. The dual-beam oscilloscope, then, really has a dual trace generated from a single electron beam by rapid switching between them. Figure 6-6 shows an electron switch that provides two traces from a single beam. If the switching signal has a high enough frequency, then the viewer will see what appears to be two completely separate beams!

There are actually two types of dual-beam systems, and both are often found on the same oscilloscope. The *chopped* mode is shown in Fig. 6-7A. The electron beam is deflected by one vertical signal on positive peaks of the switching signal, and by the other vertical signal on opposite peaks of the switching squarewave. Usually, the chopped mode is used on low frequency signals.

As the frequency of the vertical signal increases, however, the *alternate* mode shown in Fig. 6-7B might be more useful. In this system, the electron beam will sweep across the screen, displaying one channel, and then sweep again displaying the other signal.

OSCILLOSCOPE SPECIFICATIONS

There are several specifications on the oscilloscope that are critical to evaluating its performance. The specs to look for will, of course, depend upon the job to be done. The frequency response required by a car-radio servicer or a Hi-Fi technician are a lot less than that required by a digital-computer servicer who deals with 10-MHz squarewave clock pulses!

Sensitivity

The vertical sensitivity, or vertical deflection factor as it is sometimes called, is the specification that tells us how much the electron beam will be deflected on the screen of the CRT by a given signal voltage. The CRT screen has a grid pattern called a graticule, and the grid patterns have divisions 0.75 to 1.3 centimeters apart, although most are 1 cm apart.

The vertical attenuators will be calibrated in terms of volts per division or volts per cm. If the marking is, say, 100 mV/cm, then this means that 100 millivolt signal applied to the vertical input will deflect the beam one centimeter. This is the vertical-deflection factor. The maximum vertical sensitivity is usually quoted in the spec sheet, and is the smallest deflection factor marked on the vertical attenuator knob.

Bandwidth

The manufacturers of oscilloscopes will specify the maximum frequency response of their product, but, like Hi-Fi amplifier power

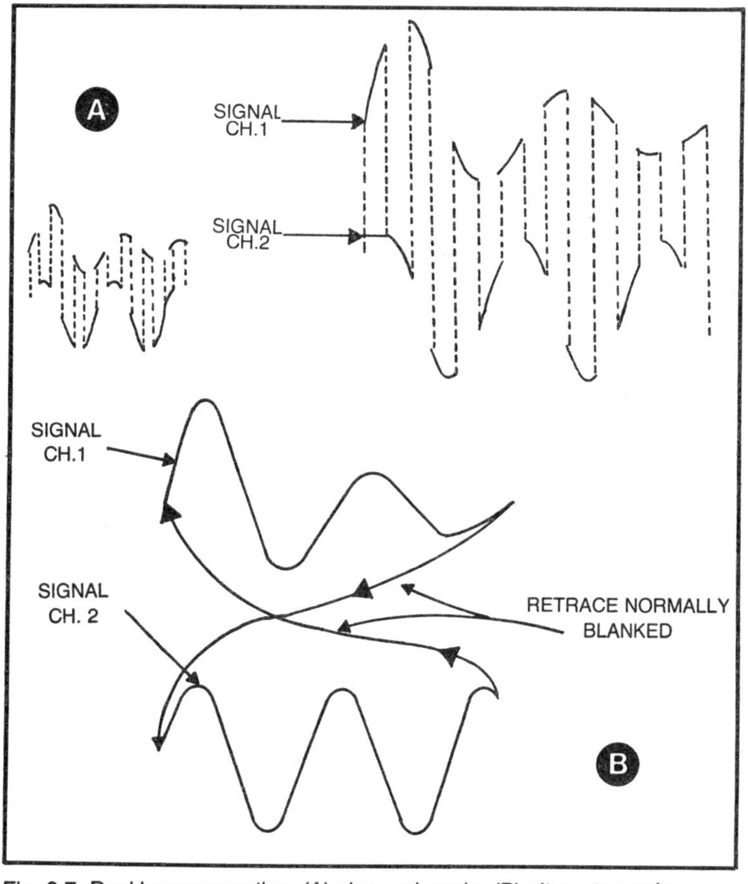

Fig. 6-7. Dual beam operation, (A) chopped mode, (B) alternate mode.

specs, this area has been subject to a little, shall we say, "creative spec writing" by unscrupulous manufacturers.

A typical oscilloscope bandwidth may be 500 kHz, 10 MHz, 35 MHz, etc. Most reputable manufacturers of scopes specify the frequency response as the frequency at which the gain of the vertical amplifier falls off −3 dB from its 1-kHz gain. It is wise to beware of oscilloscopes in which the gain is specified to some other point, i.e., 3.5 or 4 dB, because it is very difficult to make valid comparisons between these models and the others. They may very well be good oscilloscopes, but it is difficult to tell how they stack up against H-P and Tektronix models that are specified to −3 dB.

Also, be a bit wary of the scopes that are specified in the form "down 3 dB at 10 MHz." Some manufacturers will purposely peak up the response of the vertical amplifier at some frequency in the 8 to 20-MHz region so that they can claim the response is 3 dB down at that frequency, therefore implying bandwidth to that frequency. The truth may be quite different. In my color-TV days long ago, we bought a popular oscilloscope that had a supposed bandwidth of 8 MHz, clearly more than we needed (4.5 MHz) for TV work. But the spec was actually "3 dB down at 8 MHz," and we found out the expensive way that the response had been peaked up at 8 MHz to allow the claim. The response at 5 MHz was about 5 dB down! Most reputable manufacturers shun this type of cheating.

Rise time

The rise time of pulse waveforms is defined as the time required for the pulse to rise from 10 percent of its total amplitude to 90 percent of its total amplitude. An oscilloscope used to view that pulse needs a vertical amplifier rise time equal to, or greater than, the pulse rise time. If the manufacturer of your scope does not publish the rise time, use the formula below to calculate the approximate rise time from frequency response.

$$T_r = 0.35/F$$

Where:

T_r is the pulse rise time in seconds

F is the frequency response of the scope in Hz (microseconds and megahertz may also be used. Whenever the magnitude of the units is changed, however, *both* must be changed by the same factor.)

Chapter 7
Signal Generators and Tracers: Which to Use; When; and Where

In a previous chapter we noted that there were two basic approaches to troubleshooting equipment such as audio amplifiers and radio receivers; i.e., signal tracing and signal injection. In signal tracing, a signal was applied to the input, and then a signal tracer or an oscilloscope was used to find the signal at various points in the cascade stages. In the signal-injection method, an output indicator (oscilloscope, AC voltmeter, loudspeaker, etc.) was connected to the output of the receiver or amplifier, and then a signal from a signal generator was applied to each stage in succession, until one was found that would not pass the expected level of signal.

In the signal-tracing approach we troubleshoot from the input and work toward the output, while with the signal-injection technique we work from the output toward the input. The two techniques are opposites of each other. Not surprisingly, there are adherents to both methods. Even amongst the majority of professional servicers who use both methods, any given servicer will show a preference by which method is selected *first*.

In this chapter we will consider both methods, signal tracing and signal injection, and try to give you some insight on which to use when and where.

The first step in the process of learning the two techniques is to become more familiar with the various types of instruments used in each. Some instruments are commercially available at reasonable prices, or may appear on the Amateur auction and hamfest market in used, but usable, condition from time to time. In still other cases, it is possible for the enterprising Amateur or electronic hobbiest to homebrew some of the devices (especially signal tracers).

SIGNAL TRACERS

There are a number of different types of signal tracers. About the simplest and cheapest of these is probably a simple, old fashioned, pair of high-impedance earphones. Although not too common in consumer electronics shops (radio, TV and stereo) or two-way radio shops, they are sometimes considered standard equipment in shops that service intercom and public-address amplifiers.

While they are certainly the simplest, and possibly the cheapest, type of signal tracer, a set of earphones does have limited usefulness. For example, the drive requirements are enormous, compared with other types of signal tracer. Certain low-level signals, the output of some AM detectors, for example, might not be audible in the earphones unless amplified.

Another difficulty is that the usual range of impedances for "high impedance" earphones is on the order of 1000 to 10,000 ohms. This may be considered "high" impedance to a loudspeaker circuit (typically 4-16 ohms), but is almost a short circuit to a MOSFET gate or vacuum-tube grid.

One way to eliminate the problems of the earphone signal tracer is to drive the earphones with a high-gain audio amplifier which has a very high input impedance. However, a loudspeaker can also be driven by such an amplifier, thus eliminating the need to don the earphones before beginning work.

If the amplifier used in the signal tracer is a wideband type, then it may also be used as a general-purpose oscilloscope preamplifier. Most commercially available signal tracers, whether ready-built or in kit form, are little more than an audio amplifier with certain mechanical adaptations. Many units offer additional features such as external access (via banana jacks or binding posts) to the loudspeaker voice coil and the output transformer primary winding.

Figure 7-1 shows the block diagram of a highly satisfactory "homebrew" signal tracer made by a technician in an automobile radio shop in Virginia. It was made from a "surplus" Philco-Ford auto radio. The unit is cheap, simple, and, I can add from personal experience, highly effective.

The same idea as in Fig. 7-1 can also be implemented by using a table model radio, many of which sell for peanuts at hamfests and

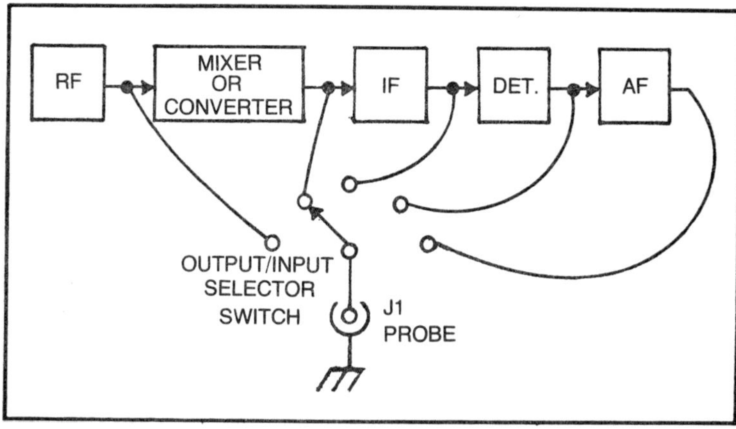

Fig. 7-1. Simple signal tracer made from a converted radio receiver.

auctions. You are cautioned, however, that these types of radios contain a potentially lethal trap for the unwary! Many of them (most, in fact!) are AC/DC models. This means that the AC mains power is rectified to form the B+ operating voltages. The AC/DC technique is even used in some transistor models, the audio output transistor being a high-collector-voltage type, and then the low voltages are derived from a Zener diode and resistor. AC/DC radios are lethal because the ground is one side of the power line. If you are 100 percent sure that the AC power plug is always plugged in the right way, and that all AC power receptacles are *correctly wired*, then there is not any danger because the AC neutral wire will always be on the ground side of the power supply in the radio. But this CANNOT EVER be guaranteed!

Only those radios that operate from batteries, such as portables and car radios, or those table models that operate 100 percent from transformer power (some sets used a low voltage transformer for all stages except the audio output, and that was AC/DC powered!), are candidates for making a homebrew signal tracer.

The technique or idea shown in Fig. 7-1 was not new when that Virginia radio shop built the device. No indeed! The same basic configuration was used in signal tracers made before and immediately after World War II. A similar set, which, by the time that I saw it, was old enough to vote, was marketed under the brand name of *Autolyzer*. In more modern times, the *transistor radio analyst* by B&K Dynascan of Chicago, is a modern version of the autolyzer idea, paired with a transistor checker and a signal generator for the AM broadcast band and common intermediate frequencies.

Unfortunately, this technique seems to be largely forgotten by modern test gear manufacturers. Perhaps in our zeal to adapt highly sophisticated new technologies we tend to forget older ideas that still work nicely. After all, what could be simpler? The signal tracer of Fig. 7-1 is merely an AM radio equipped with a probe that can be moved to any stage in the radio. It can be used as either a signal source or signal tracer, depending upon whether the servicer decides to disable the stages preceding the one selected!

SIGNAL GENERATORS

Signal generators also come in a wide variety of models and types. Unlike the signal tracer, however, there is a wide variety of quality in signal generators. In fact, the range of quality is so broad that you are cautioned against many of the "low end" models. . . they are essentially useless for servicing (or much else, for that matter).

The main hallmark of the useless signal generator is poor shielding. Poor models use only a loosely fitted cabinet as "shielding," while all high-grade models use good, tight cabinets and a technique called double (or triple) shielding. In these models, the oscillator and any on-board amplifiers used are contained within an rf-tight shielded box inside of the cabinet. Generally, the models that use a microphone connector as the rf output are poor, while those that use a BNC or SO-239 UHF connector are better. Note, however, that these rules of thumb are far from absolute!

The principle problem with signal generators that are not properly shielded is that rf "leaks" around cabinet flanges. In fact, with sensitive 2-meter FM rigs, there might be enough signal from leakage to drive the receiver into hard quieting! Hardly a condition conductive to troubleshooting and alignment!

One of the simplest to use, and least expensive, signal source (one hesitates to call it a signal *generator*, although that is what it is) is the so-called *noise generator*. These devices are nothing more than a 1000-Hz squarewave or sawtooth generator that has harmonics strong enough to be heard well into the high-frequency range. They are not to be confused with the white-noise generators discussed in later chapters, in which a wideband signal is created by reverse biasing a pn junction of a transistor or semiconductor diode.

The 1000-Hz oscillator will have harmonics easily discernible in AM broadcast band receivers and I-F amplifiers into the megahertz region, but generally fall off in usefulness as the frequency of the receiver of I-F amplifier approaches 10 MHz.

Another simple type of signal generator consists of a transistor or IC oscillator that uses resonant quartz crystals to set the frequency. Simple types, in which the technician plugs in crystals for the desired frequency, have been perpetually popular as construction projects in amateur radio and hobby publications that cater to the newcomer. One such unit I once built had crystals at 100 kHz, 262.5 kHz (the standard AM car-radio I-F), 455 kHz, 460 kHz (used on European AM radios), 500 kHz, 1000 kHz, 9 MHz, 10 MHz, 10.625 MHz, 10.675 MHz, 10.7 MHz, 10.725 MHz, and 10.775 MHz. The 100-kHz crystal can be used as an alignment marker, while the other VLF crystals were for I-F amplifiers often encountered in those days. The 500- and 1000-kHz crystals provided markers well into the shortwave spectrum, even up to 10 meters (30 MHz). The 500-kHz, 1-MHz and 10-MHz crystals can also be used to zero beat against WWV to form relatively precision frequency sources. The 9-MHz crystal serves two purposes, it is the standard I-F used in many Amateur SSB transceivers, and it also provides (via harmonics) convenient markers in the FM broadcast band. The harmonics fall at 90, 99, and 108 MHz, providing both midband and bandedge points for rapid alignment of the FM local oscillator!

The 10.7-MHz crystal is, of course, the most common I-F frequency in both broadcast and communications FM receivers. The center of the FM passband is 10.7 MHz. The other frequencies clustered about 10.7 MHz (i.e., ±25 kHz and ± 75 kHz) are used instead of a sweep generator in the alignment of FM broadcast receivers (see also Chapter 25). Of course, for amateur and commercial two-way radio receivers which use 5-kHz deviation, crystals closer to 10.7 MHz are needed.

Variable-frequency signal generators can be classified as audio, rf, and function. Note that the function generator possesses many of the attributes of both audio and rf generators, but more nearly resemble audio generators except as to frequency range.

Audio generators are generally classified as those that produce output frequencies between 20 and 20,000 Hz, although some operate as low as 1 Hz and as high as 100 kHz. The typical audio signal generator produces both squarewaves and sinewaves, and will have a precisely metered or calibrated output attenuator. The function generator produces sinewaves, squarewaves, and triangular waves in almost all models, and some also produce nonsymmetrical square pulses and sawtooth waveforms. Most function generators operate from less than 1 Hz to at least 100 kHz, with many models available that produce output frequencies up to 1 or 11 MHz. Both audio and function generators have 600-ohm standard output impedances, although the function generator does not usually have a well calibrated output control.

Most rf signal generators cover a wide frequency range, except, of course, for those models intended for a special purpose. For simple troubleshooting, a so-called "service grade" rf signal generator is sufficient, but for alignment and certain critical troubleshooting applications a "laboratory" or "communications shop" grade instrument is required. I personally prefer the better instruments for all applications because there are too many false results with the low cost instrument. Service and hobby grade instruments sell new for less than $200, while the higher end service-grade instruments sell for $200-300.

But, no matter how you cut it, or how much money you pay, the specifications of the typical service-grade instrument do not stack up well against the specifications of the laboratory-grade instrument. Such generators offer almost no leakage around the equipment flanges or cabinet seams, and the output calibration is a bit more accurate than the calibration of service-grade instruments. These instruments sell new for prices in the $500 to $5000 range, with most being in the "over-1-kilobuck" end of the scale.

I personally object to the designation "laboratory" grade, because it conjures up a picture of instruments that only the engineering staffs of large manufacturing and R&D companies can afford, or know how to use. This is far from the actual case. The instruments shown in Fig. 7-2 are so-called laboratory-grade signal generators, but may often be found in commercial two-way radio shops.

It is quite possible to buy older vacuum tube lab grade signal generators for very reasonable prices at hamfests, auctions, and through the classified sections of Amateur Radio publications. Look for models such as those by Hewlett-Packard, Boonton Laboratories, Marconi, and Measurements Corporation. The Boonton 202 (later H-P 202 when H-P bought out Boonton) and the Measurements model 80 (and the military TS-497) are exceptionally good buys. The model 80 covers from 1 to 400 MHz, with a calibrated 50-ohm output attenuator covering output levels of 0.1 to 100,000 microvolts.

OTHER TYPES OF SIGNAL TRACERS

Before moving on to the rest of our discussion for this chapter, we must spend a moment considering two other forms of "signal tracer" instrument: oscilloscopes and multimeters.

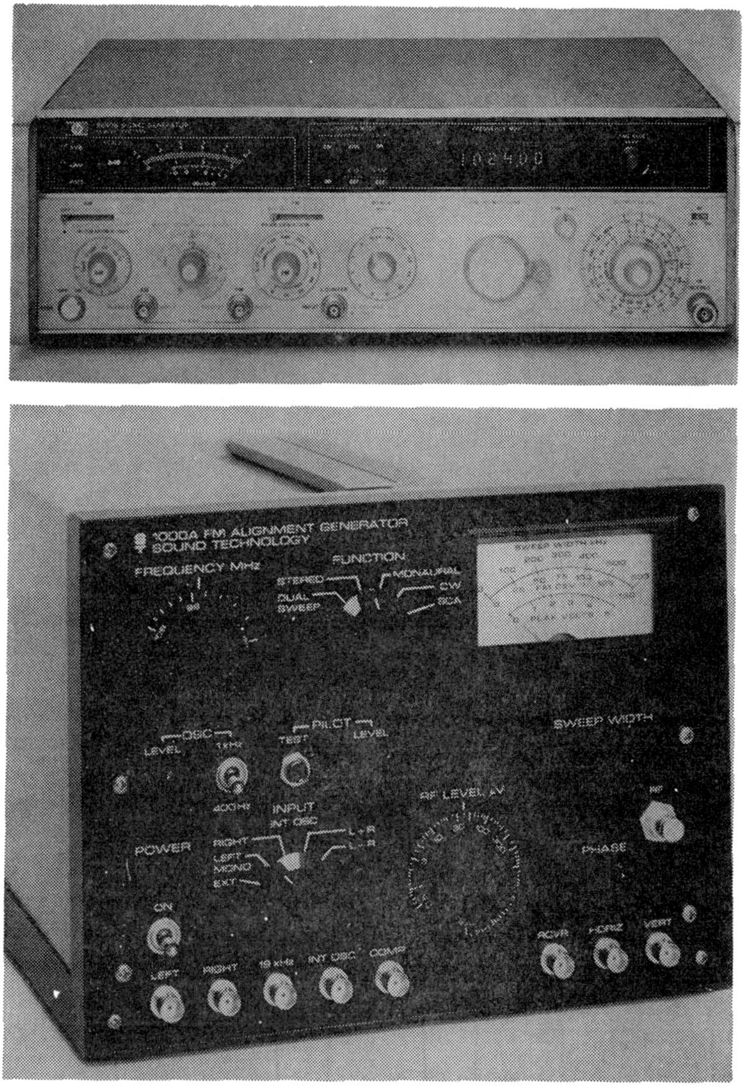

Fig. 7-2. Commercial high-grade rf signal generators.

The oscilloscope is one of the most powerful signal tracing instruments available. In fact, the scope is generally considered among professionals as the single indispensible instrument! It is a shame that more amateurs will not invest in even a used oscilloscope. The oscilloscope can view the rf signal directly if the vertical preamplifier bandwidth is sufficient, and will view the modulation if a demodulator probe is used.

The multimeter can be used to trace both AC signal voltages and DC levels through the stages of a receiver or transmitter. In fact, the famed Motorola "test set" used to troubleshoot two-way radio VHF/UHF FM equipment uses a dc microammeter to measure the activity at the grid or base of the stages in the transmitter and receiver sections. Any stage that derives all or part of its dc bias by the grid- or base-leak methods can be checked by measuring the DC level at the grid or base. If every stage in the device uses a grid-leak bias, then it is quite possible that only a VTVM or related instrument will be needed to troubleshoot the equipment!

If the VTVM is equipped with an rf demodulator probe in place of the normal DC/AC/ohms probe, then we may use the instrument to measure rf voltages, and trace them from stage to stage. As a matter of fact, some CB transceiver manufacturers specify such a probe and VTVM combination for both alignment and troubleshooting operations. One manual, for example, calls for 3 volts p-p at point "X" with 1 microvolt of unmodulated, on-channel, signal applied to the antenna terminals. Manufacturers who bother to publish this type of specification immeasurably aid the troubleshooter, especially those who use instruments that are not terribly effective at the frequencies involved.

Troubleshooting

Signal Tracing. A signal tracer is used to sample the signal at various points in the equipment being serviced. Start at one end of the radio and work in logical sequence toward the other end. Most texts state that signal tracing should start at the rf amplifier. Isolation of the dead stage is accomplished by noting at which point the signal no longer exists. This will tell you which stage is no longer capable of passing a signal. DC analysis of the voltages within the stage, or perhaps some other technique, is then used to find the defective component.

Signal Injection. When using the signal-injection method, start at or near the output stage and work backwards toward the antenna. The idea is to substitute a signal from a signal generator for the normal signal. Again, as in signal tracing, the point where the signal fails to go through indicates where, if not what, the difficulty is.

Combination. A combination method involves the use of a signal tracer at the output stage, or a chain of stages, and a signal generator at the input. Such a set up, as shown in Fig. 7-3, can be used both to measure stage gain and the frequency response of the set, as well as for troubleshooting.

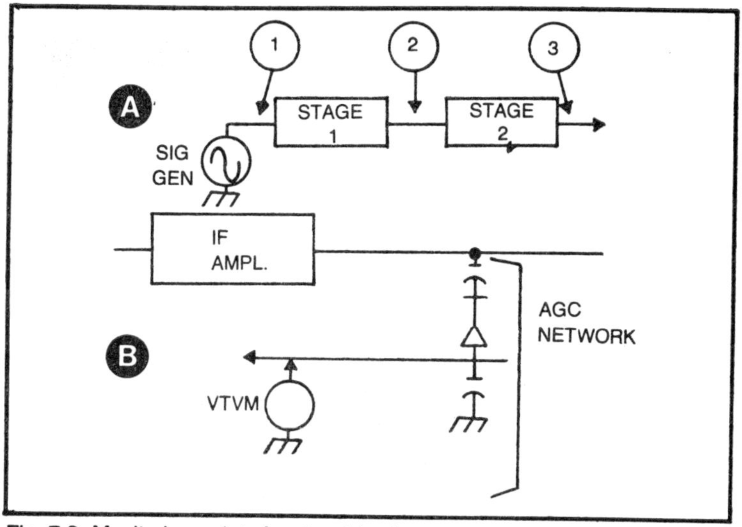

Fig. 7-3. Monitoring points for signal tracing.

Which? At first glance, these procedures appear deceptively simple and universal. An inexperienced person might ask whether or not either might be used, at the servicer's will. The only factor that seems to govern this choice is the availability of the instruments. In many cases this is true. Just as often, however, there is a good reason to prefer one over the other.

In some circuits, for example, the normal signal-voltage levels may be too low to use a signal tracer. The input of an rf amplifier in a receiver, especially the rf transformer tap leading to the base of a transistor, is such a point. Another is the tap from the input I-F transformer to the base of the first I-F transistor. I suppose that an awful lot of first I-F transformers have been replaced over the years because of this phenomenon.

Another effect of low signal levels is that the residual hum and noise pick up might be sufficient to drown out even some relatively healthy signals. The high gain needed to amplify the signals to a usable level will also amplify the hum and noise. The unshielded probe found on many tracers does little to alleviate the situation, and in fact, often serves to aggravate it.

At this point you may be on the verge of thinking that it would be wise to just go ahead and use the signal generator at all points in a receiver. It would certainly be a lot easier if we only needed one, universal, test procedure for all cases. Unfortunately, though, all signal generators have a low impedance output. They range from 50 ohms for rf-signal sources to 600 ohms for audio-signal generators and function generators. There are some circuits that simply cannot be tested with a low-impedance instrument. In certain others the test results are less valid because of the low-impedance loading by the generator. The 1 megohm (or higher) input impedance loading by the generator. The 1 megohm (or higher) input impedance of a signal tracer, on the other hand, produces significantly less circuit loading.

The main criteria, therefore, seems to involve impedance levels and signal strengths. In most cases these two parameters can be used to determine which technique and instrument to use.

APPLICATIONS

In this section I will draw upon many years of troubleshooting experience, and try to show by example some different situations. Note that not all will be strictly Amateur Radio equipment, but some consumer equipment as well. This policy is followed on the theory that an Amateur who can troubleshoot a shortwave communications receiver or SSB transceiver could also service the family stereo.

Tape Players and Recorders

Many tape units are self contained, while others are designed for use with external audio amplifiers. If self-contained models are being used, then the choice is pretty much up to the servicer, except back in the early stages near the pickup head. The output signal of the tape head is very small, on the order of millivolts, so many tracers will not be able to "see" the signal properly. If the tape player is a type that must be connected to an amplifier, then either signal tracing or the combination method discussed earlier is indicated, because there is no output device (i.e., loudspeaker) to indicate the presence of the injected signal.

FM Stereo Alignment

Alignment of an FM stereo multiplex circuit can be accomplished off the air from a strong station, but only in the cheapest of receivers. In anything like a "hi-fi" receiver, a proper stereo signal generator is needed. If an off-the-air signal is used, then the alignment will probably be less than ideal. Fig. 7-4 shows the equipment set-ups typically used for stereo alignment.

Audio Amplifier Stages in Radio Receivers

The audio amplifier stages in most radios are not particularly bothersome to most service technicians. A good understanding of

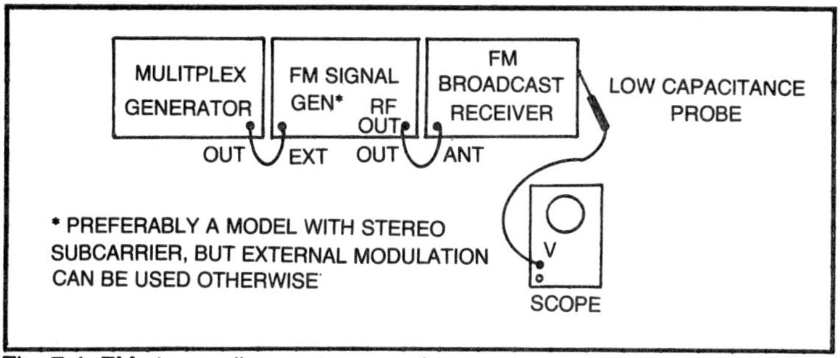

Fig. 7-4. FM-stereo alignment test-equipment set-up.

test equipment and procedures, however, can make the job a lot easier. Even an inexpensive signal tracer can usually pick up any signal that might be present, even at the input of the first stage in the amplifier chain. These signals, fortunately, are almost all of high level. At the plate or collector of this stage, however, it is possible for the signal tracer to produce a false indication. Because this stage provides power amplification instead of voltage amplification, it is possible for the signal to sound strong from the tracer, but weak from the loudspeaker.

If the stage is equipped with an emitter (or cathode) bypass capacitor (Fig. 7-5), then you can use the tracer to determine whether or not it is open. An open bypass capacitor can reduce the gain of the stage without significantly altering the dc voltages on the tube or transistor. The symptoms of an open bypass capacitor are reduced signal at the plate or collector, but a relatively strong signal at the cathode or emitter. This can be a little misleading. In at least one case, the same symptoms occurred with zero plate voltage. It seems that the screen grid of a pentode power amplifier tube acted as the plate, and the cathode bypass was too small. As a result, there was a sufficiently strong signal voltage on the capacitor to sound very loud in the tracer.

Be careful when interpreting the results of signal injection in the output stage. If you depend on the audio jack of a standard rf-type generator, it might be insufficient. Many audio stages require several volts of signal before they will produce any significant amount of output power. Because of this requirement, it might be wise to have on hand one of the audio signal generators which can produce an output of up to 10-volts rms.

Either the signal tracer or oscilloscope is a good choice for the audio driver or preamplifier stages. A signal generator also can produce good results. On the output side of the preamplifier, how-

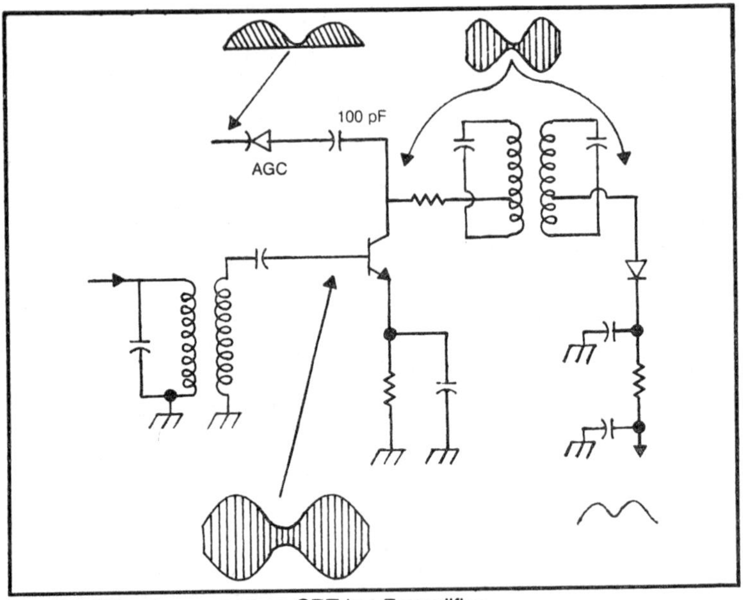

Fig. 7-5. Waveforms seen on CRT in I-F amplifier.

ever, it will, in effect, be driving the final power amplifier. A low-output-level signal generator will, as discussed above, produce insufficient output signal, and this leads to false negative results.

I-F Amplifiers

There is often good reason for preferring the oscilloscope over other instruments when troubleshooting the I-F amplifier stages in many solid-state receivers. For example, the voltage at the tap (base connection) of the secondary of the first I-F transformer is too low to drive the standard audio amplifier/demodulator probe type of signal tracer. At the plate or collector of most I-F amplifiers there is usually plenty of signal available for the tracer.

To provide plenty of sensitivity, a scope should have a —3 dB bandwidth of at least the frequency being viewed. This means 500 kHz for most radios with I-Fs in the VLF region, and up to 15 MHz in shortwave and communications receivers. Some authorities maintain that the scope bandwidth should be twice the frequency being viewed, but I feel this is excessive.

Another good reason to use an oscilloscope is that the automatic gain control (AGC) detector, as well as the AGC sampling capacitor, can be checked for proper operation. The pattern will be the same as in Fig. 7-5.

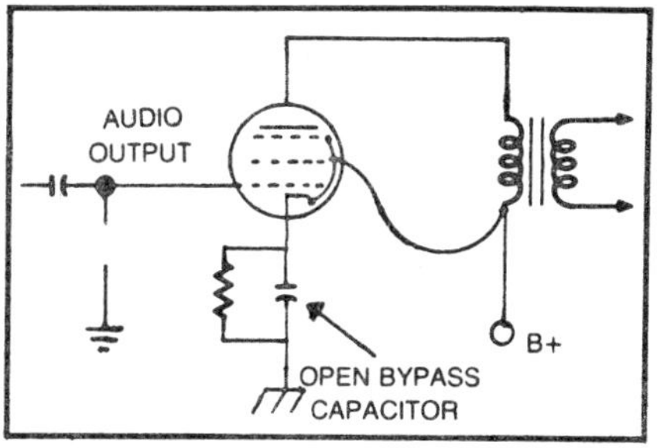

Fig. 7-6. Open cathode bypass capacitor causes weak or dead output.

A signal generator used in the I-F stage must either be of the noise generator variety or it must be dialed to a specific frequency. While the accuracy of the test isn't affected by having to "dial up" the correct frequency, it does take a little more time than does using the noise generator.

Converters and Oscillator/Mixer Combinations

The converter or an oscillator/mixer combination can cause more than its share of headaches when troubleshooting. An audio amplifier type of tracer will show whether or not a modulated signal is passing through the stage, but that is about all that it will show! An oscilloscope, on the other hand, will show a lot more. It can, for example, show whether or not the oscillator is working. If modula-

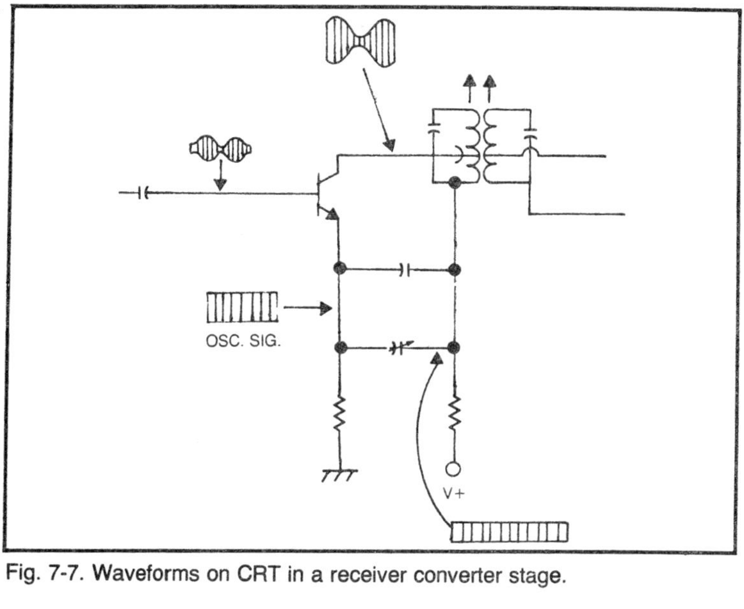

Fig. 7-7. Waveforms on CRT in a receiver converter stage.

tion shows up on the waveform at the output, we also know that proper mixing action is taking place. If it is properly calibrated, the scope will also tell the precise amplitude of the input, output, and oscillator signals (Fig. 7-7). Again, if the scope is properly calibrated, you can make rough frequency measurements without the signal generator by comparing the number of horizontal divisions per displayed cycle with the time-base calibration.

Rf Amplifiers

The weak signal levels in the rf amplifier often will cause a standard signal tracer to be useless; it will produce many false indications in which a stage is pronounced "dead," when it is actually in good working order. For this reason, it is often necessary to resort to either a signal generator or an oscilloscope (with very good sensitivity). Do not be too surprised if your signal generator actually shows a slight loss when switched from the output of the solid-state rf amplifier to the input. This is due to the difference in impedance, and is quite normal in stages employing bipolar transistors.

Chapter 8

Probes and Connectors

The best test equipment is of little use unless it is connected to the equipment or circuit being tested or measured. A probe is a device used to connect the measurement instrument, such as an oscilloscope or voltmeter, into the circuit where the measurement is being made. The simplest probe is a pair alligator clip leads, and certain complex probes may have VHF and UHF amplifiers in-line. In short, the probe may be very simple, or very complex.

We will also consider in this chapter certain pieces of hardware used to interconnect equipment. Most of these fall into the general category of connectors. We also will cover when and where the different types of probe and connector are used.

SIMPLE TEST LEADS

A test lead may be a length of regular hook-up wire with alligator clips on each end, or it might also be a special probe with a costly end-tip. These are often found on oscilloscope probes, and are used to probe in and around live circuits. A hook-tip probe will make it safer when you work on live circuits, both for you and for the components in the circuit. Alligator clip leads tend to fall off, or slip, and short-circuit when being used, and that can destroy transistors and integrated circuits in a hurry.

You can buy a special type of "test lead wire" from electronic wholesale outlets. It is well insulated (500 to 1000 volts, depending upon type), yet is flexible enough to allow easy use in practical

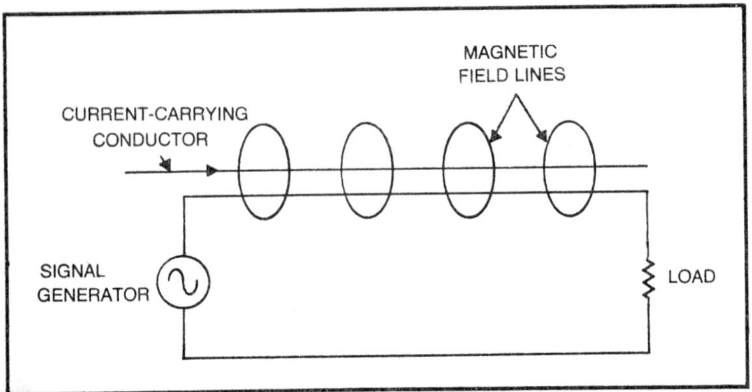

Fig. 8-1. Nearby current-carrying conductor induces currents into test-lead wire.

situations. Although red and black are commonly stocked, several other colors are also available.

The simple test lead is found most frequently in AC and DC voltmeter applications, and in certain other cases where the AC frequency is low. At higher frequencies, the distributed capacitance and noise pick-up are too great to allow easy measurement.

Figure 8-1 shows how interference can occur when open test leads are used, especially in high-impedance circuits. Note that one of the test leads is grounded, while the other carries a signal from a generator or other source. A nearby current-carrying conductor generates a magnetic field, and the flux from this field cuts the signal wire, inducing a signal. Note that nearby wire need not be actually in close proximity to the test leads. The power wiring in a building, for example, will easily induce quite a large signal into the test leads. If you doubt this, try placing your finger on the input terminal of an oscilloscope. Note that up to several *volts* of deflection can be obtained at times!

Figures 8-2A and 8-2B show the result of 60-Hz interference on a sinewave signal. These waveform photographs were taken in a circuit similar to Fig. 8-1. Figure 8-2A shows a 100 kHz, 100 mV signal as it should be, while Fig. 8-2B shows the same signal with superimposed 60-Hz interference.

We must also be aware of two additional factors when using test leads; signal amplitude and circuit impedance. If the amplitude of the signal is sufficiently high, then the distortion caused by interference may be small enough that we can ignore it. But the source and load impedances in the circuit affect the amplitude of the interfering signal. We can observe several volts of signal on the CRT screen of the oscilloscope, as mentioned above, because the circuit impedance is very high, typically 1 megohm. In the example of Fig. 8-2, the signal amplitude is relatively low, but then again, so is the impedance (typically 600 ohms from a 100 kHz signal generator). A much smaller signal, therefore, is displayed on the CRT.

Note that it is a current that is induced into the signal wire from the field surrounding the current-carrying wire. There is a rule of thumb that hum pick up will amount to approximately 10 mV per inch, but this is a bit unrealistic because of the strong dependence of the phenomenon on impedance.

A solution to the hum problem that sometimes works relatively well is the use of a twisted pair of lead wires, as shown in Fig. 8-3. These test leads may be purchased already twisted, or be made from ordinary hook-up wire. In the latter case, the two ends of the pair of wire are inserted into the chuck of an electrical drill, and then the opposite ends are anchored. When the drill is turned on, its rotation will start the twist, and make it as tight as you desire.

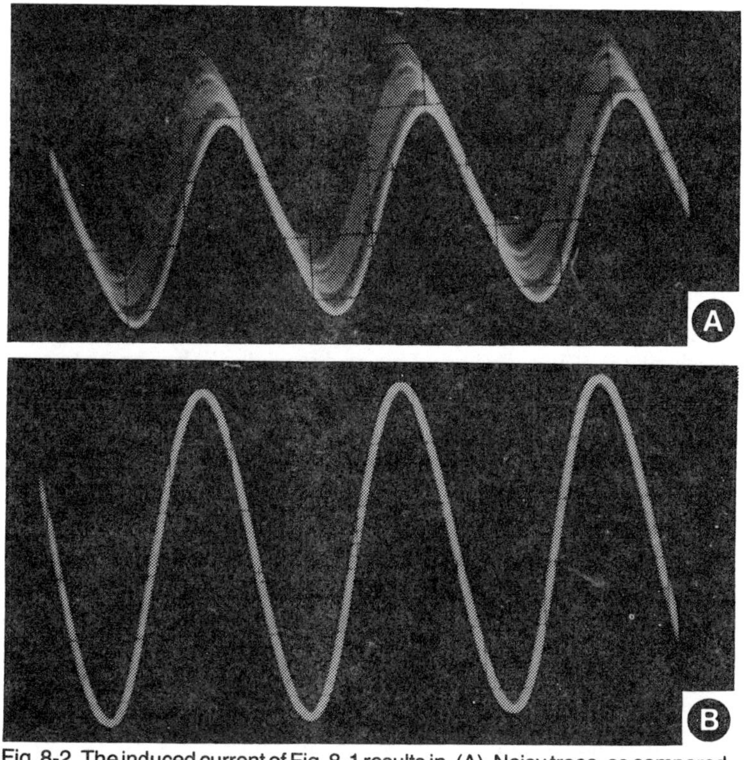

Fig. 8-2. The induced current of Fig. 8-1 results in, (A), Noisy trace, as compared to, (B), Normal trace.

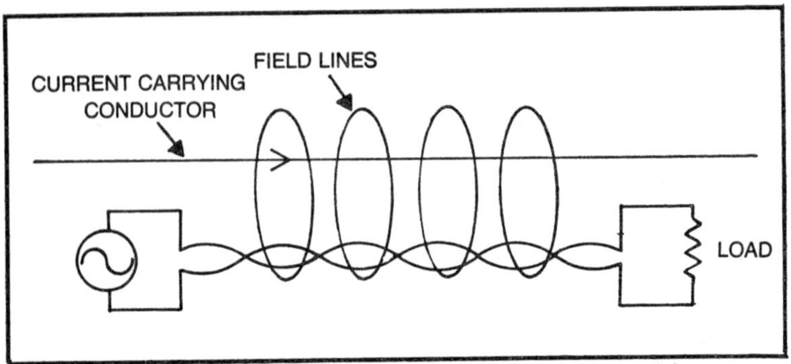

Fig. 8-3. Twisted wires offer some immunity to noise and hum pickup.

Some care is required, however, for this job to be done in relative safety. If one end of the wire comes loose, it will whip around and cause a painful sting. Damage to eyes is also a possibility.

The reasons why the twisted pair are effective in reducing hum are a) the magnetic field affects both conductors equally, and b) self-shielding takes place. The self-shielding effect is noted particularly in unbalanced circuits in which one of the two conductors is already grounded.

The criteria for using either open test leads or a twisted pair are the same, namely: high signal amplitude, low source and load impedances, and low AC frequency or DC signals.

SHIELDED CABLES

Shielded cables are both the best solution, and the worst cause of trouble, in certain hum-prone circuits. A shielded wire is a conductor that is surrounded by, and insulated from, another conductor; in most cases the outer conductor is in the form of a braided cylinder. Coaxial cable, often used as a transmission line in Amateur Radio receiver and transmitter antennas, meets this description, so is often used as shielded wire. Note, however, that shielded wires are not always usable as transmission lines!

The coupling between close-proximity conductors is both inductive and capacitive. Capacitor C1 in Fig. 8-4 is the capacitance between the shield and a nearby current-carrying conductor, and represents a path for interfering currents. Capacitor C2 in Fig. 8-4 represents the capacitance between inner and outer conductors.

There are two main reasons why shielding will reduce interference in some systems. One is that C1 and C2 are effectively in series with each other, so the total capacitance of the system

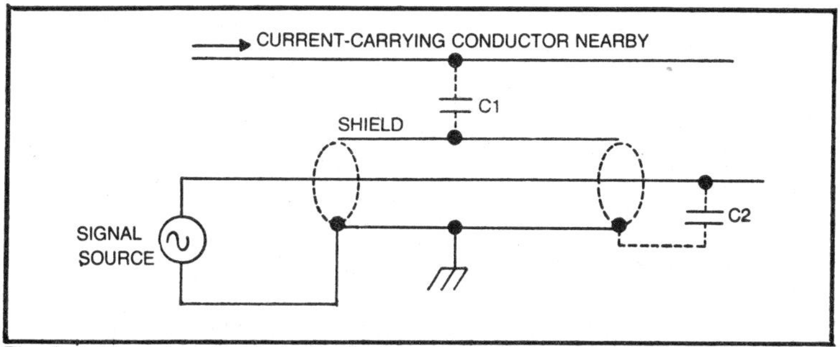

Fig. 8-4. Use of a shielded wire for hum protection.

becomes quite small. In unbalanced systems, such as Fig. 8-4, we also find that one end of each capacitor is grounded, i.e., at a point of zero electrical potential. See Fig. 8-5. The interfering potential does not usually get into the signal circuit.

CONNECTORS TYPES

It seems that there is an almost infinite variety of connectors used to interconnect instruments. It always seems that the probability of finding two pieces of equipment that will interconnect with the same connectors is approximately zero. As a result, all electronic repair shops and laboratories seem to have a large stock of adaptor cables used to interface various instruments and pieces of electronic bric-a-brac.

Figure 8-6 shows several different types of connectors often encountered in troubleshooting electronic equipment. In Fig. 8-6A

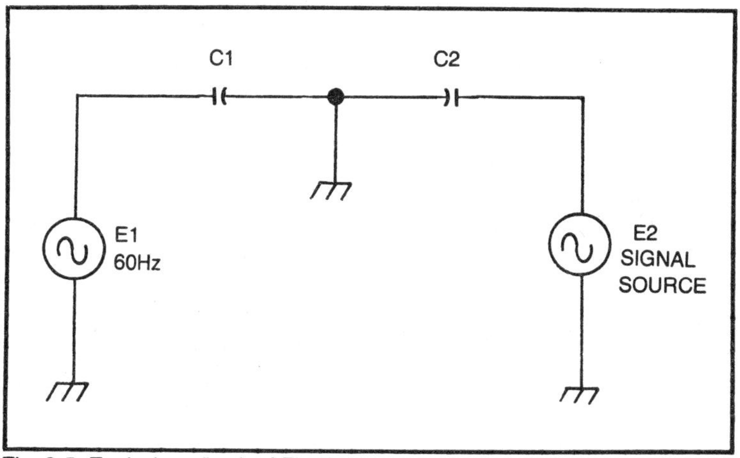

Fig. 8-5. Equivalent circuit of Fig. 8-4.

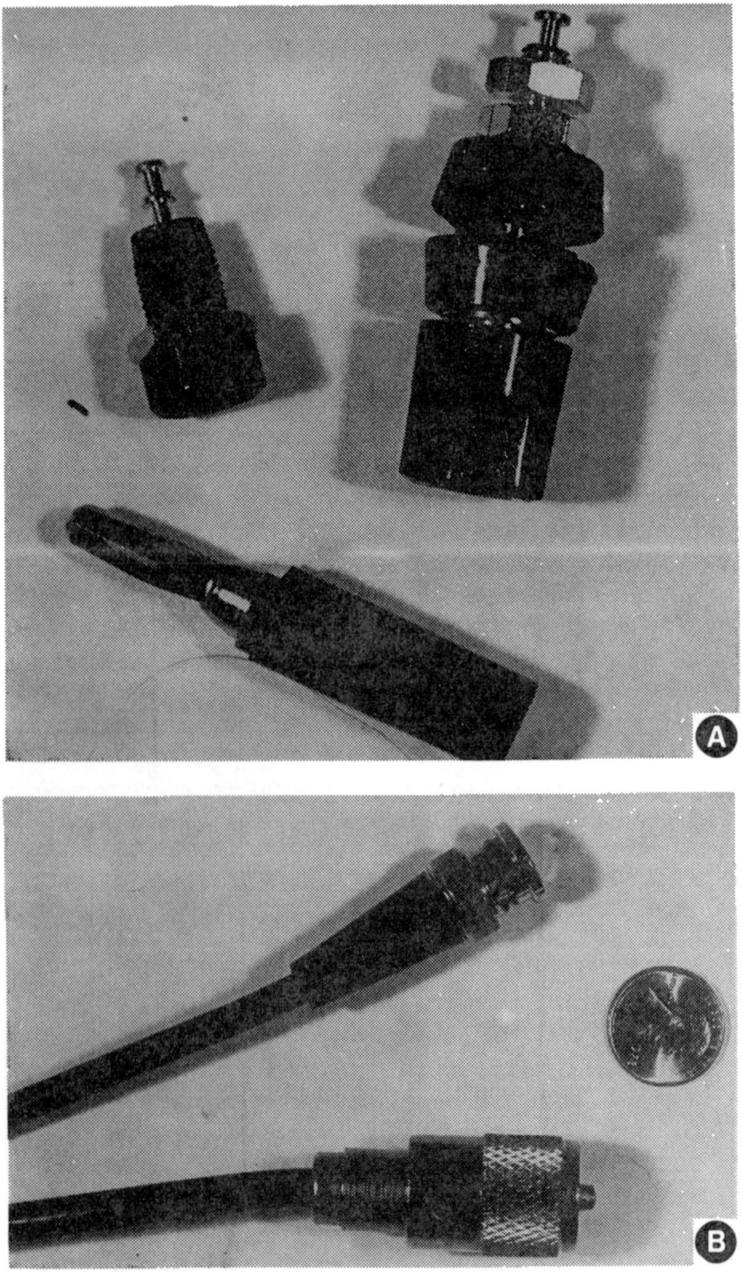

Fig. 8-6. Some common plugs and connectors. (A), Banana plugs and jacks; (B), BNC and UHF (PL-259); (C), Two types of BNC Connector; (D), BNC female connector-to-alligator-clip adapter.

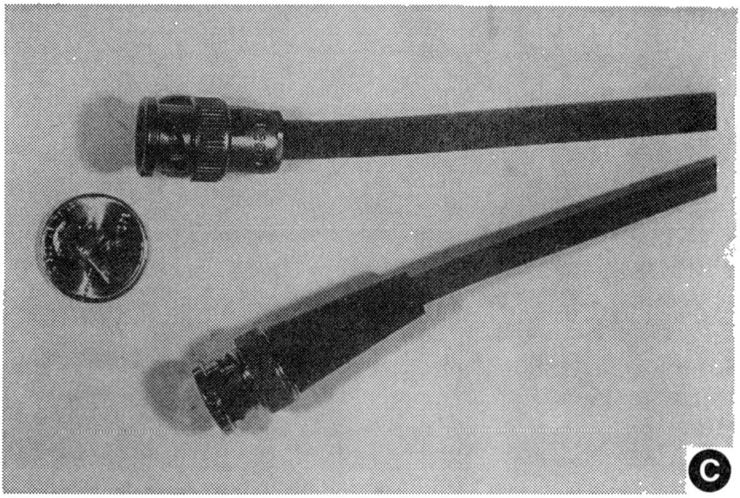

C

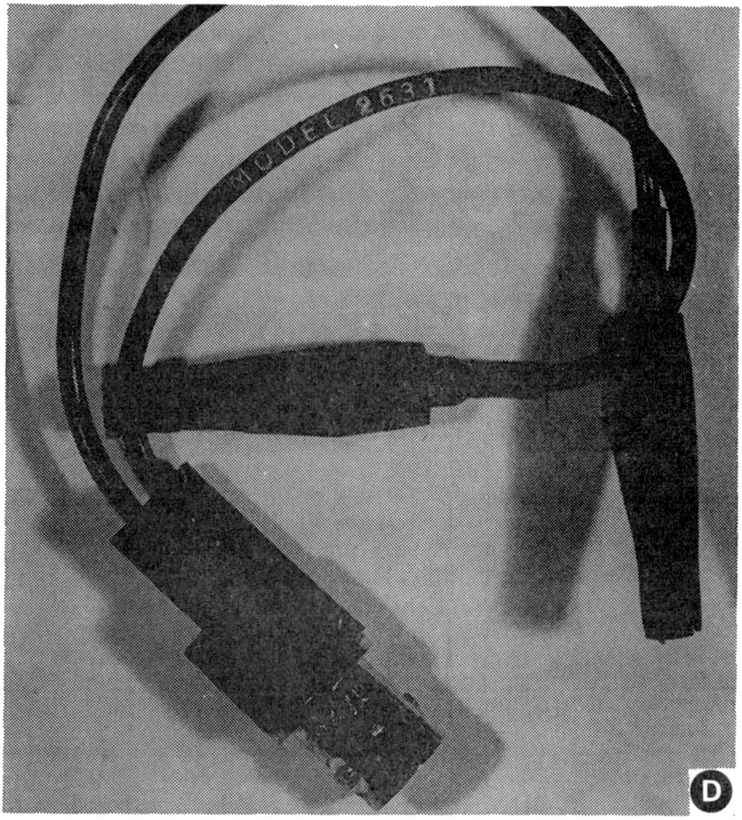

D

we see the popular banana plug and two different types of banana jack, i.e., a chassis flush-mount and a five-way binding post. These connectors are used in some audio generators and on certain types of multimeter. They are useful only to frequencies of a few dozen kilohertz.

The PL-259 UHF connector is shown in Fig. 8-6B. This connector is used as the rf-output connector on many, perhaps most, Amateur Radio transmitters. In older times it was also the connector of choice on rf signal generators. A large number of signal generators now falling into amateur hands from commercial sources are equipped with this connector.

Most modern test equipment, however, uses the BNC connector shown in two versions in Fig. 8-6C. This connector is possibly the closest thing we have to a "standard" in modern equipment. If you think installing an SO-239 on a piece of coaxial cable is a real dog, then wait til you try a BNC! You will learn rather painfully why so many professionals prefer to buy patch cords already made up with BNC connectors! They tend to be a little on the expensive side, but are well worth it in saved time and nerves.

Incidentally, most users of equipment originally equipped with an SO-239 connector have long since gone to an Amphenol dealer to buy a BNC-to-UHF adapter, so that modern cables will fit the instrument. A collection of some of the dozens of adaptors on the market is shown in Fig. 8-7.

Shielded Cables: Some Problems

Besides ground loop problems, we must also be on guard against the stray capacitance of shielded cables. That very property of shielded cables that affords protection from one type of trouble will provide equal opportunity for another. In some cases, the capacitance of the cable will severely attenuate high-frequency signals.

Although the figures vary from one type of cable to another, we are safe in assuming a value for most small-diameter coaxial cables of approximately 30 pF per foot. A four-foot section of cable used as a test probe will have a capacitance of 120 pF! Moreover, all electronic instruments have an input capacitance and an input resistance, typically 30 pF and 1 megohm, respectively. Add this to the cable capacitance, and we now have 150 pF shunted by 1 megohm.

The equivalent circuit is shown in Fig. 8-8A, and is redrawn in Fig. 8-8B. C1 and C2 represent the cable and instrument input capacitances, respectively. Together they total 150 pF. Resistor R1 represents the oscilloscope input resistance. This circuit is a

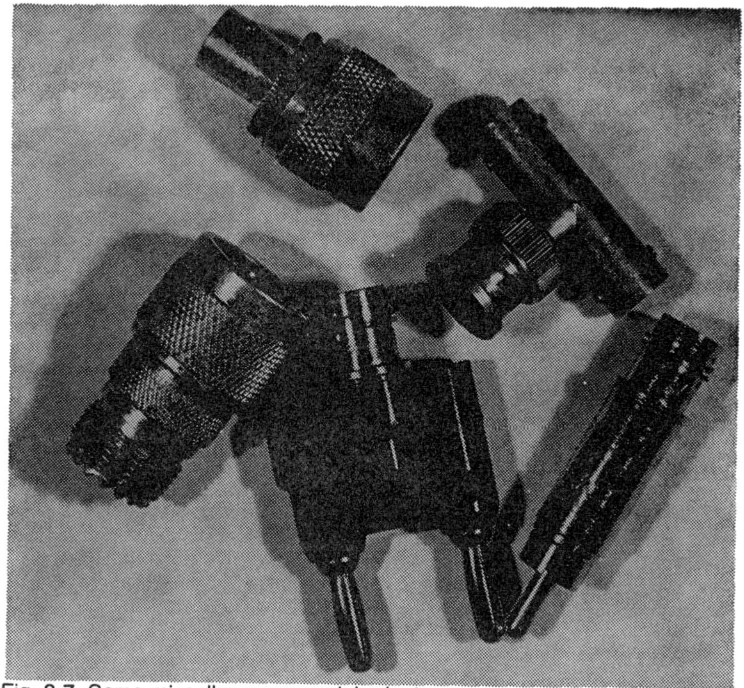

Fig. 8-7. Some miscellaneous coaxial adapters.

voltage divider, and E_0 is the effective potential applied to the input of the instrument. To make the arithmetic simpler, let us set E at 10 volts. At DC, the error due to voltage division is negligible, approximately 0.1 percent. But consider the situation at higher frequencies, where the capacitance can no longer be safely ignored. The reactance of a 150-pF capacitor is approximately 10 kohms at 100 kHz, and drops to 100 ohms at 10 MHz and 10 ohms at 20 HMz. Clearly, the capacitance would cause serious errors at higher frequencies. In addition, adding insult to injury, the capacitance conspires with the resistive portion of the impedance to create a phase-shift error. These errors are not terribly important on most sinewaves, but can spell disaster on many nonsinusoidal waveforms. The solution is the low capacitance probe of Fig. 8-9.

Shielded leads provide lessened susceptibility to interference over ordinary, open test leads and twisted pairs. But they should still be regarded as not the best solution in many cases. In general, the shielded lead is used in the same situations as open leads and twisted pairs, with the exception that they are generally capable of accepting lower level signals without too much interference. An

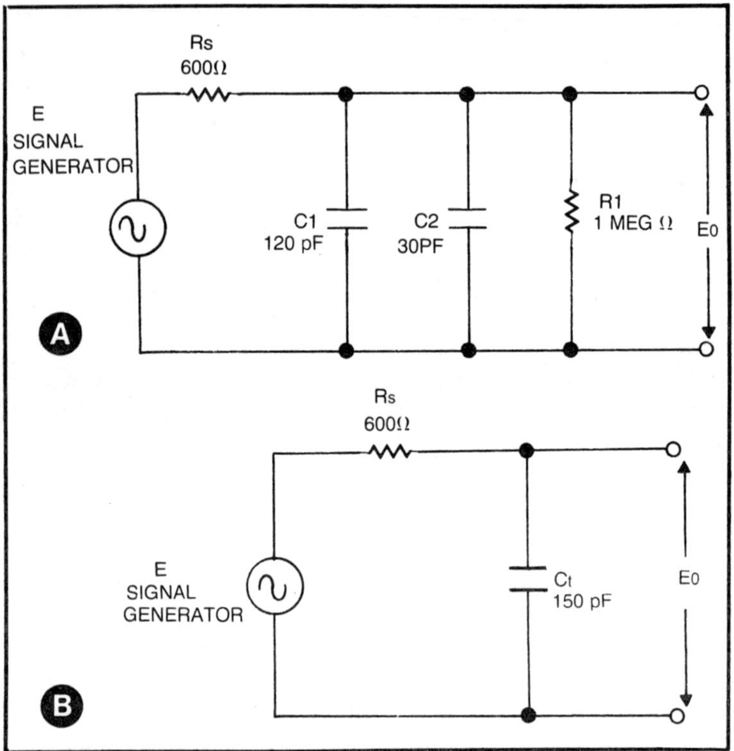

Fig. 8-8. An input-circuit for a scope or meter, (A); and a simplified equivalent circuit, (B).

audio microphone cable in a radiotelephone transmitter is an acceptable use of shielded cable.

LOW CAPACITANCE PROBES

The basic low capacitance probe (Fig. 8-9) contains a parallel RC network consisting of a high value resistor and a small value capacitor. The capacitor is sometimes variable. This type of probe is called a passive probe because there are no amplifying devices in use. The probe shown in Fig. 8-9 presents a capacitive load that is one-tenth of the load presented by the shielded cable, and also provides a 10:1 voltage division ratio (the output voltage is 1/10 the input voltage). The voltage division aspect may at first appear to be a disadvantage, but it has the advantage of extending the upper range of the instrument by a factor of ten. To read the correct voltage, then, it is necessary to multiply the indicated voltage by ten. Figure 8-10 shows a commercial oscilloscope probe.

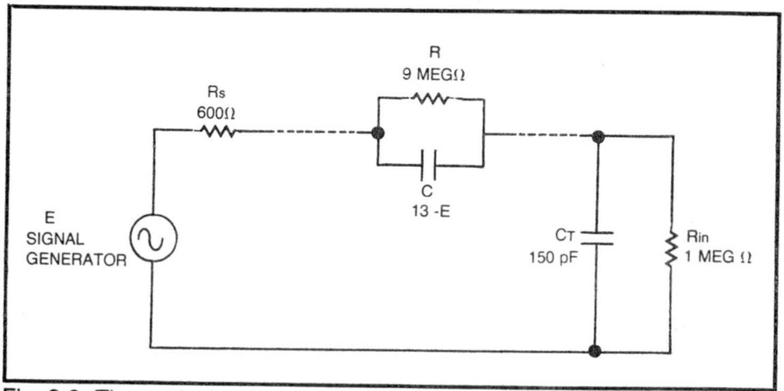

Fig. 8-9. The equivalent input circuit when a low-capacitance probe is being used.

The *RC* combinations in the circuit cause a phase shift, so some manufacturers will place a variable trimmer capacitor in the circuit. In low cost instruments the capacitor inside of the probe itself is made variable, while in higher-cost instruments the variable capacitor will be part of a phase-correction network inside of a molded BNC connector housing (see Fig. 8-11). This network will eliminate most phase-shift problems. The network housing and adjustment point is shown in Fig. 8-11, while the results of misad-

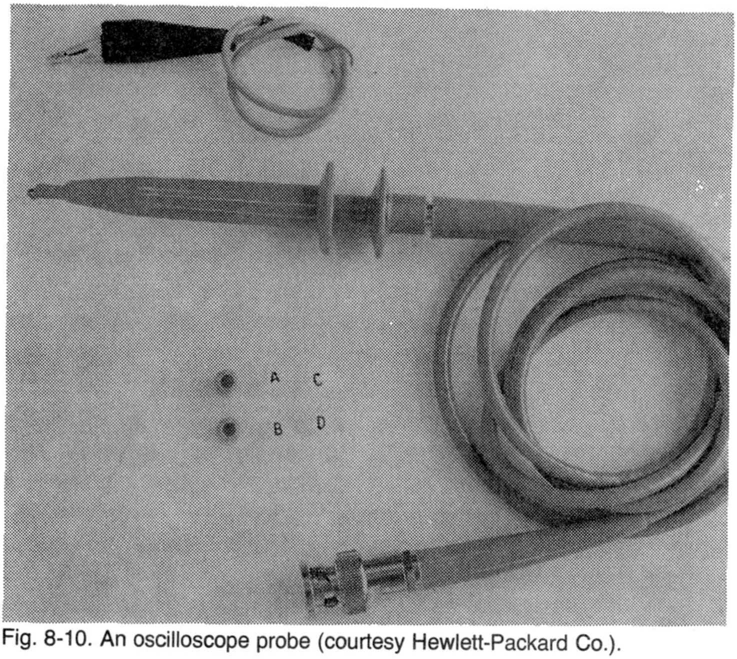

Fig. 8-10. An oscilloscope probe (courtesy Hewlett-Packard Co.).

Fig. 8-11. The connector end of a low-capacitance probe. The adjustment capacitor is located inside of hole in housing.

justment are shown in Figs. 8-12A and 8-12B. The capacitor is usually adjusted with a 1000 Hz squarewave applied, and the idea is to shoot for best "squareness." In some cases, the capacitor will overcompensate, causing the squarewave to overshoot.

Probe Use

The low-capacitance probe is very easy to use, especially if you remember the role played by phase shift. But there are some problems: You should, for example, be mindful of the manufacturer's recommendation as to the frequency range over which the probe may be used. It is also necessary that you make sure the phase-compensation capacitor is adjusted properly. The fact that the probe had been compensated on another instrument is irrelevant; the input-impedance components of the present instrument are probably different.

Some probes have a 1X-10X switch. In one position of this switch, the probe is *direct*, so the voltage indicated on the voltmeter or oscilloscope is 1 times the applied voltage. In the 10X position, the indicated voltage is 1/10 the applied voltage. If you are not certain as to the position of this switch, then it might cause an order of magnitude of error in your measurement! Some later Tektronix oscilloscopes have built-in circuits to sense the probe position, and automatically adjust the indicated vertical scale voltage factor. But in lesser-models the user is expected to know what is going on!

The ground wire is on the probe for a very special reason and should always be used. Most modern instruments use a three-wire

AC power cord. The third wire is connected to the equipment chassis, and is returned to ground, making the equipment chassis and cabinet at, or near, ground potential. Some people try to use the AC power mains ground between instruments to make the ground connection. This practice causes several problems, one of which is the old-fashioned ground loop. Current flowing in the ground wires will create a slightly different potential at the respective chassis. The difference between these potentials is sometimes seen as a valid signal by measuring instruments and other electronic equipment. At the very least, it will introduce noise (both 60 Hz and

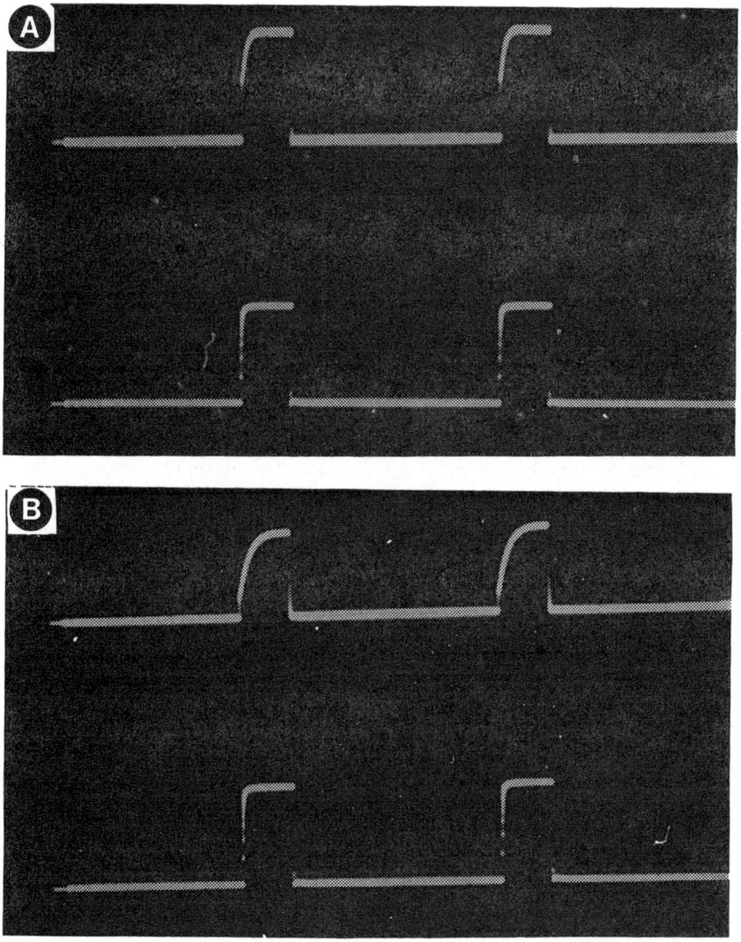

Fig. 8-12. Input and output traces when 10X is selected (A), and input and output traces when 1X is selected (B).

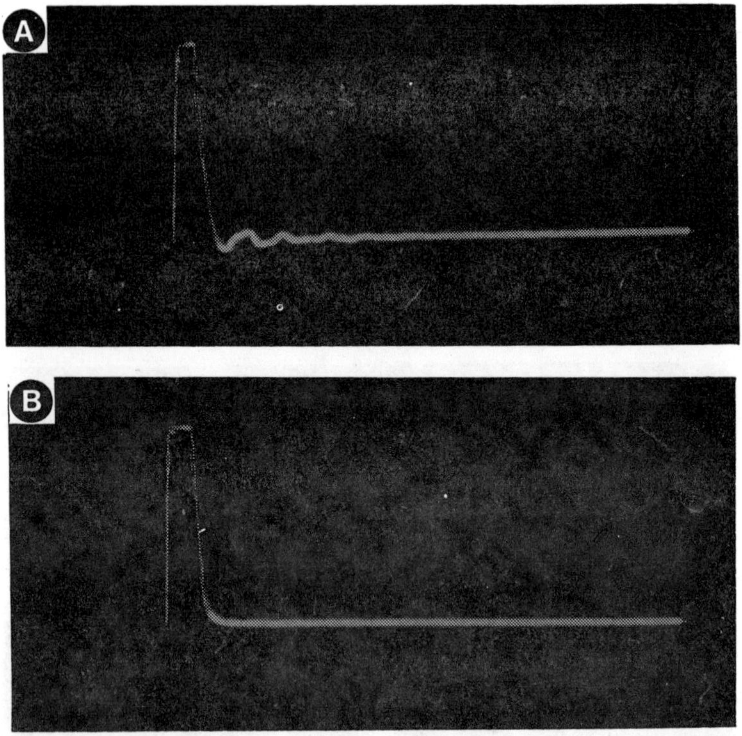

Fig. 8-13. A proper pulse display (A), and ringing when the ground wire on probe is disconnected (B).

random impulses), and may easily lead to significant measurement errors when low signal-levels are being used.

Another problem is *ringing*, the effects of which are shown in Fig. 8-13. The waveform in Fig. 8-13A was taken with the probe ground-wire connected. Note that the trailing edge of the pulse drops low, and falls off the rest of the way to zero smoothly. But, in Fig. 8-13B we see the effects of simply removing the ground wire. Note that the trailing edge is no longer smooth, but is ringing exactly like a shock-excited *LC* network. The phenomenon shown here is due to the stray inductance and capacitance in the log AC power mains ground path. These *L* and *C* components cause the circuit to behave as if it were a resonant circuit. So, when a pulse is applied, it will ring exactly like any other tuned circuit. This type of problem is eliminated by connecting the ground wire supplied by the probe manufacturer. Most companies do not add any feature to a probe just for looks, there is usually a valid reason for its inclusion, so use it as intended by the manufacturer.

HIGH VOLTAGE PROBES

A lot of different techniques to measure high voltages have been invented over the years. One amusing method used a pair of spheres connected across the unknown potential, and separated from each other. One of the balls is on a movable rack and pinion calibrated so that the distance between them can be measured. The operator moves the knob controlling the rack, bringing the balls closer together, until the air between them ionizes and a large arc occurs. The operator then measures the atmospheric pressure and relative humidity which, along with the separation, are plotted on a set of nomograph charts to find the actual voltage. While this nonsense is somewhat more spectacular than other methods, I recommend that you use an ordinary VOM, VTVM, or other electronic voltmeter with a special high-voltage probe. An example of such a probe is shown in Fig. 8-14.

Most high voltage probes are of one of two popular designs. In one case, there is a voltage divider network inside of the probe, consisting of a high-value resistance (many megohms), and a low-value resistance (less than 1 megohm, often down into the hundreds-of-ohms range). The output voltage across the lower value resistor is a fraction of the applied voltage. The advantage of this type of probe is that it is universal, i.e., independent of the type of meter used to make the measurement.

The second type of probe is simpler, containing only one resistor, but is dependent upon the input impedance of the voltmeter used to make the measurement. In most cases, the resistor

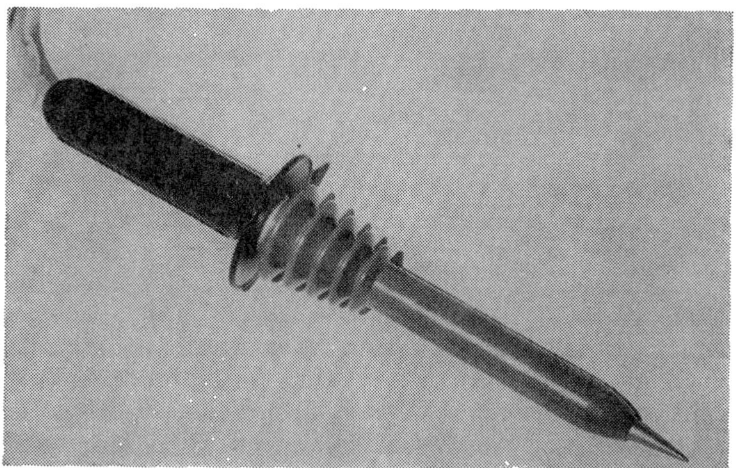

Fig. 8-14. A high-voltage-measuring probe.

inside of the probe is 900 megohms. For electronic voltmeters that have a 1 megohm input impedance, the voltage-reduction ratio is 1000, while for those with a 10 megohm input resistance, the reduction is 100:1. Those FET-input types that boast an input impedance of 100 megohms will see a reduction ratio of only 10:1.

The construction of the high voltage probe (see Fig. 8-14) is important for the safety of the operator. First, it is built of thick, insulating, plastic to prevent electrical shock. Secondly, there is a fin-like structure on the body of the probe that is designed to effectively lengthen the surface path between the operator's hand and the tip of the probe. Remember, the tip will be at some high potential, and if it is a transmitter power supply, or some other supply that is capable of delivering a lethal current, then coming in contact with it can kill you. High voltage probes are considered safe to use if you get smart and follow the regulations:

1. Use the ground wire attached to the probe, regardless of whether the voltmeter also has a ground wire.
2. Never depend upon power mains grounds.
3. Never use a probe that is dirty, or caked with film.
4. Never use a probe that is wet, or has other surface contaminants.
5. Never use a probe that is either broken or in poor mechanical condition.

RF DETECTOR/DEMODULATOR PROBES

Radio-frequency voltages cannot be directly measured on most lower-cost voltmeters. Indeed, they must be measured on specially designed rf voltmeters. Similarly, you might find that the frequency of an rf signal is too high for the vertical amplifiers on your oscilloscope, but modulation is well within its range. Or, in other cases, it might be the modulation that is of interest in the first place. In these cases, you might want to use an rf detector or demodulator probe. Two examples are shown in Figs. 8-15 and 8-16.

The probe in Fig. 8-15 is purely passive, because it contains no amplifying devices. Capacitor C1 serves to isolate the circuit from any DC potentials, or low-frequency AC potentials, passing only the rf signal. Diode D1 will rectify the signal, producing a pulsating DC that is smoothed in an RC filter network consisting of R1, R2, and C2. If the input RF waveform is sinusoidal, then the voltage across capacitor C2 will be the peak rf voltage. The resistor network R2 and R3 will reduce this to 0.707 times the peak voltage, yielding the rms value. This holds true, however, only when the input signal is sinusoidal. For any other waveform all bets are off!

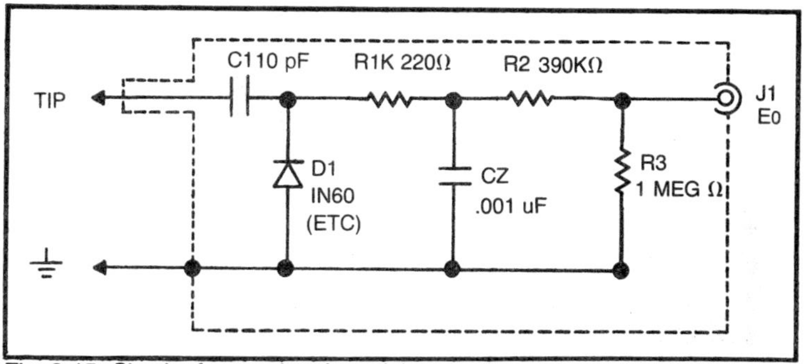

Fig. 8-15. Circuit of a passive demodulator probe.

Both minimum and maximum potentials that this probe can accommodate are set by the properties of diode D1. If a silicon diode is used, then the lowest potential is 0.6 volts (the junction potential of the silicon pn junction), while if a germanium diode is used, potentials down to 0.2 volts can be read.

The maximum voltage that can be read is set by the peak-inverse-voltage rating of the diode used at D1. For most VHF and UHF germanium diodes this will be in the 40-to 100-volt range. This level can be increased by connecting several diodes in series, but only at the expense of a higher minimum-readable-voltage.

An example of an active RF demodulator probe is shown in Fig. 8-16. This probe contains a two-stage transistor amplifier that will raise the level of the signal sometimes as much as 30 dB. This type of probe is used to amplify weak RF signals before applying them to the rectifiers.

Transistors Q1 and Q2 should be selected to have medium to high *beta* ratings, and a gain-bandwidth product of 700 MHz, or more. If proper attention is paid to good VHF/UHF layout practices, then the probe will be usable to 170 MHz, and will be almost "flat" to well over 100 MHz. The circuit of Fig. 8-16 should be constructed inside of a small metal shielded probe can (or housing).

PROBES FOR IC CIRCUITS

It is a scenario as old as the first integrated circuit. Someone is troubleshooting an IC circuit, and the probe being used slips. A momentary short circuit between IC pins, or to other circuit elements, will often kill an IC instantaneously! Digital and linear ICs are not very forgiving of accidental shorts . . . and these devices can be very costly. A certain digital-counter IC sells for $30, while some microcomputer devices are still in the $100 range!

The close packing of devices on modern printed circuit boards makes the possibility of an accidental short even more likely when troubleshooting. We had this problem in ordinary transistor circuits, but it was nothing like it now is in IC circuits. Additional care is required.

The solution is to use a special IC DIP plug that fits snugly over the IC package, making contact with the pins. A example is shown in Fig. 8-17. Pay no attention to the expensive professional scope in the picture, only to the connection devices on the PC board. These probes are attached when the power is turned off, so when the power is finally applied the signal will be going to the instrument. Alternatively, one could use unattached dip clips (AP Products,Inc., Paynesville, OH) that also clip over the IC. A regular probe can then be used, because the terminals on top of the dip clip are spread out far enough to allow a careful user to make connections.

TRANSMISSION LINE INTERCONNECTIONS

In many RF applications, the interconnection cable between two circuits and/or instruments acts like an RF transmission line. The characteristics of this simple piece of cable become very much like those of your antenna transmission line. The main problems are interrelated, and are: standing waves, characteristic impedance, and the ability of the line to transfer power to the load from the source. If you are not familiar with VSWR, and these other matters, then let me refer you to Channel 26 on antenna instrumentation, or almost any Amateur Radio or commercial antennas book.

An important implication of high SWR on an interconnection line is that the voltage along the line varies as a function of line length. This means that the actual signal supplied to a circuit from a signal generator, or measured on a voltmeter or oscilloscope, will be more or less than the signal indicated, depending upon the relative phases of the forward and reflected waves.

A principal cause of SWR on the line is mismatched impedances. The impedance at both ends of the line must be equal to the surge (characteristic) impedance of the coaxial cable used for the interconnection. This is true even if a matching pad or transformer must be used in the line.

The same problem can cause ringing on the line. The transmission line will pass both incident and reflected waves, so may create extra pulses on the line, as far as a frequency counter is concerned. Once again, the solution is to terminate the line in its correct impedance.

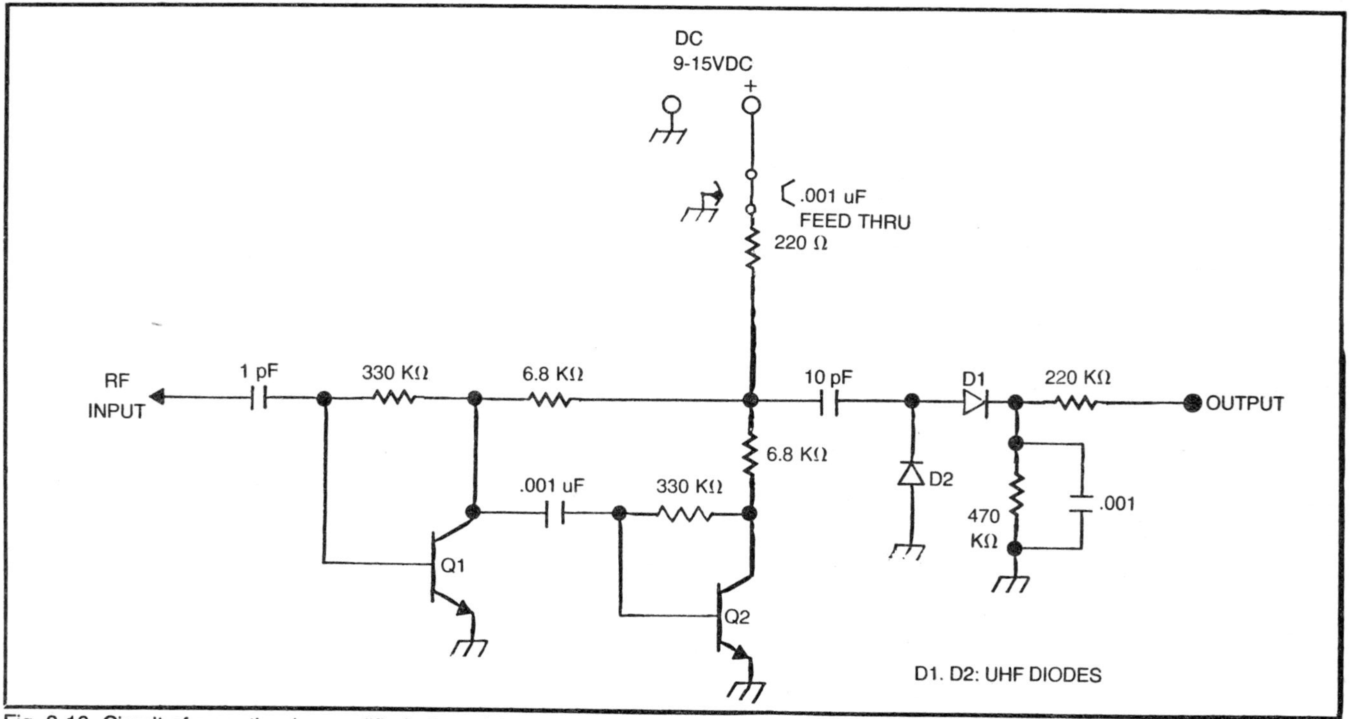

Fig. 8-16. Circuit of an active (or amplified) demodulator probe.

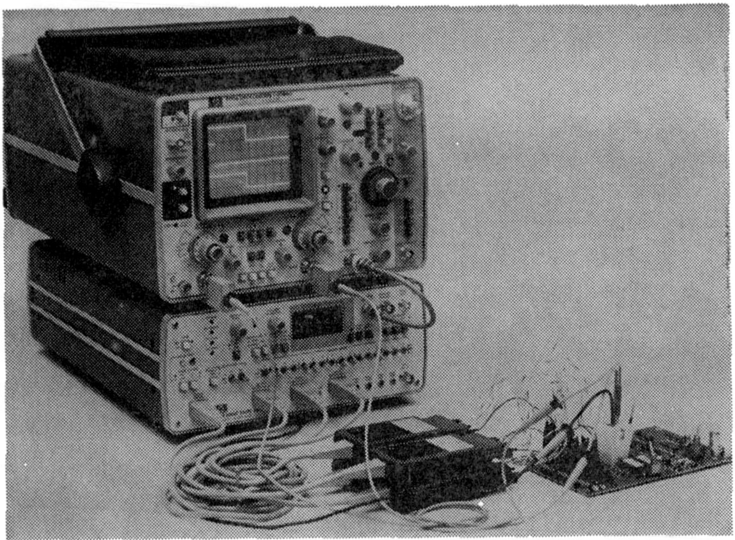

Fig. 8-17. An oscilloscope and logic-analyzer with DIP IC adapters (courtesy Hewlett-Packard Co.).

If the coaxial cable has a length that is electrically an appreciable fraction of a wavelength of the applied signals, then all of these problems are magnified. We can observe nodes and antinodes along the line, exactly as in antenna transmission lines, and this indicates that the impedance along the line varies considerably. For example, at 144 MHz, an electrical half-wavelength of polyethylene cable is

$$L_{(ft)} = \frac{(492)(0.66)}{144} = 2.26 \text{ feet}$$

If this length of cable is terminated in a signal-generator output impedance of 50 ohms, and a load of, say, 150 ohms, then the impedance that would be measured 2.26 feet from the load end will be 150 ohms. We may, therefore, use a short length of coaxial cable as an impedance tranformer at VHF and high HF frequencies. In fact, it is standard practice to use a quarter wavelength coaxial cable in instrument interconnections. The length of such a line will be given by:

$$L = \frac{246V}{F_{MHz}}$$

Where:

L is the length in feet (ft)

V is the velocity factor of the trannsmission line*

F is the frequency in megahertz (MHz)

*V is usually taken as 0.66 for regular polyethylene coaxial cable and 0.80 for foam-dielectric coaxial cable.

It is necessary to find a coaxial cable with a characteristic impedance Z_0' equal to the square root of the product of the source and load impedances:

$$Z_0' = (Z_0 Z_L)^{1/2}$$

Where:

Z_0' is the characteristic impedance of the coaxial cable used in the transformer section.

Z_0 is the characteristic impedance of the coaxial cable from the signal generator, or the output impedance of the signal generator if no additional coaxial cable is used.

CURRENT PROBES

The measurement of electrical current by a current meter requires that the circuit be broken so that the meter can be connected in series with the load. However, this is not always either possible or desirable. An active probe, such as one of the operational-amplifier current-to-voltage converters, will allow the measurement of DC and low-frequency AC signals on instruments

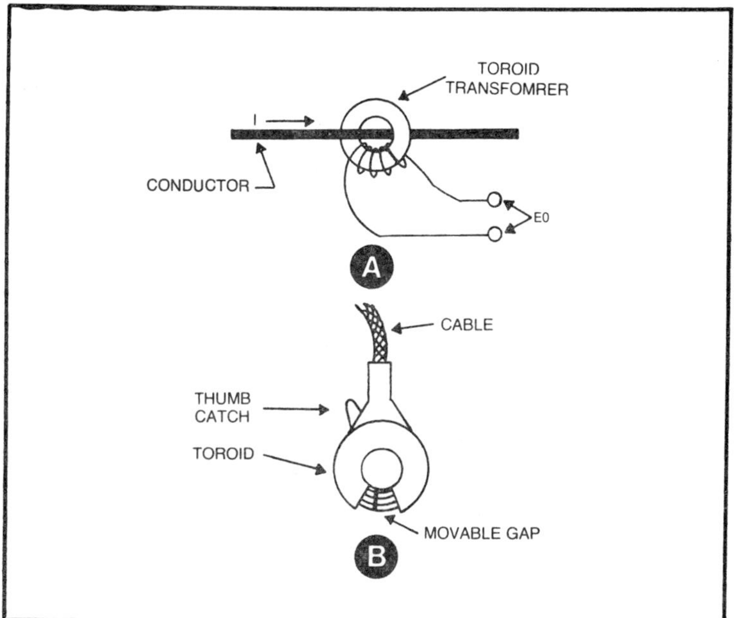

Fig. 8-18. Simplified circuit of a current-probe (A), and its construction (B).

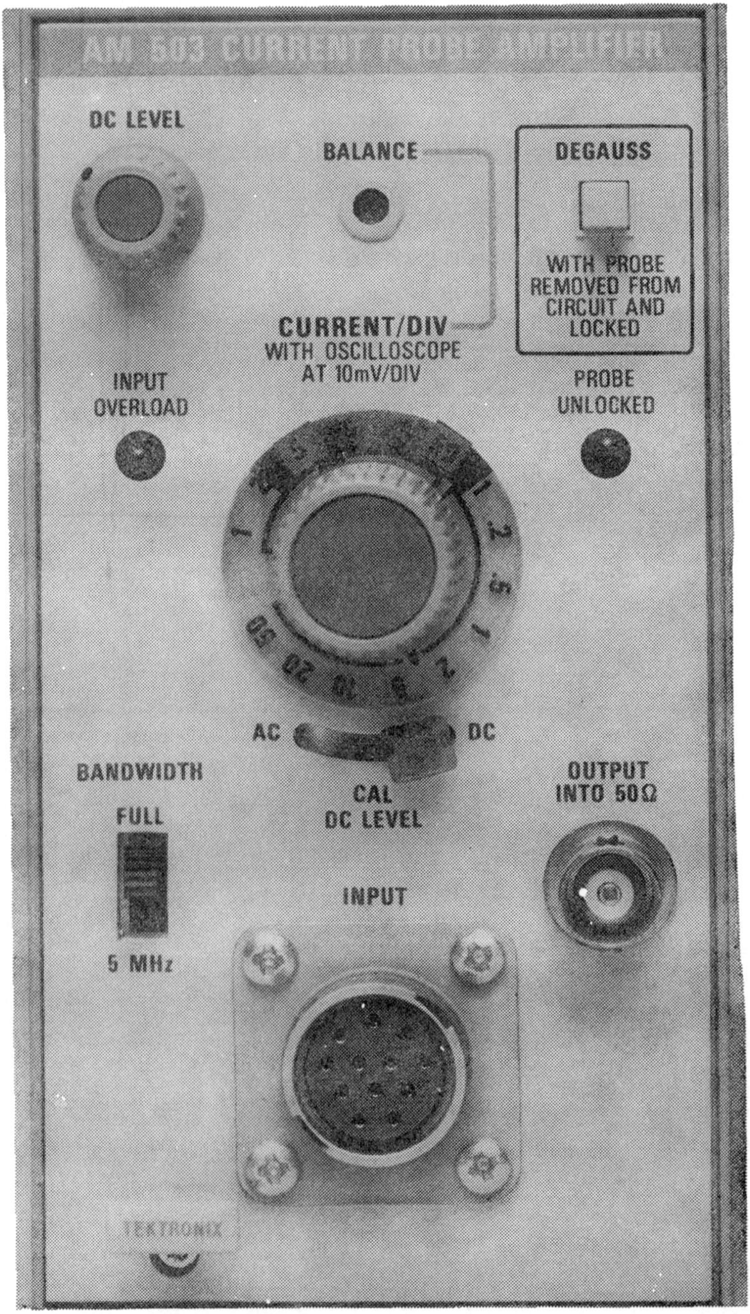

Fig. 8-19. Current-probe amplifier (courtesy Hewlett-Packard Co.).

such as voltmeters and oscilloscopes. The requirement for breaking the circuit, however, still exists.

A passive current-probe uses the magnetic field surrounding a current-carrying conductor to measure the strength of the current flowing in the conductor. An example of such a probe is shown in Fig. 8-18. The coil is wound on a toroidal form. There is a gap in the toroid. A latch on the side of the core allows the user to spread the gap to admit the conductor. Most such probes operate over a frequency range of 100 Hz to several MHz, but require an output amplifier (Fig. 8-19) that is scaled in millivolts per milliampere (mV/mA).

In operation, the probe is slipped over the conductor. This will produce an output voltage that is proportional to the current flowing in the conductor. This same principle is used in AC power-mains ammeters.

Chapter 9

Digital Test Instruments

Digital test equipment is becoming almost the "standard" in most areas of electronics. Just a few years ago, the digital test equipment that was available (only a few pieces) was very expensive; the digital readouts cost in the neighborhood of $200/decade. An eight-digit frequency counter, therefore, used $1600 worth of decade counters. Frequency counters typically cost $1800 - $4800, so only commercial R&D labs, some communications servicers, and the government could afford them. Less affluent shops used analog instruments, or other techniques.

Today, digital instruments have taken on a certain mystique that causes people to have almost unlimited faith in their accuracy. But, why is this thing called "digital?" What is a "digital instrument," and does it confer any advantage on the user? Some people would consider the last question to be a matter of heresy, but it seems valid once you begin investigating the available digital.

Some manufacturers consider an instrument to be digital if the readout device is digital, i.e., it uses something like seven-segment digital-display devices. Such a readout will present the output in the form of lighted digits, such as found on a calculator or wristwatch, but may actually be an analog circuit. It is almost standard practice in some companies to use ordinary analog circuits, and then display the result on a digital voltmeter. One commercial VSWR calculator uses in-line sensors and ordinary analog circuits (operational amplifiers) to create a voltage proportional to forward and reflected power, and VSWR. These voltages are displayed on a digital voltmeter (digital panel meters are now almost cheap!). A rotary switch selects forward, reverse, or VSWR outputs. This manufacturer uses a digital panel meter, but the rest of the circuit is purely analog. Is such an instrument really "digital?" I don't think so, because only the output device is a digital circuit. This type of instrument is not made any more accurate by being digital, although it can be successfully argued that the output indication is less ambiguous. Those lighted digits are a lot easier to read than an analog meter scale, especially from across the room. In the case where a reading must be made from odd angles, across a distance, or by unskilled operators, then the use of digital readouts is justified. In other cases it is advertising hype.

Such digital instruments can, however, prove misleading if too many digits are displayed. For example, if an instrument produces an analog output voltage of 0 to 2 volts DC, with 10 mV resolution and a digital panel meter capable of reading to 100 microvolts (a 4½-digit model) is used as the readout, then clearly no more than two of the digits to the right of the decimal point are valid. In that case, a DVM capable of reading to 1.99 (instead of 1.9999) volts is needed. A 2½-digit meter is reasonable, but the use of the 4½-digit type leads to false sense of accuracy. The company would not lie, mind you, but may be guilty of creative spec writing.

A real digital instrument uses binary logic elements such as gates, inverters, flip-flops, etc., to perform most of the measurement function. It may or may not use a microprocessor, almost unheard of a few years ago, but now more of a reality.

Digital electronic circuits operate with only two voltage levels (see Fig. 9-1), and these represent the two digits permissible in the binary number system (0 and 1). In the VSWR meter mentioned

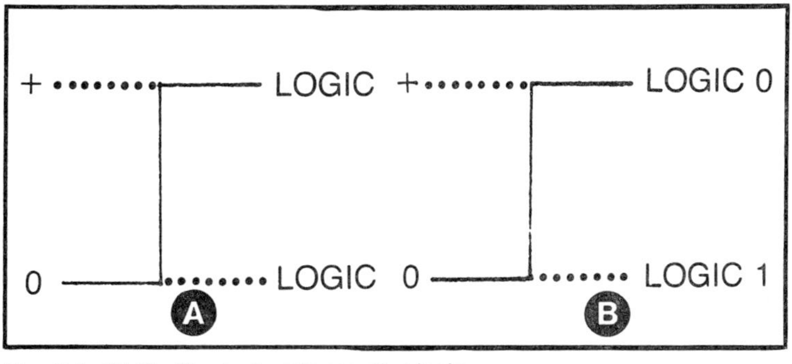

Fig. 9-1. (A) Positive logic, (B) negative logic.

above, it would become digital, in my opinion, if the analog voltage from the sensor in the transmission line of the antenna were given some initial amplification, and then converted to a binary word (that is, a digital signal) immediately thereafter. The rest of the processing would then be in digital circuits.

Although being digital does not automatically assure improved accuracy or resolution, the use of certain digital-design techniques does often improve the instrument. Digital techniques are especially appealing where the measurement is made, or held, over a period of time and analog circuits would drift or "droop."

Do not automatically think that an array of pretty digital readouts assures superior quality, accuracy, or resolution. It does not! If digital supporting circuitry is not used, or is of poor quality and/or design, then all those digital readouts do is increase the cost. A poor digital voltmeter is a lot less accurate than a good quality VOM . . . even if the VOM design is old enough to have voted for Roosevelt!

BINARY CIRCUITS

I assumed that you are familiar with basic digital-logic circuits, and the various IC digital-logic elements that are available in CMO and TTL. If you are not familiar with this area, let me refer you to the TAB books catalogue (available on request), or your nearest TAB book dealer. Any study of digital electronics, incidentally, is made easier if you understand binary arithmetic (the arithmetic of base-2), and the definitions of the various types of gates and flip-flops commonly used.

A binary counter ordinarily consists of a chain of clocked flip-flops that divide the input frequency by some integer N. In other words, there will be one output pulse, for every N input pulses. In frequency counters and other digital test equipment the counter is usually a decade or decimal counter, meaning that the division ratio is ten. The standard decade counter uses four divide-by-two flip-flops and suitable gating that resets the flip-flops when the eleventh count occurs, see Fig. 9-2.

A four-bit binary counter, as described above, normally counts from 0000_2 to 1111_2 (i.e. 0_{10} to 15_{16}), while the binary coded decimal (BCD) counter used in test equipment only counts from 0000_2 to 1001_2 (0_{10} to 9_{10}). Table 9-1 shows the BCD code, a subset of straight binary coding.

DECIMAL COUNTING UNITS (DCUs)

The heart of any electronic counter, or other digital instrument, is a circuit called the decimal (or decade) counter unit, DCU.

Table 9-1. The BCD Code.

Decimal	BCD	Decimal	BCD
0	0000	5	0101
1	0001	6	0110
2	0010	7	0111
3	0011	8	1000
4	0100	9	1001

All decimal counting units consist of a decade counter circuit, a code decoder, and a display device. Some also include a circuit called a data latch to make operation easier.

An example of a DCU is shown in Fig. 9-3. This circuit has been widely used, and is constructed of TTL integrated circuits. The decade counter is a type 7490 device which uses a chain of flip-flops not unlike the circuit of Fig. 9-2A, although in a little different arrangement. The 7490 is a *biquinary* counter, meaning that it contains a single divide-by-two stage, and a single divide-by-five stage. An external connection (between pins 1 and 12) connects the output of the binary stage to the input of the quinary stage. The resultant division ratio is 2 × 5, or 10. The 7490 will sense the tenth count, and will overflow on the eleventh and reset itself to the 0000_2 state.

Very few circuit designers use actual gates and flip-flops anymore. They prefer to use the "prepackaged" IC devices, like the 7490, to achieve their goals. Both TTL and CMOS lines contain

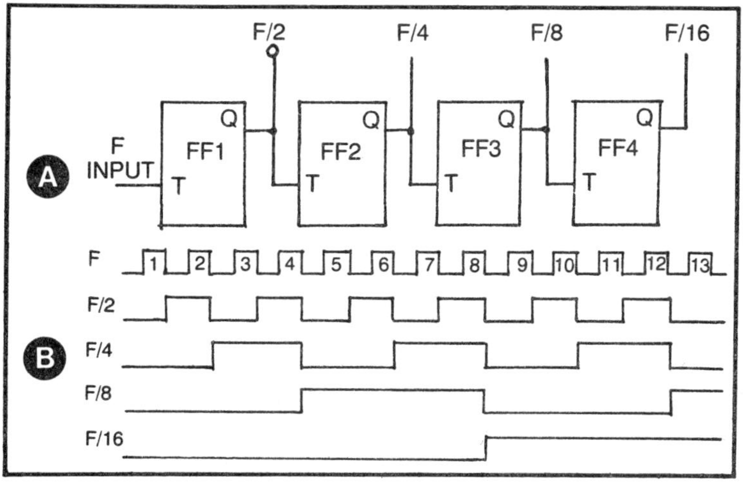

Fig. 9-2. (A) Binary counter chain, (B) waveforms.

several examples of decimal counters. There is even an LSI (large scale integration) CMOS device that contains a complete 8-digit decimal counter that operates to 10 MHz.

The four output lines from the counter in Fig. 9-3 are coded in the 1248 format common to BCD instruments. These lines change state with every input line to reflect the new count data. There are also two other lines, *reset* and *data input*. The input line accepts the pulses being counted, while the reset line will cause the counter to reset to condition 0000_2 when the reset line is brought HIGH momentarily. This line provides a means for clearing the counter.

The decoder, in this case a TTL 7446 or 7447 IC, is used to convert the four-bit BCD code, produced by the counter output, to the code needed to correctly drive the readout or display device. Several different code converters are used, but more about that later. The 7490 counter output may be used to directly drive the inputs to the decoder, but this results in a difficult to read display presentation. The "rolling effect" obtained when using the counter directly is created because the display device will follow the count. If the count is to be "7" for example, the display device will show "0-1-2-3-4-5-6-7" before coming to rest on the proper count.

The rolling effect is eliminated by using a quad-latch circuit (TTL device 7475, for example). The purpose of this latch is to hold the last previous data while a new count is taking place, and then to transfer the new output data to the display when the count is completed. A latch is actually a bank of four type-D flip-flops, connected so that their clock terminals are all shorted together. This common clock terminal becomes a *strobe* line for the latch circuit. When the strobe line is LOW, then the data on the four Q outputs will remain the same as it was at the instant the strobe line dropped low. If the strobe line is brought HIGH, then the data appearing on the four inputs will be transferred directly to the output lines.

In normal counter operation, the strobe line is kept LOW while the count is accumulating, but when the count is finished, the strobe line is brought HIGH for an instant. This transfers the new data to the output lines so that it may be displayed by the decoder/display combination.

DISPLAY DEVICES

Early electronic counters, many of which are still finding their way onto the amateur radio auction/hamfest circuit, used a column of ten incandescent or neon lamps (mostly incandescent) to represent the ten digits of each decade. If you had an eight digit counter,

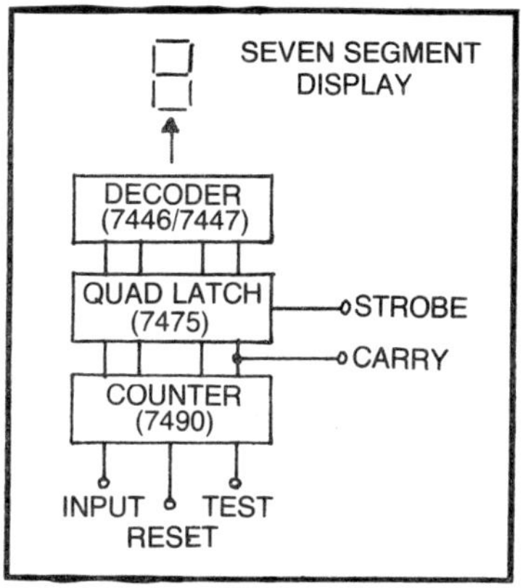

Fig. 9-3. Simple decade (÷10) counter with quad latch and decoder.

then, you would have not less than 10 × 8, or 80 lamps to read. If the counter reads 8257, for example, we would read the "8" from the eigth lamp in the thousands column, the "2" from the second lamp in the hundreds column, the "5" from the fifth lamp in the tens column, and the "7" as the seventh lamp in the units column . . . a pain in the neck, at best.

The Burroughs Nixie ® tube was a step forward. In this device a series of wire filaments were fashioned to the shape of the ten digits 0 through 9. These were connected together inside of a glass envelope which was filled with low-pressure neon gas. The ten digits then became cathodes. A high potential (150-200 volts DC) was applied to the common anode. To light up a single digit required a decoder that would ground the appropriate cathode see Figs. 9-4 and 9-5. If cathode "6" were grounded, then the potential between the common anode and cathode "6" would be 170 volts, so the gas molecules in the vicinity of the wire formed in the shape of a "6" would glow. The color produced was a dull orange.

The basic format of the decoders used for both lamp columns and Nixie® tubes is a single-pole rotary switch such as Fig. 9-4. In the example shown, we are dealing with incandescent lamps. The decoder switch is set to "4", so the fourth lamp is lighted. Of course, the actual decoder was either an integrated circuit designed for that duty, or a collection of ICs and discrete transistors that would examine the BCD code at the output of the latch (or counter) and convert it to a "one-low-out-of-ten" output: i.e., one line of decoder output would drop LOW when the count is completed.

Neither the Nixie® tube, nor the column of lamps, is used in modern equipment. Most modern equipment uses either the seven-segment bar readout, or the 5 × 7 matrix readout.

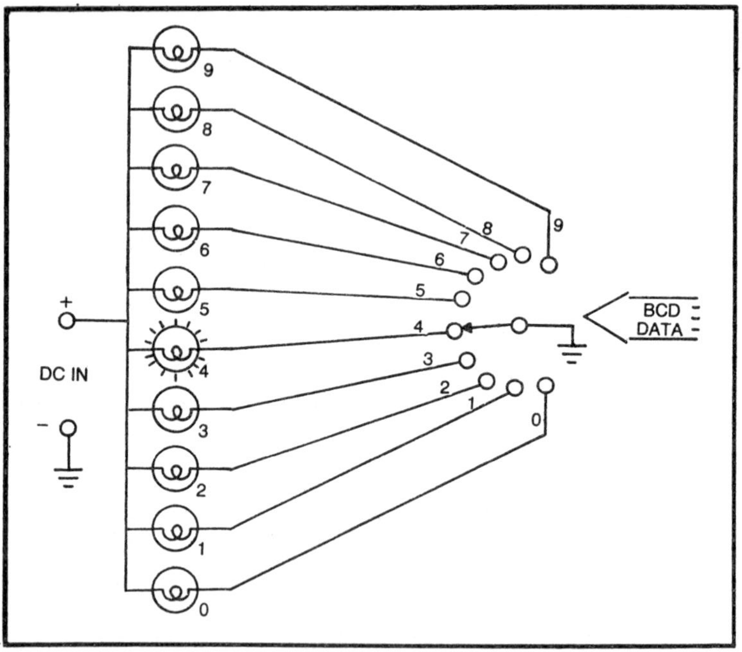

Fig. 9-4. Equivalent circuit of a one-of-ten decoder.

Seven segment readouts (Fig. 9-6) are available in a variety of different forms, using several different types of light source. One device by RCA, for example, uses incandescent wire filaments to form the segments. This type of readout is constructed inside of a modified seven- or nine-pin vacuum-tube envelope.

Probably most common amongst the seven-segment readouts is the light emitting diode (LED) type. The use of these displays is so common, in fact, that many people erroneously call all seven segment readouts a "LED display." Most LED displays are red in color, although a few are yellow or green. There are, however, several other types of seven-segment display. One, for example, is the fluorescent. It produces a greenish-blue display, and usually requires about 25 volts of DC for proper operation. The gas discharge display, also seven segment, requires about 150-200 volts DC, and is used where brightness and readability across the room is required.

The liquid-crystal is also a seven-segment display, although it needn't be. Two advantages of the liquid crystal display are that it can be formed to produce any letter, number, character, symbol, Logo, message, or whatever is desired by the equipment designer,

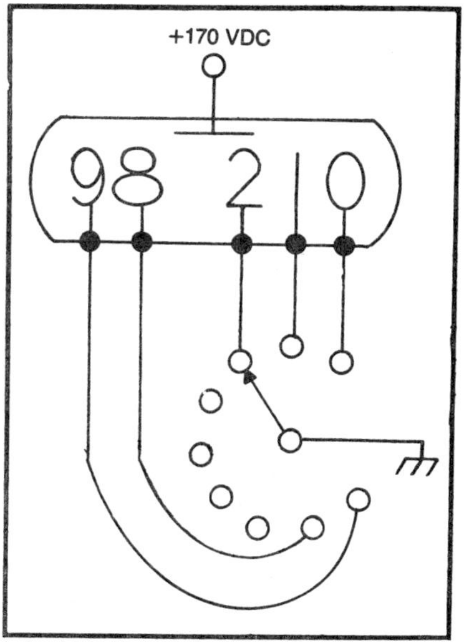

Fig. 9-5. Nixie® tube equivalent circuit.

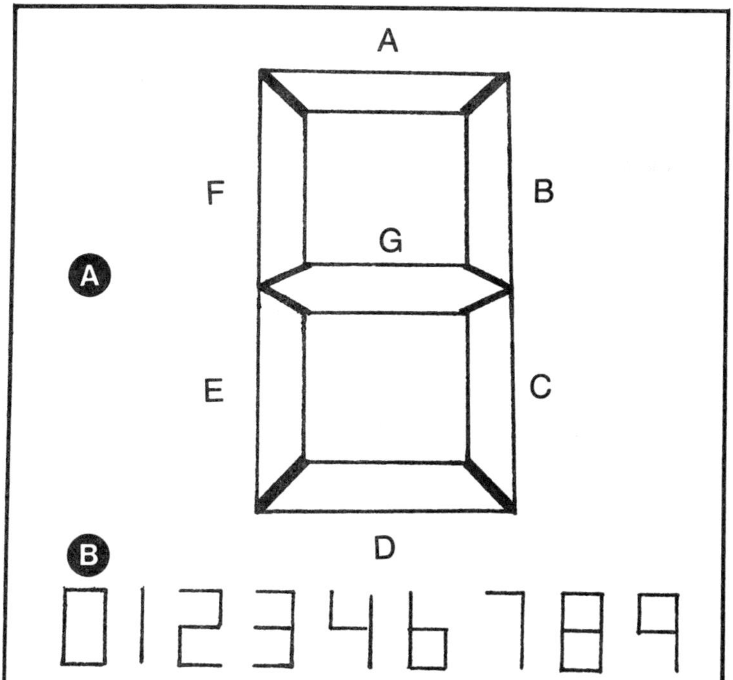

Fig. 9-6. (A) Seven-segment display, (B) digits 0-9 formed on a seven-segment display.

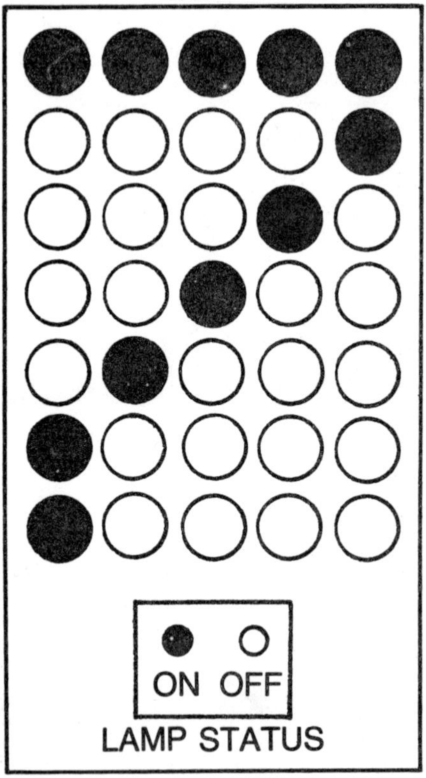

Fig. 9-7. 5×7 dot matrix display. Dark dots represent lighted lamps.

and that it requires very low current. The current requirements of the liquid crystal are so low because it is an electrostatic device, i.e., it will draw current (like a capacitor) only when it is changing state (charging, in the capacitor analogy). One brand of pocket calculator uses CMOS circuits and a liquid crystal display to provide as much as 1500 hours of operation from a pair of hearing aid batteries.

One last type of display is the 5 × 7 dot matrix shown in Fig. 9-7. This type of display was used long before the era of electronic digital displays, in relay-controlled athletic score boards and outdoor time/temperature billboards. It consists of a matrix of light sources (lamps, LEDs, etc.) arranged in five vertical columns of seven lamps each. The 5 × 7 matrix can be illuminated, with proper decoding, to produce all the alphabet and numerical characters in the English (and several other) language. Furthermore, the characters look more "natural" than those produced by the seven-segment readout.

The 5 × 7 matrix, is also used in television typewriters, such as computer CRT readouts, to produce the characters on the screen. In that case, the light points are unblanked points on the CRT raster (which is normally turned off). Each horizontal row of dots at each location represents a single horizontal sweep of the CRT by the deflection circuits.

Decimal Counting Assemblies (DCA)

An individual DCU can only count from 0 to 9. But if two DCUs are connected in cascade, so that the "D" overflow of one becomes the input to the second, the pair together form a two-digit decimal counting assembly (DCA) that can count from 0 to 99. We must add one more DCU to the DCA for each order of magnitude desired in the counter (see Fig. 9-8). This means that four are needed if the counter must go from 0 to 9999, and six if it must operate from 0 to 999,999.

The *strobe* and *reset* lines from all DCUs in the DCA are tied together to form common strobe and reset lines. This allows all three DCUs to be affected by a single reset or a single strobe pulse.

The *carry* output of any DCU is the "D" line (i.e., the line weighted 8), and it is used as the input source for the next most significant DCU in the cascade chain. Recall, from digital electronics, that the J-K flip-flops used in decade counters change state only on the negative-going transition of the input clock pulse. The

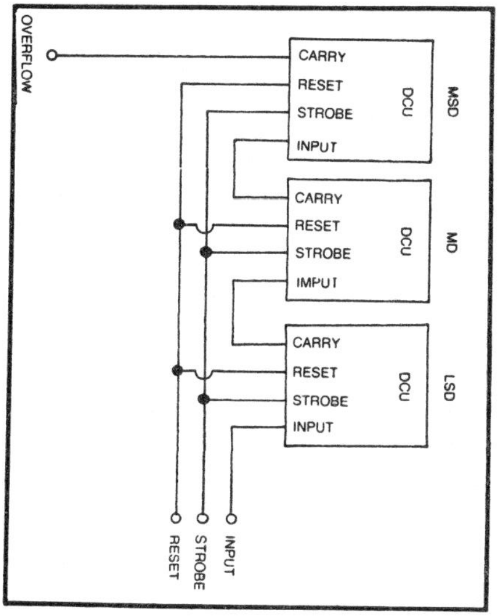

Fig. 9-8. Decimal counting assembly (DCA).

BCD line weighted "8" is HIGH only on the "8" and "9" counts, and drops LOW when the counter overflows from the 10 count. It will, therefore, toggle the input of the next most significant flip-flop on the trailing edge of the tenth count. . . it is a *decade* counter. This same tenth count also returns the input DCU to the zero state.

The original signal to be counted, then, is applied to the input of the DCU that is in the least-significant-digit position in the DCA. The carry output of the LSD DCU is applied to the input of the next most-significant-digit.

DCU, and the output of this DCU is applied to the input of the following. The output of the last, or most-significant-digit, stage is used as an overflow flag.

A counter that is constructed in the manner of a basic DCA will be a *totalizer,* which means that it will count the total number of pulses applied, and will continue to accumulate counts until the power is shut off. A totalizer is used whenever you want to know the total number of events without reference to time.

FREQUENCY COUNTERS

Figure 9-9A shows the block diagram for a frequency counter. The main sections are the *DCA, main gate, main-gate-control flip-flop, time base,* and a *display clock.*

The DCA is the totalizer counter of Fig. 9-8. The overflow stage is a flip-flop that is set when the MSD carry occurs. The overflow flip-flop turns on a lamp to make the operator of the instrument aware that the counter overflowed, and not to trust the data.

A frequency counter must measure events-per-unit-of-time (EPUT), i.e., cycles per second. This means that the DCA must be turned on only for a certain period of time (one second). The main gate, main-gate flip-flop, and the time base conspire together to allow this to happen. They will allow input pulses into the DCA for only some given period of time, such as 0.1, 1.0, or 10 seconds.

The time-base section consists of a crystal oscillator that produces pulses at a precise rate such as 1, 2, 4, or 10 MHz. In many lower-cost instruments, the time base will be 1 MHz. A chain of frequency dividers, such as TTL 7490 chips connected as divide-by-10 stages, is used to reduce the time base frequency to a lower frequency. Commonly used are 10 Hz for 0.1 second, 1 Hz for 1 second, and 0.1 Hz for 10 seconds.

The timing diagram for one complete measuring period is shown in Fig. 9-9B. Pulses T1, T2, and T3 are output from the time-base section. When pulse T1 goes HIGH, the control logic section generates a short pulse to reset the DCA to zero. When T1 goes low again, the Q output of the main gain flip-flop will go HIGH. The main (AND) gate has one input tied to the Q output of the flip-flop, and the other to the signal being counted. As a result, the main gate passes input pulses to the DCA *only* when the Q terminal of the flip-flop is HIGH.

The main gate flip-flop remains set until the negative transition of T2 occurs. At that time the Q output of the flip-flop drops LOW, turning off the flow of pulses into the DCA and causing the control-logic section to generate a *strobe* pulse. This pulse tells the DCU latches to transfer the data from the counter output terminals to the decoder input terminals. The latches will hold this count, and display it to the world until the next count is completed.

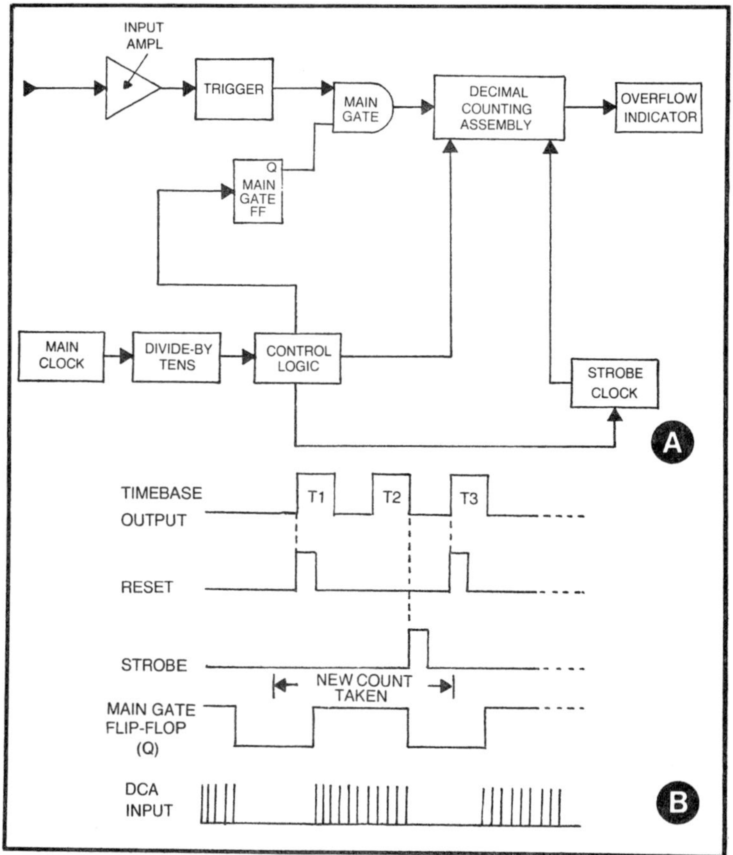

Fig. 9-9A. Block diagram of a frequency counter. B. Timing diagram for Fig. 9-9A.

The frequency counter of Fig. 9-9A counts frequency because the DCA is enabled for only certain specified periods of time. The frequency of the input signal is the number of counts accumulated in the DCA, divided by the time base period in seconds. If the time base period is one second, then the reading will be in hertz, and if a sub- or integer multiple of 1 second, then the reading will be in other units.

Counter Displays

The display used on a frequency counter will be one of those discussed earlier in this section. But there might also be additional features to make it easier to use, or to conserve battery power.

Ripple blanking is a feature of some decimal counters in which each DCU issues a signal that tells whether the count is zero or nonzero. If it is zero, then this ripple blanking output tells the ripple blanking input of the next least significant DCU to shut off the display if it is also zero. In this way, the counter is able to turn off, i.e., blank, zeroes to the left of the first significant figure. All leading zeroes are blanked. If we have a six stage DCA, for example, and want to display the number "9346", it would appear as "009346" on an unblanked counter, and "9346" on a blanked counter. This type of display is not only more easily read by the operator, but is also more conservative of battery power. The display device is usually the most power-hungry circuit in the counter. My LED calculator, for example, draws 40 mA when only a single zero is displayed, and 850 mA when all eights (888888888) are displayed. The "8" in a seven-segment readout is the most power hungry.

Another tactic that saves both power and device interconnections of the printed circuit board is display multiplexing. This technique is now very common on all forms of electronic equipment. All of the seven lines of the seven-segment display (*a* through *g*) are tied together to form a seven-line *bus*. A timing-control circuit will output the seven segment code for only one digit at a time, along with a control pulse that turns on that digit. Each digit receives its code in parallel, but sequentially to the other digits. Only one display is turned on at a time. But if the switching between the digits is made fast enough (10 to 250 kHz is commonly used as the switching speed), then the persistence of the LED and the persistence of your eye will make the display appear to be lighted constantly. The display only appears constant, because of the persistences involved; it is actually switching on and off at a high rate. If N is the number of displays in the system, then each one will be turned on

1/N-th of the time. Since current is a function of time (charge per unit time), the overall current demand made of the power supply is reduced considerably.

PERIOD COUNTERS

Period is defined as the time elapsing between identical features on successive waveforms in a wave train. The period can be calculated from frequency because period is the reciprocal of frequency:

$$P_{sec} = 1 / F_{Hz}$$

A period counter can be made by reversing the roles of the input stage and time base. The input amplifier is connected to, and controls, the main-gate flip-flop, and the time base is connected to the main input directly. This is shown in Fig. 9-10A.

The timing diagram for a period counter is shown in Fig. 9-10B. The main-gate flip-flop is set, i.e., Q goes HIGH, on the negative transition of T1. This allows the gate to pass pulses from the time base into the DCA.

The time-base frequency determines the time interval of each pulse. For example, suppose we used a time-base frequency of 1000 hertz. This yields a period resolution of 1/1000 second, or 1 mS. Similarly, a 10-kHz time base yields 0.1 mS resolution, and a

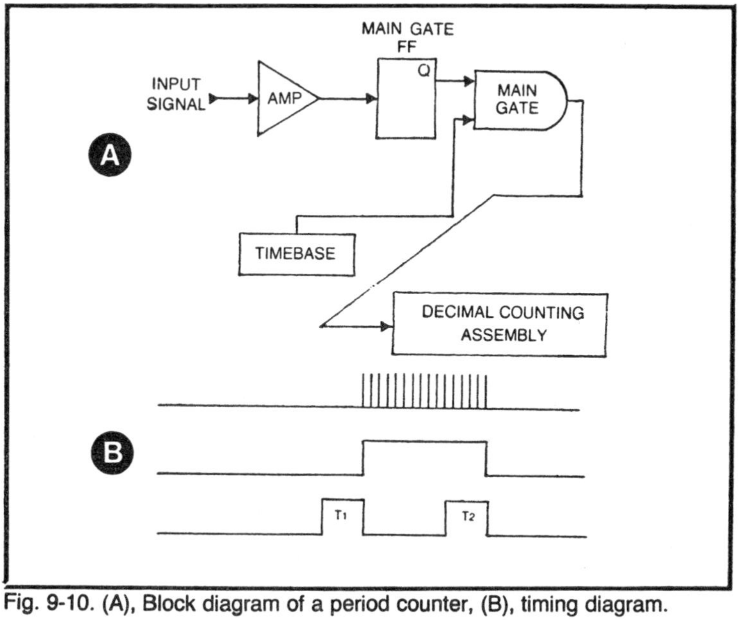

Fig. 9-10. (A), Block diagram of a period counter, (B), timing diagram.

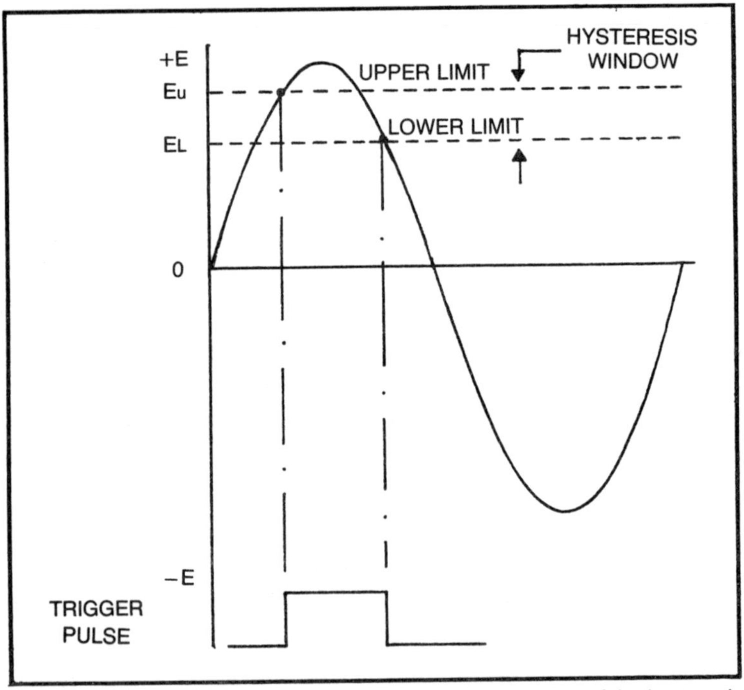

Fig. 9-11. Input signal must cross both upper and lower limits of the hysteresis window before triggering can occur.

1-MHz time base yields 1 microsecond of resolution. The period is given by the count accumulated by the DCA and the time-base frequency.

TRIGGER CIRCUITS

The input signal will probably not be the nice, fast-rise-time squarewaves needed by counter circuits, unless you are measuring the frequency output of a squarewave generator, pulse generator, or the frequency at some point in a digital electronic circuit. The waveform produced when measuring the frequency of your transmitter, for example, is probably a sinewave. Similarly, the input waveform, although a squarewave, may be too noisy to be used in the counter directly (noise on signals tends to cause false counts, but more of that later). Remember, a good rule of thumb for TTL is that the toggle point at which the device recognizes a change of state is either 1.4 or 2.4 volts, while on CMOS devices it is V+/2. If noise impulses produces a signal that greatly deviates from these norms, then false counting can occur.

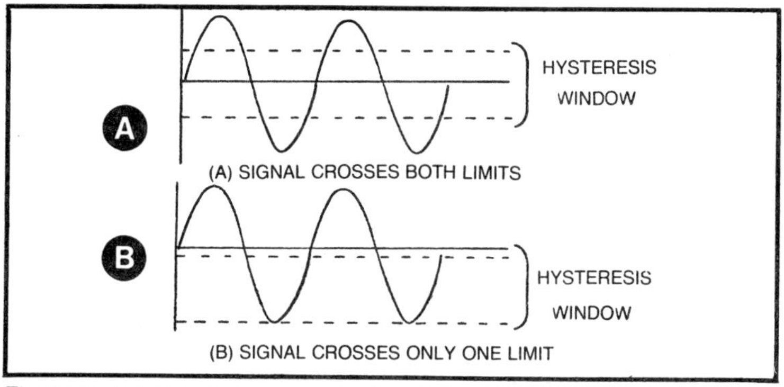

Fig. 9-12. With the window properly placed about the signal, the triggering will occur, because the signal crosses both limits (A), But, when the window is improperly placed, no triggering can occur (B).

The input signal should, therefore, be conditioned by an amplifier and a trigger stage. The amplifier will be a wideband voltage amplifier that has sufficient gain to build up low amplitude signals (usually some low figure in the 25 to 100 mV range is specified) to a level great enough to drive the trigger amplifier. The trigger stage is a Schmitt trigger, or similar, circuit with a built-in hysteresis. This type of circuit is used to clean up irregularly shaped signals by making them into squarewaves. Figure 9-11 shows the normal operation of a trigger circuit. The output snaps HIGH when the input signal crosses the lower hysteresis limit, and remains high until the input signal crosses the upper limit in a negative-going direction. The hysteresis window is the quantity ($E_u - E_L$). Note

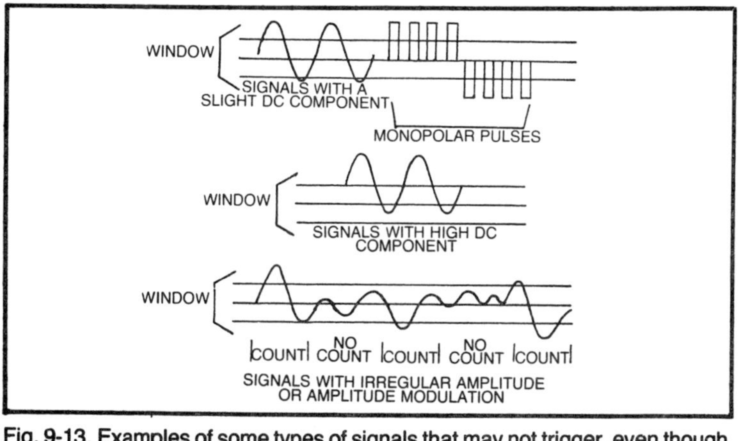

Fig. 9-13. Examples of some types of signals that may not trigger, even though the amplitude is sufficient.

that the trigger output crosses only one of the limits in Fig. 9-12B. When situations like this occur, only those counts where the input crossed both limits will be registered. All other counts will simply be lost. Figure 9-13 shows some of the types of signals that will typically *not* count, or will fail to count properly. Note that these cases are for a hysteresis window that is placed symmetrically about zero volts.

Some counters are equipped with a *trigger level* control that allows the user to adjust the position of the window over a specified range, while other models are equipped with a three position "–, 0, +" trigger control that allows the trigger window to be placed in three different locations, but is not continuously adjustable. Some of the latter use designations of "–, preset, +" instead of "–, 0, +". A continuously variable trigger control, however, allows us to position the control anywhere in the range (Fig. 9-14).

The trigger level, or trigger control, will not allow the operator to vary the width of the window, only its location. A *trigger amplitude* control, however, will allow the user to vary the width of the window.

COUNTER ERRORS

If you were to mention the possibility of digital counters producing erroneous results within earshot of certain "true believers," you would probably be stoned! But counters do have certain sources of error, and the wise user will be aware of them.

Counter errors can be grouped as *inherent errors* and *signal-related errors*.

The inherent errors are a function of the quality, age, and service history of the individual counter device. Little can be done about these unless their source is in need of a trip to a recalibration laboratory. Signal-related errors, on the other hand, are often correctable by proper manipulation of sensitivity, trigger-level, and trigger-amplitude controls.

Inherent Errors

There are two main sources of inherent errors in all frequency and period counters: time base error and a ± one-count ambiguity.

The time-base error is expressed in terms of a percentage, or in parts per million. The error from time-base inaccuracies is directly reflected in all measurements of frequency or period. For example, suppose a 1 MHz time base is off frequency by 40 Hz. This is an error of 40 parts per million, and can mean that the time base frequency is actually 1,000,040 or 9999960 Hz instead of 1,000,000

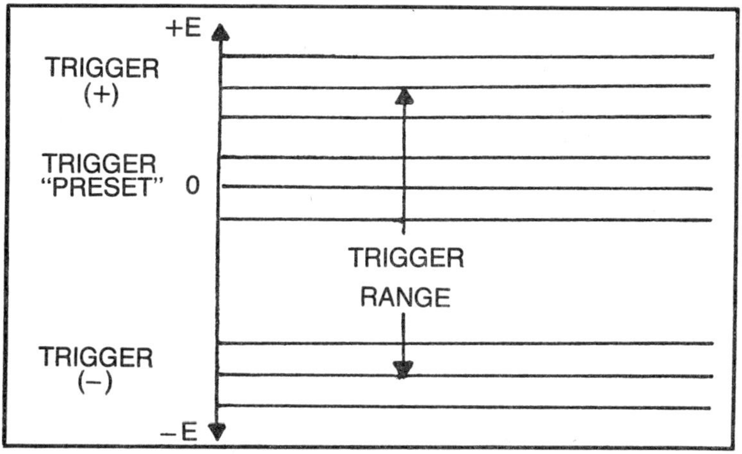

Fig. 9-14. Trigger-hysteresis-window range.

Hz. Assume that it is +40 Hz. The error in percent is given by:

$$\frac{1{,}000{,}040 - 1{,}000{,}000}{1{,}000{,}000} \times 100\% = 0.004\%$$

The measurement error from time-base inaccuracies is constant regardless of the input frequency being measured. That is to say that we would have a 0.004 percent error whether the input frequency were 100 kHz or 100 MHz. If the time base frequency is too high, then the counter reading will be too low, and vice versa.

The total time-base inaccuracy is the sum of several individual errors: initial calibration error, short term stability, long term stability, and temperature change.

The initial calibration error is a reflection of how well the factory (or homebrew builder) was able to calibrate the time base when the unit was constructed. Similarly, the initial error will be the error in setting the time base when the instrument is recalibrated by you or by a recognized metrological laboratory. Different methods are used to set the initial calibration of the time base. In years past, the principal method was to set the time base against one of the standard radio broadcast transmissions of the National Bureau of Standards (WWV, WWVH, or WWVB). The 60-kHz WWVB transmission was usually specified for most accurate work. Today, more and more metrology people prefer an atomic standard, such as a cesium beam, or rubidium cell.

The short-term stability is the time-base-oscillator frequency drift per day. Long term stability is the frequency drift per month or per year (check the specs on the individual counter.) . The long-term stability is usually called the *aging rate*.

The temperature and line-voltage stability specifications refer to a frequency change for variations over the 0 to 50 °C temperature range, and for ±10 percent changes in the primary line voltage (or battery voltage, if portable), respectively.

There are four different classes of counter time base: AC line, room-temperature crystal oscillator, temperature-compensated crystal oscillator, and oven-controlled crystal oscillator.

The use of the 60-Hz AC line as a frequency counter time base was popular in low-cost instruments, until the cost of crystal oscillators came down. Today, only the very cheapest models use the 60-Hz line as the time base source. Power companies will usually tell you that their generators keep the frequency to within less than 0.1 percent, but this is not quite the truth. They average the frequency overtime, so that the average frequency over, say, one day is within 0.1 percent of 60 Hz. This is fine for electrical clocks at bedside, or to tell when it's quitting time, but is disastrous for frequency counters. The error during the one-second-period when a count is being taken might actually be quite a lot.

The room-temperature crystal oscillator is an ordinary crystal oscillator, operated at room temperature (clever, huh?). The accuracy of such an oscillator can be good, if the room temperature varies over quite a range. The room-temperature oscillator will, however, be capable of better frequency accuracy than the AC power line.

The temperature-compensated crystal oscillator (TCXO) also operates at room temperature, and over quite a large range of room temperature, but is specially compensated to allow for these temperature excursions. Most TCXOs are encapsulated, and contain special circuit elements that cause the oscillator frequency to remain stable over wider ranges of temperature. The typical high-accuracy TCXO is less expensive now than in the past, so even medium-priced counters can boast of having a TCXO.

Those counters equipped with an oven-controlled oscillator are, perhaps, the most accurate on the market. A crystal, or component, oven is used to keep the temperature of the oscillator circuit at some constant temperature, usually 75 - 80 °C. These oscillators are capable of at least an order of magnitude better accuracy than are TCXOs.

Table 9-2 lists the typical stability specifications for several models of counter, by several different companies. Note that the

Table 9-2. Stability Specifications for Types of Counters.

Condition	Xtal	TCXO	Oven
Long term	5×10^{-7}	2×10^{-7}	5×10^{-10}
Short term	-	-	10^{-10}
± 10 voltage	10^{-7}	10^{-8}	10^{-9} *
0 - 50 °C	10^{-6}	10^{-7}	10^{-9}

*After 24-hour warm-up period

short term stability is given only for the oven type of counter time base.

The TCXO and the crystal oscillator must be operated for not less than 24-hours straight before the accuracy can reach the specified level. At operating times less than 24 hours, the stability is poorer. Some models, designed for mobile-radio and on-site customer service, have a power switch that turns off the power to the main unit, but keeps power applied to the time-base oscillator circuit. Rechargeable batteries are sometimes used for this purpose, so that the TCXO is not turned off while the counter is being transported from one job site to another.

The ± 1 count ambiguity mentioned earlier is caused by a lack of synchronization between the input signal and the time base. This phenomenon is illustrated in Fig. 9-15. During period T1 eight pulses are gated into the DCA, while during period T2 only seven pulses reach the DCA. One fundamental rule for all digital electronic instruments is that there is an error of ± 1 count of the least significant digit. In other words, a counter that reads 1000 Hz, is actually telling us that the frequency is 1000 Hz ± 1 Hz, or something between 999 and 1001 Hz.

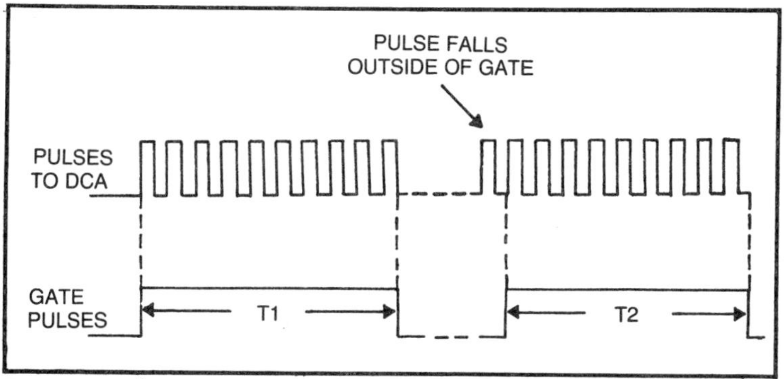

Fig. 9-15. Origin of the last digit ambiguity in digital counters.

The ± 1 count ambiguity produces an error that is inversely proportional to the frequency being measured and the gate time:

$$\text{Error (\%)} = \frac{\pm 100}{F \times T}$$

Where:

F is the frequency being measured, in hertz (Hz)

T is the time base gate-open time in seconds (s)

Example:

Find the percentage of error due to the ±1 count ambiguity at a) 500 Hz, and, b) 50 MHz. Assume a 1-second gate time

Solution:

a) Error (%) = (100)/(F × T)
   Error (%) = (100) / (500 Hz × 1 s)
   Error (%) = (100) / (500) = (1/5) = 0.2 percent

b) Error (%) = (100) / (F × T)
   Error (%) = (100) / (5×10^7 Hz × 1 s)
   Error (%) = $(10^2) / (5 \times 10^7) = 0.2 \times 10^{-7} = 0.00000002\%$

The error is ± count regardless of the frequency being measured, so the error reduces proportionally as the input frequency increases. Compare the results of parts a) and b) in the example above. The difference in percentage between the two is staggering!

SIGNAL-RELATED ERRORS

Poor-quality input signals can introduce errors that add to, or subtract from, the true count. Most of these errors result from misadjustment (or inability to adjust) the hysteresis window by the trigger control.

Most trigger errors occur because the input signal crosses the hysteresis window limits *too few* or *too many* times for each input cycle. We saw in Figs. 9-12 and 9-13 that a signal will fail to increment the DCA is it does not cross both limits of the window. This causes a too-low count.

Figure 9-16A shows how severe ringing on a signal can create a count that is too high. This problem is caused by the fact that the counter operator set the trigger level control too high, so that the ringing sections of the waveform can cross the limits. This results in extra outputs from the trigger circuit, one for every two crossing (if proper). In the case shown, there are two spurious output pulses gated into the DCA. The cure for this problem is to readjust the trigger level control until it is operating over the lower portion of the

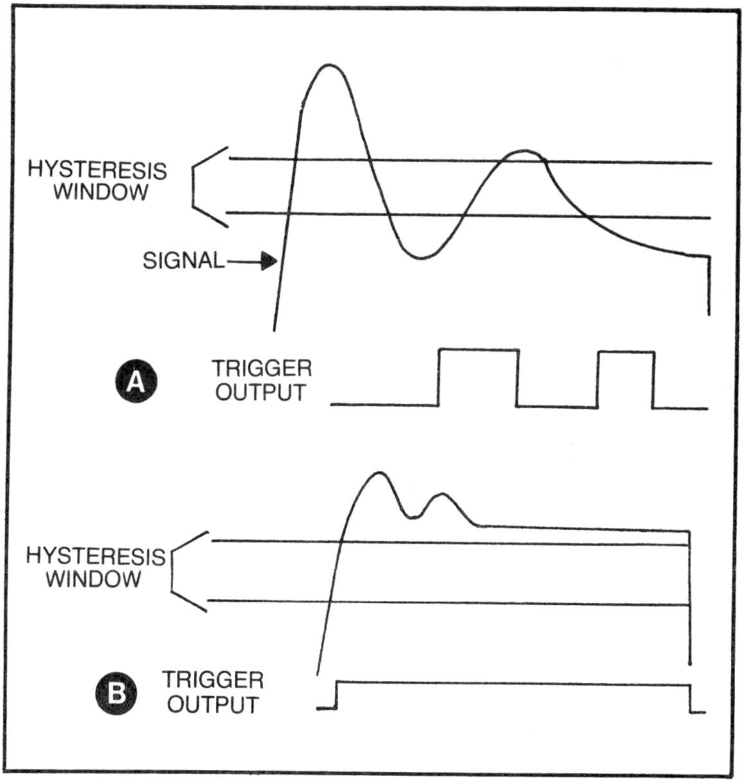

Fig. 9-16. (A), Improper placement of the hysteresis window causes extra, erroneous triggering; (B), proper placement allows correct triggering.

input signal, so that the ringing oscillations on the top cannot cross both limits. This is shown in Fig. 9-16B.

The same problem can exist on sinewave inputs (which normally won't be subject to ringing) if there is a large amount of second-harmonic distortion. The cure, however, is the same: readjust the trigger level control to cover a protion of the waveform that is lower than the affected section. This problem is illustrated in Fig. 9-17.

Impulse noise riding on the input signal can also have sufficient amplitude to cross both limits of the misadjusted hysteresis window. An example of this phenomenon is shown in Fig. 9-18, in which a pulse in a symmetrical wavetrain is carrying some impulse noise. In the case shown, the two noise bursts cross the window limits and thereby force the trigger output to create three pulses instead of just one.

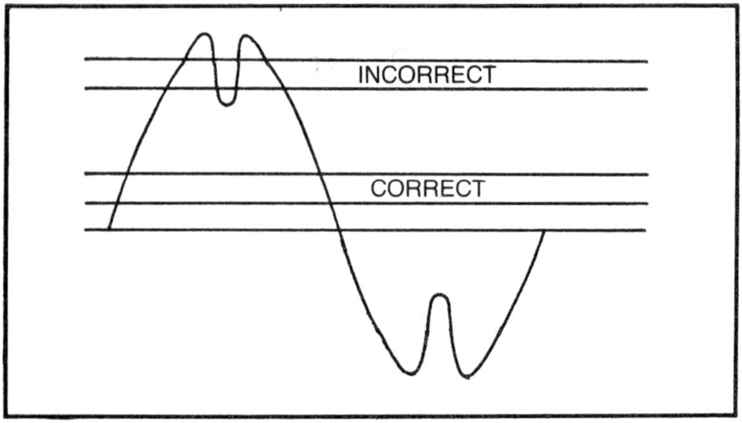

Fig. 9-17. Second-harmonic distortion can cause triggering errors.

Once again, the corrective procedure is to readjust the trigger-level control to a point further down the waveform. In the case of a non-square wave, however, the noise may appear on the leading or trailing edges, and still cause the problem. Fig. 9-19 shows the proper and improper positions for the window on such waveforms.

Note that the filtering of the noise is not usually feasible because of the bandwidth requirements of the input amplifier. A few expensive counters, however, allow you to trigger on either high or low frequency signals, thereby permitting the rejection of some noise.

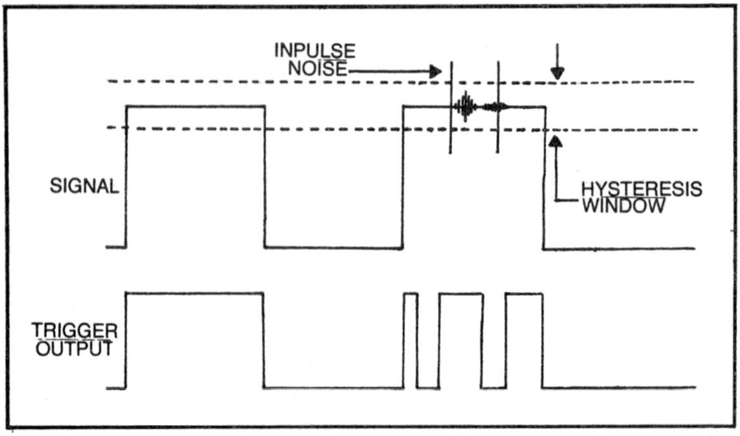

Fig. 9-18. Impulse noise in the hysteresis band will cause erroneous trigger outputs.

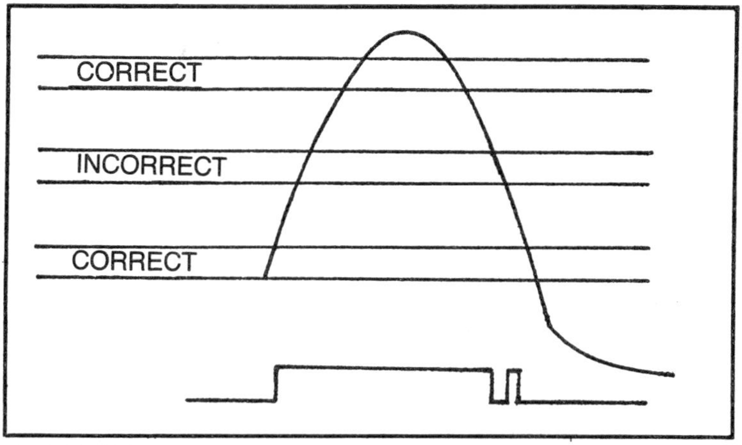

Fig. 9-19. Noise on a sine-wave can cause false trigger.

Figure 9-20 shows a type of error, also caused by noise, that is particularly troublesome on period measurements. In this example, noise rides on a signal that has a shallow slope, so it creates a band of uncertainty around the signal. The trigger circuit should produce a high output when the signal crosses the upper limit, and drop low

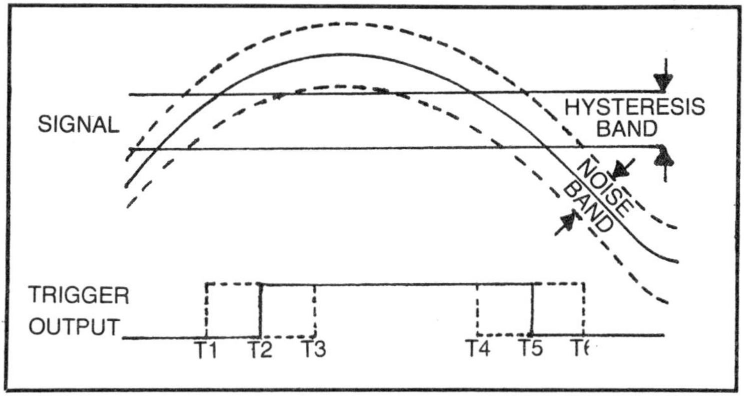

Fig. 9-20. Noise band around a slow-rise-time signal causes errors in triggering.

again when the signal crosses the lower limit in a negative-going direction. But, noise impulses, adding to the signal algebraically, could provide premature or inhibited trigger transitions. The correct duration of the trigger output pulse in Fig. 9-19 is (T4 - T1), but, under worst-case conditions the actual duration may be as much as (T5 - T0), and that amounts to a considerable error.

Chapter 10
Troubleshooting the Power Supply

The DC power supply in any type of electronic equipment is probably the most common seat of failure. And, with the exception of some of these new switching-regulator types, are generally the simplest (and most dangerous) circuits to troubleshoot. One piece of advice that I always give newcomers to the servicing industry is to check the power supply in any equipment *before* you try anything else. This habit is ingrained from many years of experience (there's something about being nailed to the wall a few times . . .). For several years I kept the service records for an electronic service company. On a bet, one day when other work was slow, I tabulated the exact nature of all problems in the previous year. The problems were broken down as to the section of the equipment involved. Invariably, regardless of the type of equipment, the dc power supply counted for one-third to one-half of all service problems! This suggests you should never overlook the supply in your *initial* investigation.

Proper evaluation of any DC power supply requires a voltmeter and an oscilloscope, but if you lack the scope, then it is still possible to evaluate the supply in most cases.

THE BASIC FULLWAVE SUPPLY

I do not want to vex you too much with trivial power supply theory; such would be an unnecessary rehash for most readers. But, I will cover the basics in abbreviated form.

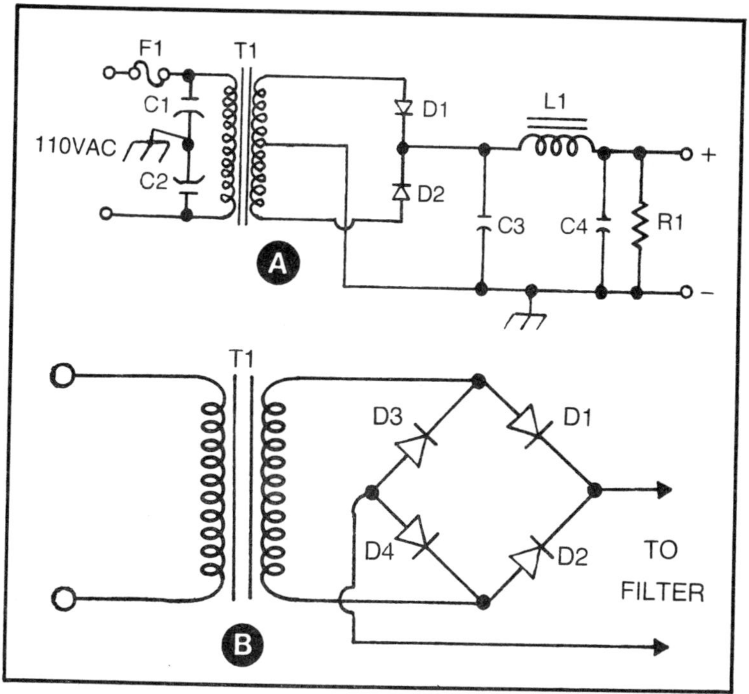

Fig. 10-1. Full wave power supply using center-tapped transformer (A), and a full-wave bridge rectifier (B).

Figure 10-1 shows two forms of elementary DC power supply found in Amateur equipment. Interestingly enough, these general schematics cover almost all Amateur equipment in existence, from small 100 mA supplies in an SSB outboard speech processor to a supply for a multi-hundred watt rf power amplifier. Of course, the inductor filter circuit of Fig. 10-1A is more common in transmitters, while the resistor filter is used in smaller equipment. Some equipment might use only C3 in the smoothing filter, deleting the other components.

Figure 10-1A shows a regular fullwave power supply using two rectifier diodes and a center-tapped secondary winding on the power transformer. The fullwave circuit will deliver the entire rated secondary current of the transformer, but only one-half of the end-to-end secondary voltage. Center-tapped transformers are usually rated as so-many volts "C.T" or, in the form "xx volts-0-xx volts." in recognition of the fact that the center tap is the zero reference point.

The bridge rectifier of Fig. 10-1B is also a fullwave circuit, but does not require a transformer with a center-tapped secondary.

This type of circuit uses four rectifier diodes, and establishes a zero reference point at the anodes of two joined diodes. The zero reference point is labeled "-" in Fig. 10-1B.

The bridge rectifier delivers the full end-to-end secondary voltage, but is cabable of delivering only one-half the tranformer's rated secondary current without exceeding the primary *V* X *A* rating. The exception is when the transformer is designed specifically for use with bridge rectifiers.

The output of a rectifier is not pure DC, but is a pulsating DC such as shown in Fig. 10-2A. This type of current is unidirectional, like pure DC, but is not very useful in electronic circuits. Connecting a small amount of filter capacitance (C3) across the output of the rectifier reduces this output *ripple component*, but does not eliminate it (see Fig. 10-2B). A large amount of capacitance in C3, or a multi-section filter circuit, will produce a smaller ripple, as in Fig. 10-2C. The ripple can be reduced to nearly pure DC, especially if the filter is followed by a voltage regulator. More about that later.

COMMON PROBLEMS

Although there are actually quite a few different problems that can afflict the DC power supply, they can generally be lumped into the following categories:

a) Hum or ripple on the DC output
b) Fuse blows
c) Fuse does not blow, but there is still no DC output
d) Poor regulation

Ripple appears in any number of forms, depending mostly on the type of equipment in which it occurs. In a receiver, for example, the ripple appears as a 60 or 120-Hz hum in the earphones or loudspeaker. In a transmitter it is a hum that will modulate the output signal, or cause a T-zero note on CW. On television receivers of standard construction the ripple appears as "hum bars," i.e., black horizontal bars (two bars for 120-Hz, or one bar for 60-Hz ripple).

In almost all cases where ripple or hum originates on the power supply, it's reasonably safe to blame the filter capacitors. These components are the most frequent sources of failure, although in regulator circuits the regulator itself can also produce these problems.

Be well aware, however, that not all hum or ripple problems are due to power supply defects. There are actually several different sources for these troubles.

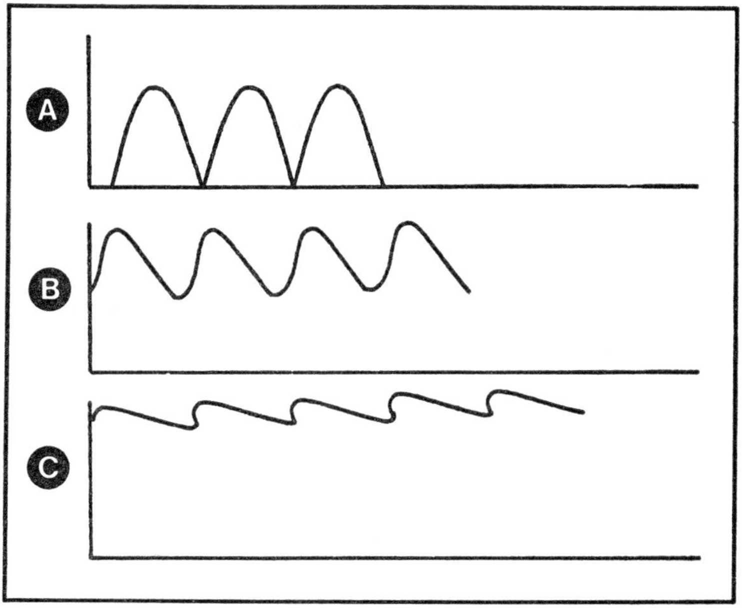

Fig. 10-2. Output waveform of fullwave supply without filtering (A), with capacitor filter (B), and with two-section filter (C).

One of the most prevalent in vacuum-tube circuits is a resistive short circuit or leakage path between elements inside the tube. The filament (or heater) is sometimes heated directly from a source of low-voltage alternating current. The heater voltage can appear on the grid, or may modulate the cathode/plate current, if certain internal shorts take place.

Another possibility is a low-frequency oscillation that is relatively close to the AC line frequency. This problem may be difficult to differentiate from real, live, hum without a frequency or period counter.

In certain circuits using differential amplifiers there might also be a lot of hum produced by common-mode rejection problems. Differential amplifiers generally see 60 Hz power-line interference as a common-mode signal, so none will appear in the output. But, it is possible for certain types of defect in the input circuitry of such an amplifier to manufacture a differential signal from essentially common-mode raw materials.

We also see "ground loops" in some electronic circuits. This problem is where a 60 Hz current in a circuit creates different electrical potentials in different points in the circuit. This would then be seen as a valid signal voltage, and amplified as such.

One differentiating technique is to examine the frequency of the hum. In all equipment that uses fullwave rectification in the power supply, regardless of whether it is a bridge or C.T. transformer circuit, the ripple frequency will be *twice* the AC power line frequency. In the U.S. this would be 2 × 60, or 120 Hz. If the hum is nearer, or identical to, the AC power line frequency, then you can temporarily forget about the power supply as the cause.

The waveforms in Figs. 10-2A and 10-2B were taken from the same power supply, during the same working session. The difference is capacitor C3. This problem would show up on the oscilloscope, and will dramatically point out a defective filter capacitor.

COMPONENT SUBSTITUTION

If you do not have an oscilloscope (any cheapie model will do in power supply servicing), then you can use component substitution as a troubleshooting technique. This means that you must substitute a known good capacitor for the suspected bad capacitor. In practice, this means shunting the known good capacitor across each filter capacitor in its turn, until all are eliminated.

Hint: Always use a replacement capacitor of at least the same capacitance and voltage rating. Higher values are permitted, but not lower values. Also, *always observe polarity markings on electrolytic capacitors!*

An aluminum electrolytic capacitor may *explode* under three conditions: a) too high applied DC voltage, b) it is in an AC circuit, and c) if the capacitor is connected in the circuit backwards.

Even if the shrapnel misses you, the cleaning of a piece of electronic gear following an electrolytic-capacitor explosion is a tedious job, at best. So, be careful, and follow the rules.

Bridging a capacitor across an electroytic filter in a DC power supply can be dangerous, if not done correctly. Many professional servicers will bridge the new capacitor across the old one while the power is still applied. Even though I have sometimes been guilty of this error, let me forthwith brand it as a stupid, dangerous practice.

Discharging and Checking Filter Capacitors

1. Turn the equipment off, and remove the AC power cord from the receptacle.
2. On small capacitors, in low-voltage supplies, use an alligator clip-lead or a screw driver to ground the filter capacitor for a few seconds. Touch the grounding device to the chassis or so-called "cold" side of the supply first, then touch the "hot" terminal.

3. Use a DC voltmeter to check to "make sure" all the charge in the capacitor has been drained off.
4. Using either alligator clip leads, or solder tacking, connect a known-good capacitor across the suspect capacitor.
5. Turn supply on and measure hum or ripple.

If the substitute capacitor seems to cure the problem, then make the substitute permanent. Do *not* make the repair by soldering a tubular capacitor across the defective section of a multi-section electrolytic.

Shunting a known good capacitor across a defective capacitor is a diagnostic aid, and it is not a repair technique.

If the power supply produces no DC output, then it may be necessary to use a VOM or VTVM or DVM to find the fault. These instruments are so low in cost these fine days that it is almost a crime for an Amateur Radio person to not own one! Although the digital multimeters look pretty, they are not always the best selection for amateur radio work, for the following reasons: portability is poor, cost is high, and they are often susceptible to rf fields.In contrast, the VOM is very portable, is inexpensive, and is not generally susceptible to rf fields.

Table 10-1 shows the proper procedure to follow when dealing with a DC power supply that does not produce an output voltage. Set your voltmeter to the highest AC scale, turn the supply off, and connect the probes (carefully!) across the primary of transformer T1. Turn the power to the equipment on, and observe the reading. If the reading is too low on the scale for convenience, then turn the range selector on the meter, one step at a time, until the reading is obtained. If you get to a scale lower than 150 VAC, however, stop. Either you somehow goofed (the probes are not across the primary), or no AC is getting to the primary.

Table 10-1. Troubleshooting procedure for power supplies.

Steps to follow when troubleshooting a dc power supply when there is no dc output

Measurement across	Approximate Value	Check for open circuit if no voltage at
1. T1 primary winding	115 VAC	F1. F1 holder. S1. ac line cord
2. T1 secondary winding	(high voltage ac[1])	(T1 primary. T1 secondary. center tap
" " "	(Low voltage ac[2])	open
3. C3	(High voltage dc[1])	T1 primary. T1 secondary. center tap open.
"	(Low voltage dc[1])	center tap grounding. rectifier diodes
4. C4	(High voltage dc[1])	L1 or a resistor in its place. if used.
"	(Low voltage dc[2])	

Notes: 1. High voltage power supplies 2. Low voltage power supplies
* All checks done with a voltmeter. An ohmeter can be used as a continuity tester, however, once an expected voltage fails to be found. Be sure the ac power cord is removed from the wall socket, and that all filter capacitors are discharged. Incidentally, it may take five filter is discharged enough for an ohmmeter's safety . . . the dielectric will store a charge even after you might think it is gone!

A good reading (105-125 volts AC in the U.S.) indicates that you should progress to the next step. If, on the other hand, there's no voltage across the T1 primary, turn off the set, unplug it from the wall, then use ohmmeter continuity checks on the *R* X 1 scale to find the open component. Check those items in the right hand column of Table 10-1 for continuity. Do not become alarmed if the DC resistance of the pirmary winding seems a bit low. It's the *reactance* of the transformer that keeps it from melting the house wiring.

Similarly, you must check the voltage across the secondary of T1 if it is found that primary voltage is present. Again, work down from the highest AC voltage range on your voltmeter until a reasonable reading is obtained. If no secondary voltage is found, then use an ohmmeter to find out whether it is the primary or secondary that is open.

In steps 3 and 4 of Table 10-1, it will be necessary to use the DC scales of the voltmeter. The basic procedure is the same; start with the highest DC scale, and work down until a usable reading is obtained.

My advice to use the voltmeter from the highest scale first is calculated, not to aggravate you with all of that scale switching, but to save your voltmeter from a very expensive (for you) demise, should you make an error.

THE OPEN FUSE

Whenever a fuse blows you must assume that there was a reason for it! In many cases, amateurs are prone to making some chortling noise about "surges" on the ac power line, but this is a rarity. To be sure, there are painfully large surges on the line during thunderstorm activity, but in most residential areas few truly serious surges happen. Besides, most fuses will not blow, if even very high overloads exist, instantaneously. A passing surge will not even heat the fuse up! It is a fair bet that there is something wrong in the circuit; find it.

Advice: fuses don't cause trouble, they indicate trouble!

Going back to our first chapter advice, the job now is to find an unwanted path for current. There are three approaches to solving this problem.

One school of thought maintains that it is best to obtain a box (or several) full of the correct fuses, and keep replacing them as they blow. The idea is to check something (pull it loose, temporarily replace it, etc), and then replace the fuse and turn on power. If the new fuse blows, that wasn't it; now try something else. I have used this method, but it is awfully expensive in the long run. For a Amateur repairing a single item it may be feasible to use this method—all you would be out is a couple of bucks in fuses. For the professional, and the frugal amateur, I offer a more elegant approach . . . method 2:

The second approach, which I fondly call "method 2" is to use a TV-type circuit breaker in place of the fuse. You can connect the leads from the breaker across the fuse-holder terminals, substituting it for the fuse. Place the circuit braker inside of a well-insulated plastic box for your protection. This method can be dangerous if you are a sloppy worker, so if there is *any* doubt, burn up a couple of boxes of fuses.

A third school of thought is to use an ohmmeter to locate the short. Unfortunately, this method can be misleading, because the ohmmeter battery does not stress the power supply components at all. They may test "good" with an ohmmeter, but will not work when exposed to 1000 volts!

The ohmmeter, fuse, and circuit breaker methods are all used in a "divide and conquer" scheme. First, disconnect or remove the rectifier(s) and see if the short perissts. If it does, then disconnect the load and the filter capacitors (one at a time), until the short is located.

TROUBLESHOOTING RULES

1. Use an isolation transformer to power all instruments and equipment being used or serviced (an isolation transformer is a 1:1 transformer that keeps your bench electrical system isolated from the power mains).
2. Always use well insulated alligator clip-leads and meter probes. If they are not in good condition, then repair them before use.
3. Always unplug the power cord when connecting or disconnecting clip leads, or when solder tacking . . . despite what you might see "experienced" servicers doing.
4. Work on a bench which has a ground-fault interrupter and a master power switch that is accessible to both you *and* any potential rescuers.
5. Use the "buddy system" and inform the buddy where the master shut-off switch is located, and how to operate it in an emergency.
6. If your equipment and test instruments are AC powered, but lack a three-wire power cord, then install such a cord.

The third wire is usually green in color, and is connected directly to the equipment chassis.

7. Never work outside, or in an area with a concrete or dirt floor unless *all* of your equipment is designed for such by reason of double-insulation construction. Many people have been killed by using an indoor appliance (i.e., television set) outdoors.
8. Never defeat interlocks.
9. Never service AC/DC equipment unless it is operated from an isolation transformer. *Never* use AC/DC equipment outdoors.
10. Always do quality work.
11. Always use quality, new, components.
12. Never allow a "temporary" repair to remain and become permanent. A Murphy's law corollary states that "temporary repairs become permanent if left for more than two days, or if there are more than two screws holding the cabinet on."
13. Switch to safety, think safety, work safely, and may you live a long, healthy life.

Chapter 11
Receiver RF Amplifiers and AGC-Related Troubles

The radio receiver's rf amplifier is the first line of defense in the battle to eliminate unwanted signals that would interfere with normal reception. Although it offers minimal selectivity, and only a small percentage of the overall receiver gain, an rf amplifier is frequently the major difference between a model that works well and another whose performance is only mediocre. This chapter will analyze the operation and troubleshooting of the rf amplifier and its related circuitry, including the automatic gain control (agc).

Fig 11-1 shows a typical rf amplifier circuit. Transistor Q1 is an NPN bipolar type. A small amount of thermal stabilization, as well as a slightly higher input impedance, is provided by the 330 ohm resistor between the emitter and ground. The forward bias applied to the base of this transistor is derived from the action of the agc circuit. How this is accomplished will be covered shortly.

The rf choke (rfc) in the antenna lead is one of those components which may seem out of place until its use is explained (who wants an rf choke in the antenna feedline?). This particular choke, for example, has a very low inductance. At low frequencies (1 – 10 MHz), it has a very low reactance. But as frequency increases, however, the reactance also increases. The function of the choke is to attenuate noise pulses, such as those generated in mobile receivers by the ignition system of the vehicle, or lightning bursts. Such a choke will help reduce the amplitude of any pulse that contains a fast rise time (i.e., high frequency component).

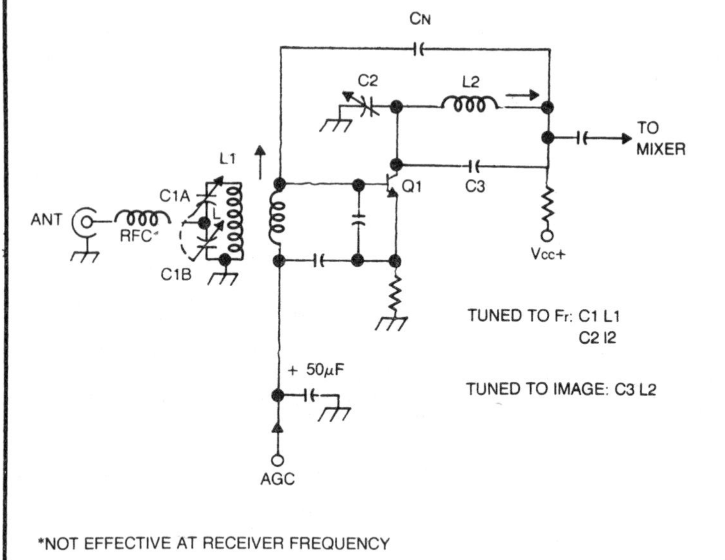

Fig. 11-1. NPN rf amplifier.

TUNED CIRCUITS

Amplifier tuning is accomplished by two *LC* tank circuits; one in the base circuit and another in the collector circuit. In car radios, and some early tunable Amateur Radio mobile receivers, the tuner is permeability tuned. That means the inductors of the rf amplifier and local oscillator are mounted so that the ferrite slugs are mechanically ganged together in a *permeability tuning mechanism* (PTM). Stations are selected by moving the ferrite cores in and out of the coil forms in the PTM assembly.

In more recent mobile receivers, the PTM has been replaced by a voltage-variable capacitance diode (varactor). These were used first in VHF receivers, because it is possible to tune the AM broadcast band (550 - 1650 kHz) with varactor diodes.

NEUTRALIZING

As with all tuned "triode" amplifier stages, the transistor rf amplifier may have some small amount of regenerative feedback. This feedback is caused by the small internal capacitance of the transistor. Theoretically, the best, or at least the most common, method for neutralizing this feedback is to introduce some negative feedback at the input. In the best case, the feedback comes from a series-resonant circuit, but this is often impractical in tunable receivers. Such a receiver must cover a wide range of frequencies, so tracking problems might exist between the main tuning and the feedback tuning.

Because a tuned-feedback circuit might present very ticklish tracking problems, it has become standard practice to use just one type of feedback—capacitance. The 33 pF capacitor (C_n) in Fig. 11-1 accomplishes the neutralization task.

The 50 μF capacitor connected to the rf amplifier agc line is used to decouple any signal from the i-f agc take-off point which manages to get into the circuit from the agc rectifier. If this capacitor becomes open, the radio will exhibit tunable squeals throughout the entire rf band. In vacuum-tube receivers, the agc bypass capacitor will have a value in the 0.01 to 0.1 μF range, while in solid-state equipment (where circuit impedances are generally lower) it will have a value between 10 and 100 μF. According to the rule usually applied, the bypass capacitor must have a reactance of not more than one-tenth the circuit impedance.

THE PNP CIRCUIT

Not all rigs will use an npn rf amplifier. In fact, most of the older receivers used PNP Germanium transistors. Figure 11-2 is a partial schematic of a typical pnp-equipped rf amplifier stage. For most practical purposes, it is essentially similar to the npn circuit of Fig. 11-1. Despite similarities, however, there are significant differences that can be used to determine which type is used in any given

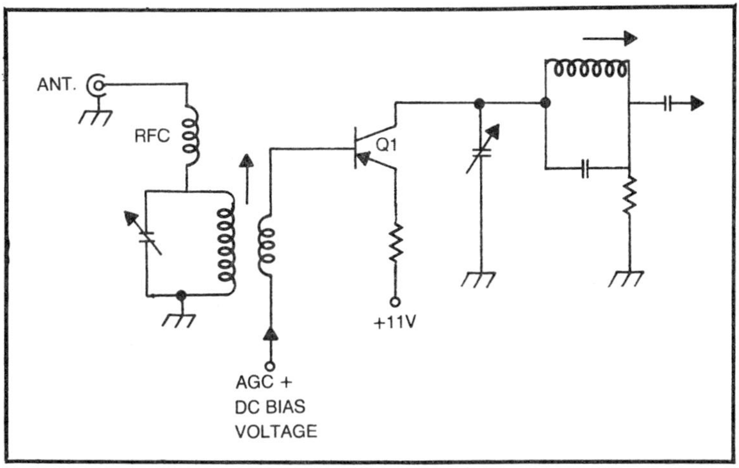

Fig. 11-2. PNP rf amplifier.

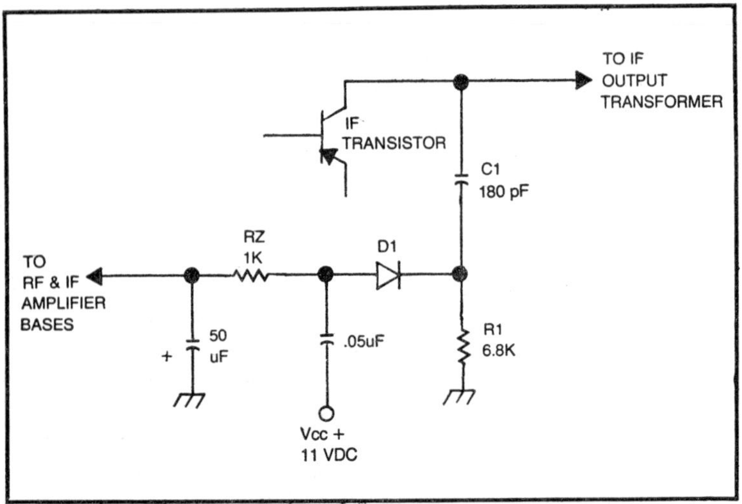

Fig. 11-3. Simple agc circuit.

radio receiver. The element voltages, with respect to ground, are one point of identification. In Fig. 11-1, for example, high voltage is found at the collector, while lower positive voltages are applied to the base and emitter. In Fig. 11-2 exactly the opposite situation exists; the emitter and base voltages are high, while the collector voltage is either zero, or a near-zero, positive voltage. These conditions are valid for most negative-ground circuits.

AGC CIRCUITS

A typical solid-state automatic gain control (agc) circuit is shown in Fig. 11-3. This circuit is typical of the single-diode circuits normally encountered. In this circuit, a sample of the signal voltage is taken from the collector of the I-F amplifier (or the output of an I-F amplifier IC), and is fed through a 330 pF capacitor to a 6.8 kohm load resistor. This signal sample has an amplitude that is proportional to the amplitude of the received signal. It can, therefore, be rectified and used as a DC correction voltage which controls the gain of the RF and I-F amplifiers. Modulation and residual signal voltage components are removed by the decoupling and filtering action of C2, R2 and C3.

A variation of this basic circuit is the two-diode circuit shown in Fig. 11-4. This variation is basically the same as the single-diode version, except that it uses two diodes to form a half-bridge rectifier instead of a simple halfwave rectifier. The exact circuit configuration of Fig. 11-4 is that of a voltage doubler. There are also

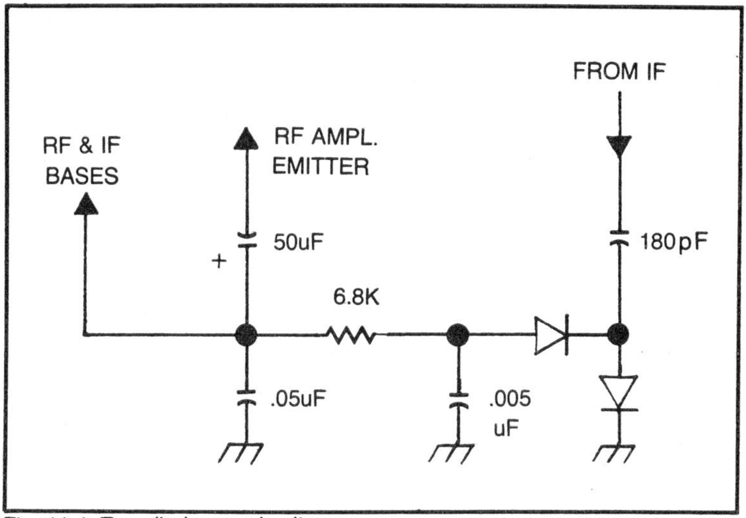

Fig. 11-4. Two-diode agc circuit.

reverse-polarity versions of this circuit for use in negative-ground pnp stages.

Figure 11-5 illustrates the relationship between the RF amplifier and the agc system. A sample of the signal is rectified by the diode. The resultant DC voltage is applied to the controlling element (base of a transistor in some cases, or a control input on IC devices) or the RF, and sometimes, the I-F amplifiers. This will reduce the overall system gain as needed to maintain a constant output amplitude level.

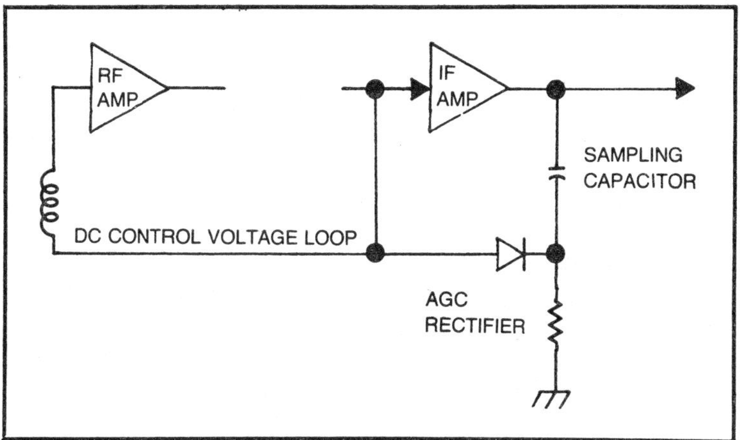

Fig. 11-5. Simplified agc circuit.

The interrelated nature of RF amplifier and agc circuits often makes the troubleshooting of certain defects much like a guessing game. For purposes of analysis, we are going to break the usual faults down into two separate categories: distortion on strong signals and the "dead" RF amplifier.

DISTORTION

Distortion on strong signals is usually caused by a defective agc circuit. The root of this problem will most likely be a bad agc rectifier diode. This diode can generally be quick-checked in the circuit by making forward and reverse resistance readings with an ohmmeter set to the Rx100 scale. The usual admonition against using a low-voltage electronic ohmmeter applies. Be sure, however, that the voltage across the open terminals of the ohmmeter is not excessive. Many older instruments (and some still recent models) use 7.5 to 22.5 volt batteries in the ohmmeter, and these will blow out most agc rectifier diodes.

Besides a bad diode, however, there are certain other defects that can cause distortion on strong signals. Included among these are broken printed circuits (and other open circuits in the agc line), and a certain type of defect in the rf-amplifier transistor.

The base bias of the rf amplifier transistor is a reliable indication of the source of trouble; if it increases and decreases as the receiver is tuned across the band, but conduction of the transistor (as indicated by the voltage drop across the emitter resistor) remains constant, then the probable cause is a leaky transistor. Substitution is the best method for checking this defect. In such cases, it seems that collector-to-base leakage is the villain, and it might not show up on ordinary transistor checkers, except occasionally as a lower *beta* reading.

In cases where it is the RF amplifier that is inoperative, you also can rely on the voltages measured on the transistor elements. Figure 11-6 shows the normal range of these voltages in a typical circuit. The emitter-to-base voltage will be about 0.2 to 0.3 volts if a Germanium transistor is used, and 0.6 to 0.7 volts if a Silicon transistor is used.

There is also one other voltage of interest in this stage: the DC conduction voltage. The voltage drop across the emitter resistor is caused almost exclusively by conduction between the emitter and collector. Consequently, you can consider the emitter voltage drop as caused entirely by the collector-emitter current (the base current contributes only 2 - 5 percent). In a pnp stage, there is also

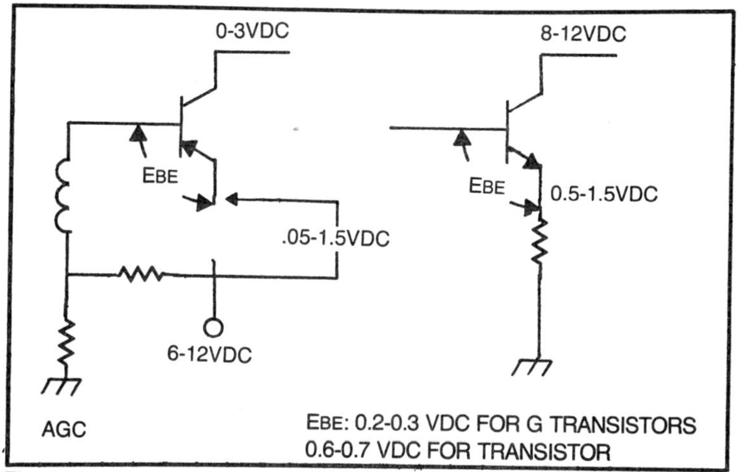

Fig. 11-6. Effect of agc on RF amplifier voltages as radio is tuned.

frequently found a small voltage (0.25 to 3.0 volts) between the collector and ground. Again we are referring to negative-ground circuits. The pnp collector voltage is directly proportional to current conduction. Since, in a normally operating system, these voltages vary according to the strength of the received signal, they can be used to judge the condition of the rf amplifier/agc circuit. Do not be confused by normal variations in these voltages. In a Germanium PNP rf amplifier, for example, a reading of 0.3 volts from emitter-to-base might indicate that no signal (or a very weak signal) is being received, while 0.15 volts might mean that a strong signal is being received. A reading in the 1.5 - 3.0 volts range, on the other hand, could indicate an open emitter-base junction. It is the large change in bias that is of consequence.

To efficiently troubleshoot the rf amplifier and agc sections of a receiver, it is necessary to determine how particular defects affect the relative bias and conduction voltages. A high emitter-to-base voltage, as already indicated, can mean an open emitter-base junction. When the bias varies normally as the set is tuned across a station, yet the conduction voltage indication remains relatively constant, you can generally bet that the transistor is leaky. These problems are discussed below in further detail in the form of a few case histories.

Case No. 1

A mobile receiver is capable of receiving only from stations that are relatively nearby, or are running substantial amounts of power.

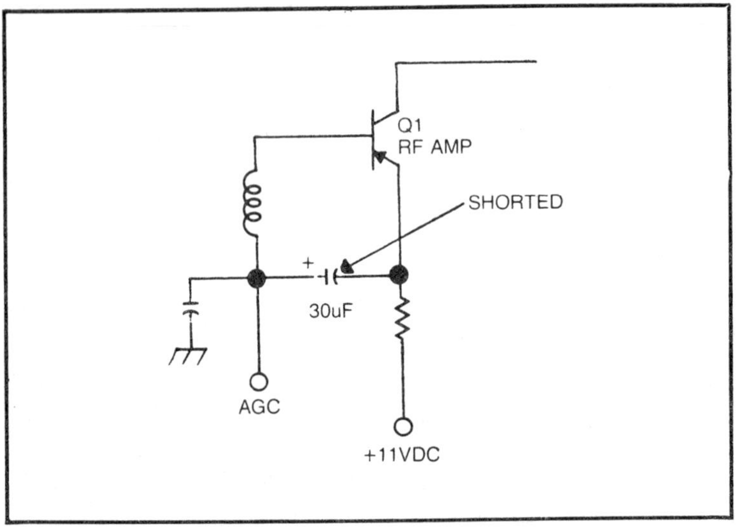

Fig. 11-7. RF amplifier is turned off when agc filter capacitor is shorted.

Stations at a distance were inaudible. The loud hiss coming from the speaker indicated that the trouble was in either the antenna or the rf amplifier (experienced people can often tell approximately which stage or section is at fault by listening to the hiss level).

A test antenna proved that the trouble was not in the antenna system (all decent shops that service mobile rigs will keep a test antenna handy to make this type of determination).

Inside on the bench, we popped the cover off the radio and used a voltmeter to check the voltages in the rf amplifier circuit. The collector-to-ground and emitter-to-base voltages both were zero. The voltages from ground to both the emitter and base were close to the normal 1 volts expected. Because a transistor in this type of circuit seldom develops a short circuit across the base-emitter junction (and even when it does there is some resistance), it was decided to look in the circuitry external to the transistor. When we disconnected the agc capacitor (in Fig. 11-7). The receiver began oscillating. The 30 μF electrolytic agc bypass capacitor was shorted.

Case No. 2

Figure 11-8 is the partial schematic of the rf amplifier in a multi-band receiver. The complaint in this case was intermittent operation on one, or a few, bands. Bandswitches are subject to build up of dirt and grease, so often become intermittent.

Bandswitches are also very difficult to replace, and this is recommended for a local professional or the factory (preferred). Before packing the rig off to the factory, however, try cleaning the contacts with a good chemical cleaner, or, if possible, the tip of a pencil eraser. Be sure to get into all corners of the bandswitch, and clean it several times in different portions of the rotor (with the switch in different positions).

Bandswitch problems can also produce sysmptoms such as an intermittent squeal, or oscillation.

Case No. 3

This case involved a type of distortion that occurs only on strong, local signals. If you tune across any relatively well used Amateur band, you will find some blockbuster signals and many not-so-strong signals. If the receiver exhibits proper reception on weak and moderately strong signals, and bad distortion on strong signals, then it is possible that the problem is in the agc circuit.

We checked the bias on the rf amplifier transistor, and the transistor proved to be completely cut off. Bridging a 1-megohm resistor from collector to base turned the transistor on, and it conducted normally. This fact led us to inspect the agc circuit. An ohmmeter set to the Rx100 scale indicated that the agc rectifier was defective. A new 1N82A cured the problem.

Case No. 4

Static was the complaint in the next case. Because the area had experienced rain storms on and off for three weeks, we suspected the trimmer capacitors, or the capacitors inside the I-F transfor-

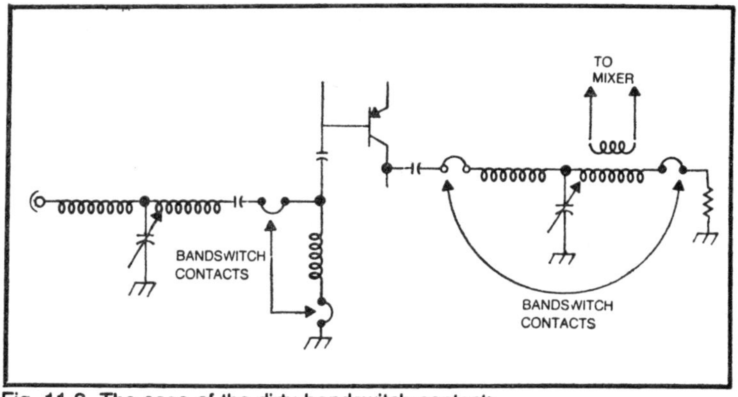

Fig. 11-8. The case of the dirty bandswitch contacts.

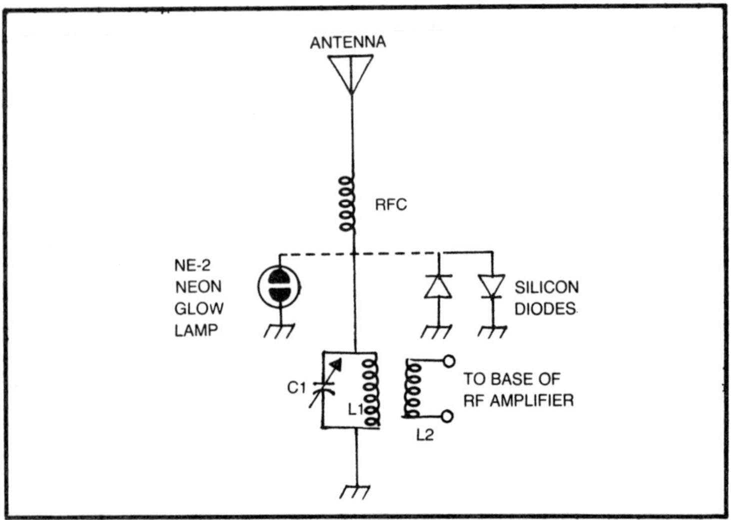

Fig. 11-9. RF amplifier input protection against static electricity.

mers. A signal tracer proved that the noise was in the rf amplifier. A voltmeter reading of the collector voltage showed some variation when static was present (this is difficult to see on digital voltmeters). We then cut the printed circuit foil to one of the trimmer capacitors, and inserted a 0.01 μF disc ceramic capacitor across the break, in series with the trimmer. The noise ceased, proving that the trimmer was arcing. A new trimmer capacitor cured the problem.

Other Problems

I recently had a fellow tell me that his solid-state high frequency mobile transceiver developed a problem when somebody *touched* the antenna! It seems that he was getting set to adjust the antenna for VSMR. The rig was on, and in the receive mode. The rig was operating normally until one of the workers went over to the antenna and touched it. Then the set went dead to all but strong local signals.

The problem was a static discharge from the worker's body entering the rf amplifier via the antenna transmission line. Figure 11-9 shows how this could be prevented.

Chapter 12
Receiver Detector Problems

The radio signal transmitted by the station cannot be heard by the human ear. The nature of the signal is electromagnetic, not acoustical. The station transmitting the signal will impress something that sometimes passes for "intelligence" onto the rf "carrier," so that it will propagate through free space. Although the audio modulation is "riding" on the rf signal, we cannot hear it. It is, after all, a part of the rf signal. Even if, as in certain medical-electronic applications, the signal was acoustical waves, we could not hear them because their frequency is many times higher than the range of human hearing.

What to do? The answer lies in *demodulating* or *detecting* the signal to extract the modulation and reject the carrier component. The exact circuits used for this service vary with the type of modulation (i.e., AM, fm, SSB, DSB, CW, etc), but all seek to accomplish essentially the same end.

AM DETECTORS

Without getting into any arguments over some of the new (and resurrected old) detecor ideas, let us point out that the AM detector is probably the simplest radio detector in use.

Amplitude modulation (AM) requires that the audio signal modulates the carrier in such a way that it varies the amplitude of the rf carrier signal (see input waveform in Fig. 12-1A). The amplitude of the carrier, then, will rise and fall in step with the audio signal modulating it. This type of signal requires an *envelope detector*, in the receiver because the rising and falling peaks and crests of the carrier form an rf envelope.

The simplest envelope detector, and the one principally used in AM receivers, is the simple diode circuit of Fig. 12-1A. The diode is used to rectify (just as in a power supply) the signal from the I-F amplifier. The rectified signal is then averaged and filtered in a low pass *RC* network to extract the audio. Note that this filter will have a cut-off frequency of 5-kHz in AM broadcast receivers, and something a bit lower (2.5 - 3 kHz) in communications receivers. The *RC* network is also sometimes called a "tweet filter" because it serves to attenuate the high frequency heterodynes from nearby stations.

Most of the problems associated with the AM detector involve the diode; it can become shorted or open. Oddly enough, in many instances the symptoms resulting from both types of failure is the same: weak, distorted audio output. There is often just enough capacitance across the open diode to pass some signal, which is then rectified by the base-emitter junction of the first audio preamplifier. Similarly, when the diode shorts, this applies the I-F signal to the same point. Of course, the tweet filter attenuates most of the I-F signal, but enough reaches the audio preamplifier input to cause at least some output. The 1.5-kohm resistor (R3) in Fig. 12-1A is a diode load, and will cause low output if it is open.

A variation on the AM detector theme is shown in Fig. 12-1B. In the former case, the diode return, through the transformer secondary to ground, is direct. In other cases, however, as in Fig. 12-1B, the transformer might return to a V+ supply. The AC return for the I-F signal, then, is through capacitor C1 to ground. A second V+ supply is used to keep the first from forward biasing the diode.

There are actually two problems associated with circuits such as Fig. 12-1B. One is that the return capacitor, C1, will open, thereby losing the AC return for the detector. This results in a weak, but usually not distorted, output. Generally, one can suspect the detector circuit whenever the S-meter reads a very strong signal, yet the output is low. Even when the last I-F transformer is poorly aligned, there is usually sufficient output signal to produce loud audio on stronger stations. Only the sensitivity will suffer.

To find a bad diode, disconnect one end, and then apply the ohmmeter tests given in Chapter 19, or use a diode tester. To find if the capacitor is bad, it is best to bridge a "known-good" capacitor across C1. If the reception returns to normal, then assume that the problem is the capacitor.

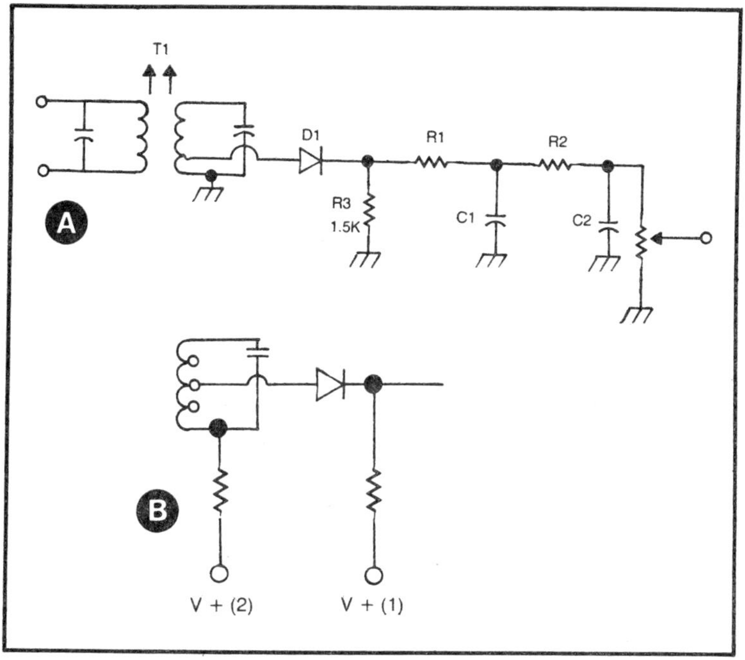

Fig. 12-1. (A), Simple envelope detector, (B), variation of the circuit.

The other unique problem in this circuit is that the diode may lose one of its V+ supplies. If V+(1) is lost, then the diode will become forward biased, and pass a DC level. The result will imitate a shorted dioded, although (due to circuit resistances), it will rarely actually burn out the diode. If, on the other hand, we lose the V+(2) supply, the diode will be heavily reverse biased, and will therefore appear to be an open diode. It will be wise to examine the schematic of the set to determine whether or not this circuit is being used.

Although not specifically "detector" problems, the I-F output transformer problems can often resemble loused up detectors. In most cases, Amateur equipment uses an S-meter in the receiver, so you can get a pretty good idea whether or not the other stages are working by observing the meter. If there seems to be normal S-meter activity, and the S-meter circuit is taken from the output I-F amplifier, then you must suspect a problem in the last I-F, the detctor, or the stages to follow. If you hear normal scratching sound as you operate the volume control (on an unused frequency, or else other noise will interfere!), then you can assume that the audio stages are all right. This leaves the detector, and the output side of the I-F amplifier.

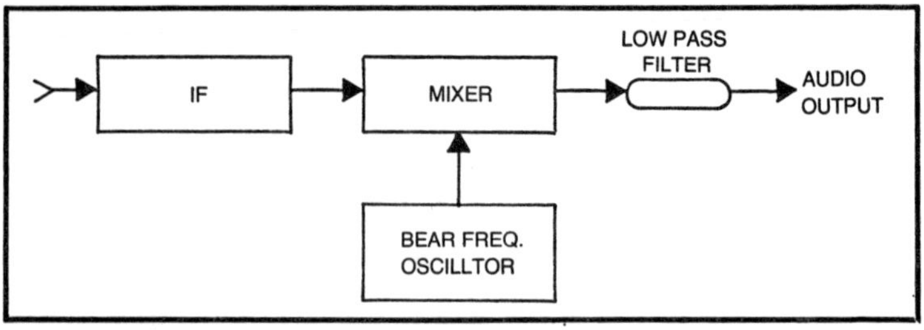

Fig. 12-2. Simplified view of the product detector.

One of the best ways to examine the circuit is to use an oscilloscope, or a signal tracer equipped with a demodulator probe, or an rf voltmeter, on the output side of the last I-F transformer. If there is a healthy signal voltage at that point, and if it seems to vary normally as you tune across the band, then assume that the transformer is good. Do not attempt to "align" the transformer as a troubleshooting technique. That brands you as a boob. The principal faults in I-F transformers are, a) open windings, b) open or shorted tuning capacitors (hard to find), and, c) winding-to-winding shorts. The latter problem will place the V+ from the I-F amplifier on the diode detector circuit, so it is relatively easy to spot.

SINGLE-SIDEBAND DETECTORS

Single sideband (SSB) is a variation of amplitude modulation in which greater efficiency and reduced bandwidth are accomplished by eliminating the rf carrier signal from the transmitted signal. At the receiver end, however, the rf carrier must be reinserted before the signal can be properly demodulated. In fact, it must be reinserted at exactly the same frequency and in exactly the same phase as the original, or distortion of the recovered audio will result.

In Amateur equipment, we cannot hope to reinsert the carrier at so precise a frequency and phase relationship to the original, so will settle, instead, for a little distortion. Typically, Amateur equipment reinserts the carrier at a frequency within ± 25 Hz of the original frequency.

The demodulator used in SSB receivers (Fig. 12-2) is a *product detector*. This means that the signal from the I-F amplifier will be mixed with a local oscillator signal from a *beat frequency oscillator* (BFO) in a nonlinear element. This will cause heterodyning that results in the difference frequency being output. The difference frequency, of course, is the audio modulation from the carrier.

SSB product detectors are subject to all of the same ills as the AM detector, plus a few related to their own little world. Principal among these is loss of the BFO signal, or having the BFO off frequency.

For an example of what a receiver sounds like when you lose the BFO, turn it off (or switch to AM), and then tune in some SSB stations. The "donald duck" sound you hear is exactly the same as will be heard when the BFO is not working! Of course, a few designs are such that the receiver simply goes dead when the BFO is lost.

An off frequency BFO causes an extremely aggravating problem: the set seems to tune . . . but never quite make it! This is not too much of a problem on older models that used a variable frequency BFO, or on more modern equipment with a "clarifier" (a small amount of variability of BFO frequency), but is a tremendous problem in receivers in which the BFO frequencies for upper and lower sideband are fixed by crystals. In general, both USB and LSB being off points to the BFO circuit, while just one being off points to the appropriate crystal.

FM DETECTORS

In past decades only two major FM detectors were seriously considered for Amateur (and commercial) communications receivers: the Foster-Seely discriminator and the ratio-detector. While both of these circuits are still with us (even in IC form!), other designs are more often seen than was previously the case. Some of these circuits are brand new designs, spurred by IC technology, while others are merely ICized versions of older ideas.

A Review of FM Fundamentals

In frequency modulation (FM) systems, the carrier frequency of a radio transmitter is varied by an audio-frequency voltage (see Fig. 12-3). The relationship between the modulating voltage and the rf-carrier frequency is shown in Fig. 12-3. When the audio signal voltage is zero, then the rf carrier will be at a frequency F_c, called the "carrier frequency." This is the frequency quoted as the channel frequency in communications equipment (for example, 146.34 MHz). When the audio-signal *voltage* increases in the positive direction, the rf carrier frequency will increase, until it reaches a maximum point (F_2) when the audio signal reaches its peak. As the audio signal passes its peak, the rf-carrier frequency will swing back the other direction, until it is once again at the carrier frequency (F_c) at the zero crossover of the audio signal. The rf carrier will then swing lower in frequency as the audio signal goes negative. It will reach a minimum frequency (F_1) when the audio reaches its negative-amplitude peak. The rf carrier frequency will then start to increase (toward F_c) as the audio signal decays back to zero.

The "true FM" transmitter described in the previous paragraph is only occasionally found. In order to meet any decent frequency-stability criterion, most communications FM transmitters use a similar modulation type called *phase modulation* (PM). From the receiver's point of view, FM and PM are practically identical. In the PM transmitter, the rf carrier freuency is held constant, and its *phase* is varied with the audio carrier. This arrangement allows the carrier to be generated in a highly stable crystal-oscillator circuit that is all but immune from frequency changes. Phase variations occur in a stage following the crystal oscillator, called a *reactance modulator*.

There are several concepts in FM and PM circuits that are sometimes misunderstood. For example, *deviation* is the amount of change in carrier frequency between its unmodulated value (F_c) and one of the extremes (either F_1 or F_2). Deviation is specified in units of frequency,i.e., hertz or kilohertz. *Frequency swing*, on the other hand, is the total frequency shift from the lowest extreme to the highest extreme (i.e., F_2 minus F_1), and is also specified in units of frequency. The relationship between the deviation and the frequency swing depends upon the modulating waveform. For a perfectly symmetrical waveform, such as the sinewave in Fig. 12-3, the deviation will be precisely one half of the frequency swing. But if the sinewave were distorted, or if another (nonsymmetrical) waveform were used, then the positive deviation ($F_2 - F_c$) would not equal the negative deviation ($F_c - F_1$) so we would not expect the frequency swing to be exactly twice the deviation, but instead, it would be $[(F_2 - F_c) + (F_c - F_1)]$.

Neither deviation nor frequency swing is affected by the modulating frequency (i.e. the audio) in a true FM system; this is not true in a PM system. The PM modulator provides a natural 6 dB/octave rising (preemphasis) characteristic, while true FM is essentially flat. In many cases, however, extra circuitry in a true FM transmitter provides a 6 dB/octave preemphasis because it improves the signal-to-noise ratio of the high frequencies.

In FM transmitters, the audio frequency determines the *rate* at which the carrier swings through its excursions. The *amount* of deviation is determined, not by the audio frequency, but by the *amplitude* of the audio signal.

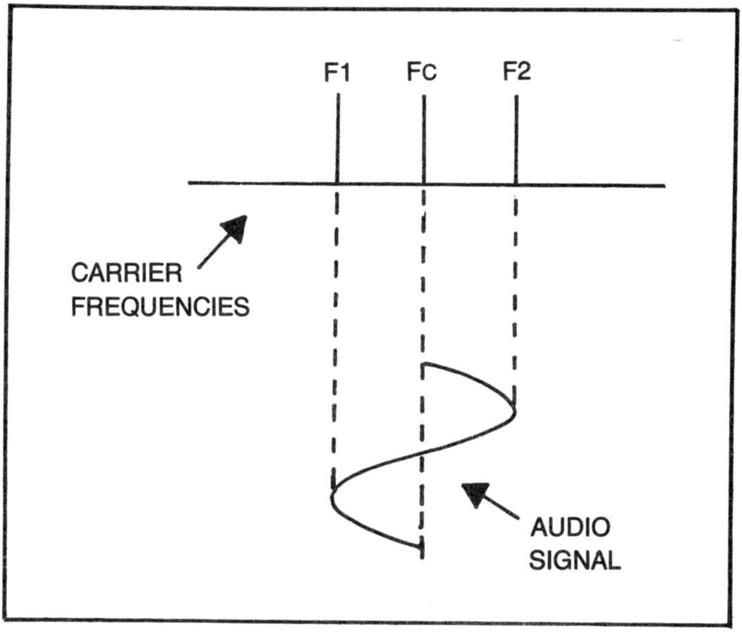

Fig. 12-3. Relationship of the audio signal to the FM RF carrier.

Full, 100-percent modulation in an AM transmitter occurs when the audio signal causes the carrier amplitude to double on positive peaks and drop exactly to zero on negative peaks. FM signals, however, have no such easily recognized physical feature on which to hang a definition for "100-percent" modulation. In FM and PM systems, "100-percent" modulation is determined by more arbitrary methods, such as allowable channel bandwidth, etc. In the FM broadcast band, for example, 100-percent modulation is a deviation (by a sinewave) of ±75 kHz. But, in television audio (an FM signal in the USA) 100-percent modulation is *defined* as 25 kHz. In two-way communications, including most Amateur systems, 100-percent modulation may be defined as a deviation of only ± 5 kHz. All of these are "100-percent" modulations, not for a physical reason, but because the F.C.C. or common practice (as in the Amateur service) has caused the definition to be set by convention.

Until recently, when specifications were changed to accommodate noise-reduction systems such as Dolby, FM broadcast transmitters gave the audio signals a great deal of high frequency preemphasis. A 75 microsecond *RC* deemphasis network was used at the output of the FM demodulator to restore the proper audio balance.

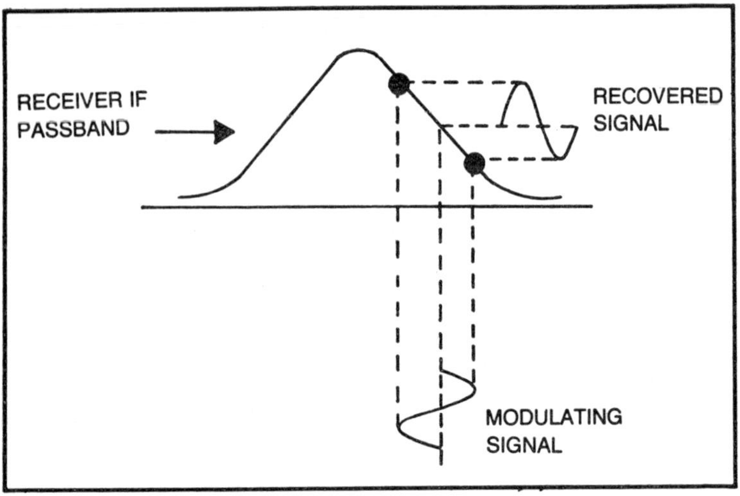

Fig. 12-4. Slope detection of FM using AM receiver.

SLOPE DETECTION

Figure 12-4 illustrates a crude, but effective, method of FM demodulation. This technique is shown here to emphasize the most critical aspect of any FM demodulator circuit: *a frequency response which varies as a function of input frequency*. This method is aptly called "slope detection." It requires a receiver with a relatively narrow passband. The center of the carrier is tuned so that it lies on the *downslope* of the I-F response curve. The incoming signal, then, sees an I-F response which varies as the carrier frequency varies.

The slope detection system is not terribly effective for most FM communications. For one thing, it is difficult to tune, and is responsive to impulse noise (as is any AM detector). The principal use, several years ago, was in using the car radio, or a transistor AM portable, to demodulate the output from a VHF converter. There were many VHF "police band" or "2-meter" converters, sold for mobile use, that plugged into the car radio. These devices depended upon slope detection. When everything was perfect, including the tuning of the AM radio, then the system worked passably well.

FOSTER-SEELY DISCRIMINATORS

Figure 12-5A shows the circuit for a Foster-Seely discriminator, used for many years as one of the principal forms of FM detector. In fact, the old discriminator is still with us in many varieties, but essentially the same form.

Note that rf choke L1 is common to both the primary and the secondary windings of the detector transformer T1. In fact, it is in series with the secondary and in parallel with the primary winding; look closely at the redrawn circuit of Fig. 12-5B.

This common conection of L1 allows the use of its voltage and current as references. When the i-f signal applied to the primary of T1 is unmodulated, it will be at a frequency equal to the resonant frequency of the T1 secondary. This causes L_{s1} and L_{s2} to be equal, and currents I_1 and I_2 to also be equal. Since these currents flow in opposite directions, however, they will tend to cancel each other, and the new output voltage will be zero.

Figure 12-6A shows the voltage and current-vector relationships in the discriminator when the frequency of the input signal increases above F_c. Because the secondary tank circuit takes on inductive properties at these frequencies, the secondary current I_s *lags* behind the secondary voltages E_{ls1} and E_{ls2} by 90 degrees. Since the voltage and current in an inductive circuit are out of phase, they must be vectorialy added to find the resultant. These are labeled E_{d1} and E_{d2} in Fig. 12-6A. In this case, the voltage applied to diode D1 is greater than the voltage applied to D2, so the

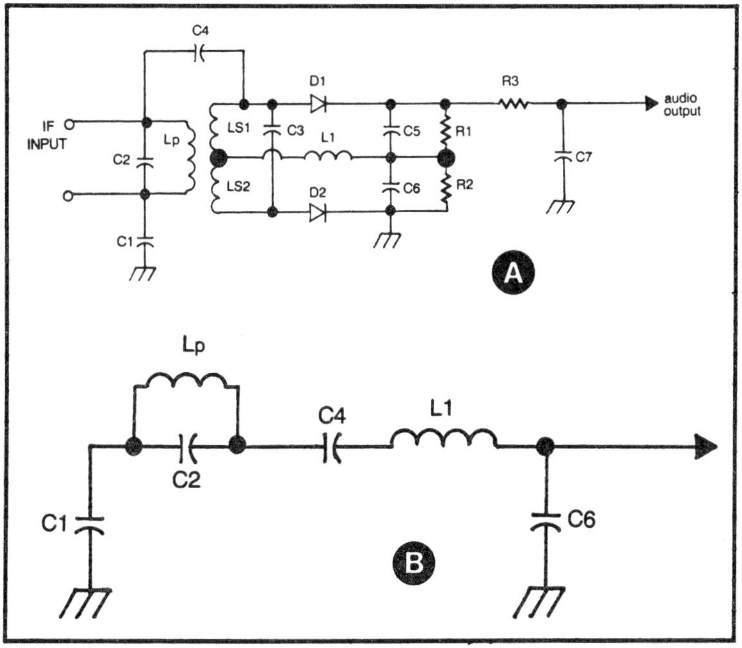

Fig. 12-5. (A), Foster-Seeley discriminator, (B), partial simplified circuit.

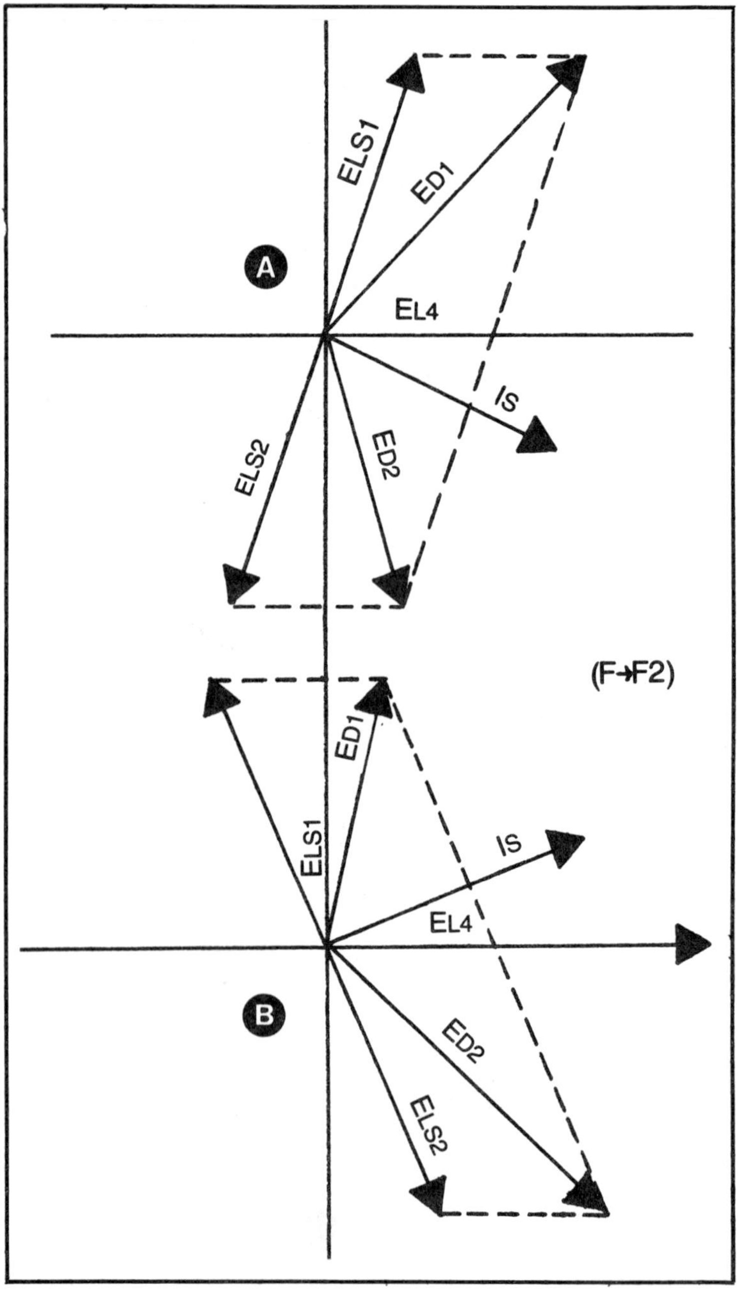

Fig. 12-6. Voltage vectors in discriminator circuit, (A), as F approaches F2, and (B) as F approaches F1.

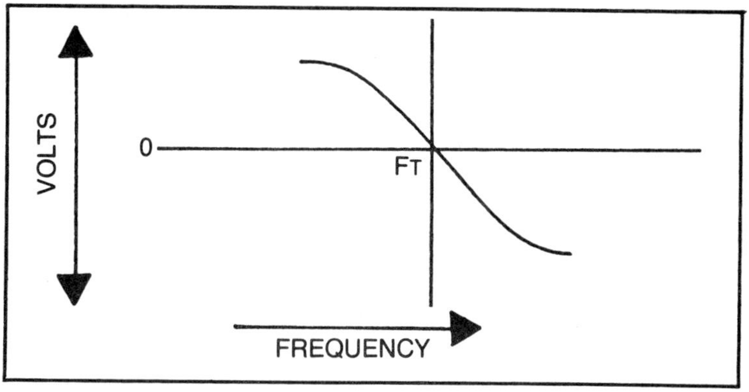

Fig. 12-7. Discriminator "S" curve.

current I_1 can be expected to be greater than I_2. Under these circumstances, the currents no longer totally cancel out, and an output voltage is generated. Similarly, in Fig. 12-6B, we see the situation existing when the carrier frequency decreases below F_c. The resultants E_{d1} and E_{d2} are reversed, so E_{d2} predominates.

Figure 12-7 shows the typical voltage-vs-frequency response curve of a typical discriminator. Part of the servicer's task in the alignment of any FM receiver is to place F_c right at the zero crossover point on this curve. The bandwidth of the circuit must be such that the expected deviation (i.e., 5 kHz, 25 kHz, or 75 kHz) will not drive the curve into a nonlinear region.

RATIO DETECTORS

Figure 12-8 shows a typical ratio-detector type of FM detector. The major difference between this circuit and the Foster-

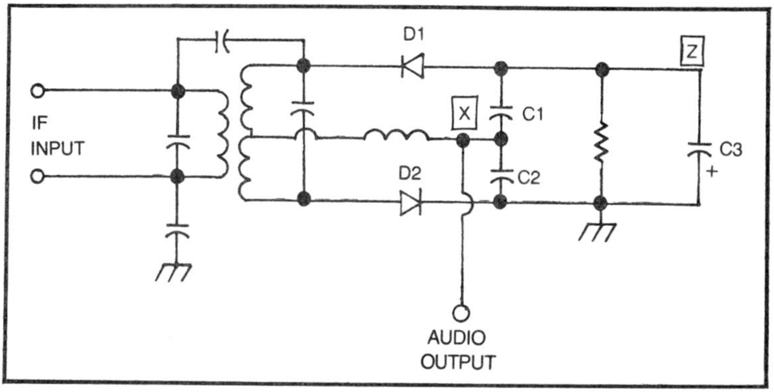

Fig. 12-8. Ratio-detector circuit.

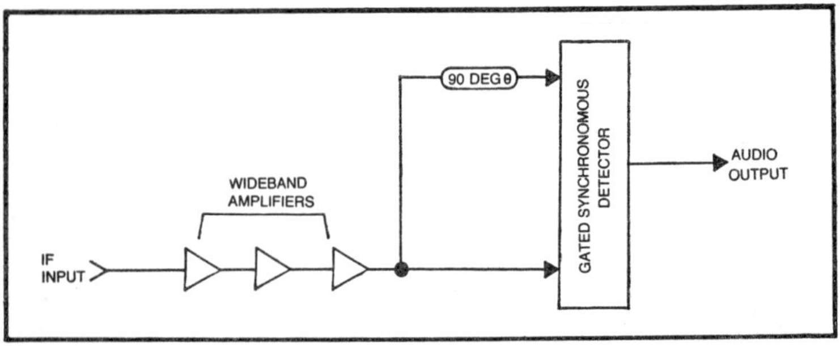

Fig. 12-9. Block diagram of a quadrature FM detector.

Seeley discriminator is that the diodes are connected in series in the ratio-detector circuit. This allows the voltages across C1 and C2 to add rather than cancel. When the input signal is at its unmodulated frequency, voltages across these two capacitors will be equal. When the carrier is deviated to a higher frequency, however, the voltage across C2 increases and that across C1 drops. Just the opposite occurs when the deviation is in the other direction: E_{c1} rises while E_{c2} drops. This, of course, results in a dc-voltage level which varies as the modulation causes the frequency of the carrier to deviate above and below the carrier frequency.

Capacitor C3 in this circuit has two main functions: 1) it stabilizes the voltage across the series combination C1/C2 so that the ratio can be taken, and, 2) it suppresses any AM including noise impulses, which may be on the carrier. It is this last function which makes it possible for ratio-detector-equipped receivers to function without a limiter stage prior to the detector (as is needed to make discriminator circuits noise free.)

IC QUADRATURE DETECTORS

Integrated circuit technology has revived a type of FM detector that was once used extensively in television receivers; the *quadrature detector*. Once popular using the 6BN6 gated-beam tube, the quadrature detector has made a substantial comeback in the form of several integrated circuits such as the ULN2111, MC1357P, and others.

Figure 12-9 shows the block diagram of a typical IC quadrature detector (ICQD). The input stages form a wideband, high-gain, limiting preamplifier whose output is a series of square waves. These are fed to two places: to one input of a gated synchronous detector, and to a quadrature (i.e., 90 degrees) phase-shift network (usually an *LC* tank circuit external to the IC.) The output of this network is brought back inside the IC, as shown in Fig. 12-10, and is connected to the alternate input of the gated detector. This detector produces output pulses with constant amplitudes, but whose periods vary with the modulating signal. These are then integrated in an *RC* network to recover the audio signal.

The ICQD is used in many applications. FM receivers for broadcast reception, Amateur communications, commercial communications, and other FM receivers are found with one of the various ICQD devices. But, be aware that the use of an IC in the FM detector doesn't automatically mean it's an ICQD. Many FM detectors are IC gain blocks that also contain the diodes needed to form ratio detector and discriminator circuits. It is the transformer (instead of a phase coil) that provides the clue.

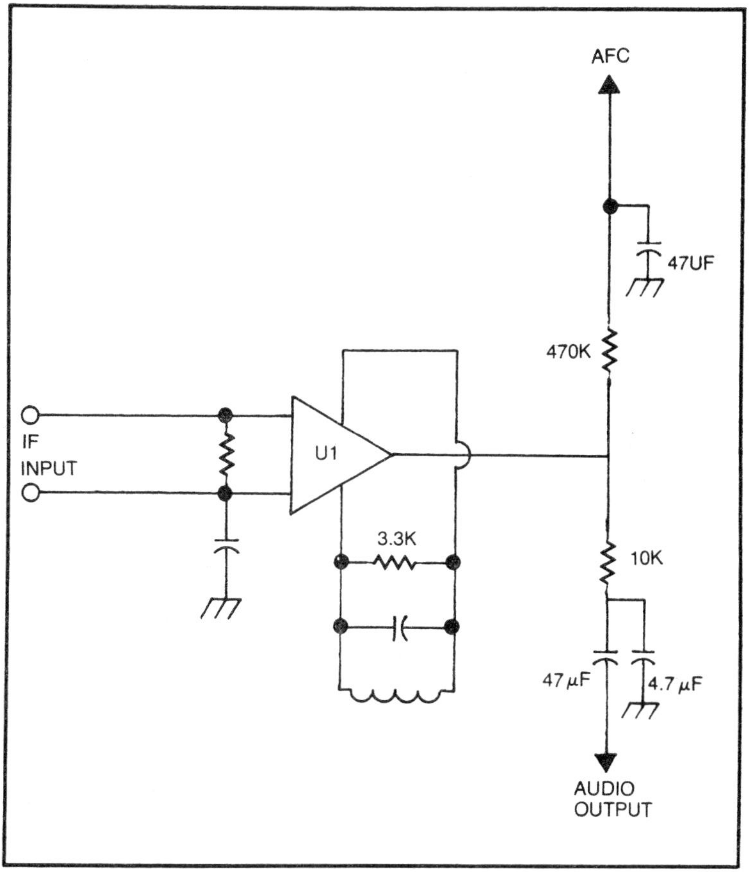

Fig. 12-10. Circuit of an IC quadrature FM detector.

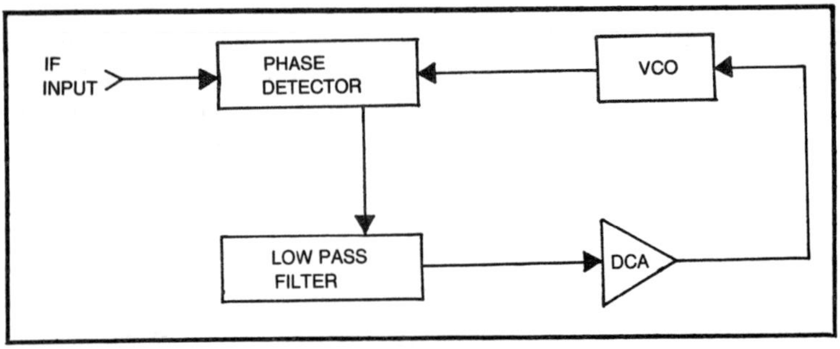

Fig. 12-11. Block diagram of a PLL FM detector.

PHASE LOCKED LOOP (PLL) FM DETECTORS

Although originally developed in the 1930's when, oddly enough, it was billed as an AM detector, the PLL has only come into its own with the recent introduction of the PLL IC. Figure 12-11 shows the block diagram to a typical PLL chip. Although used for other purposes as well, here I shall describe the circuit action as if it were exclusively used as an FM detector. (Additional information is presented in Chapter 20).

A phase detector in the chip receives two inputs: one from an internal voltage-controlled oscillator (VCO), and the other from the I-F amplifier. Under conditions where the carrier is unmodulated, the I-F and VCO frequencies are identical, so the phase detector output is zero. As the I-F signal deviates, this equality is lost, and a DC error signal is developed which is proportional to the frequency difference between the I-F and VCO signals. This is fed through a low-pass filter to remove residual rf signal, and then on to a DC amplifier (DCA). The output of the DCA controls the VCO frequency. This pulls the VCO to the new input frequency. Since the signal is always deviating about the center frequency, the VCO will always be trying to "catch up." The control voltage will be continuously varying at the rate of the audio which modulates the carrier. This voltage, therefore, may be used as the recovered audio.

Figure 12-12 shows an actual IC PLL FM demodulator using the popular Signetics 560B chip. The I-F signal is coupled to the chip via capacitor C7 and pin 12. The frequency of the VCO is set by capacitor C4. Since internal resistances normally vary ±20 percent in ICs, this capacitor will most likely be a trimmer. The input signal must be between 2 and 15 millivolts. Below 2 mV the PLL may have difficulty in tracking the signal, while above 20 mV the AM suppression is lost. Networks R1/C1 and R2/C2 form the low-pass filter between the phase detector and the DCA. The output signal is

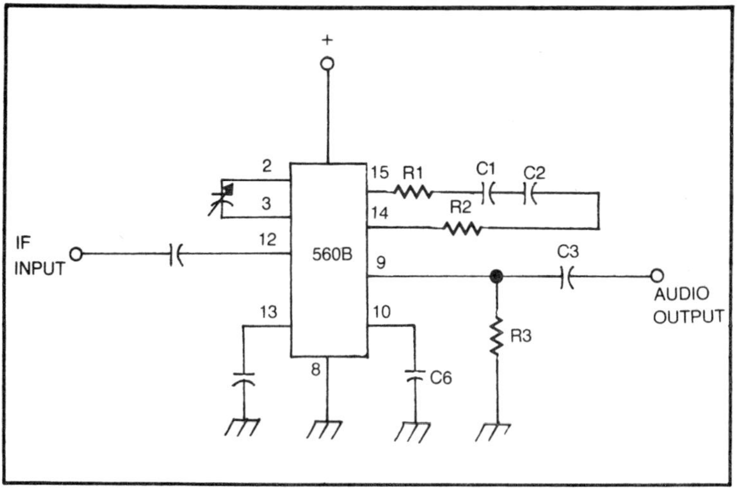

Fig. 12-12. Circuit of a PLL FM detector.

obtained from pin 9 and is coupled through network R3/C3 to the following circuits. The function of capacitor C6 is deemphasis. It has a value selected (by the PLL manufacturer) for the 75 μS deemphasis used in broadcast receivers when the internal resistance of the IC is 8 kohms. In practice, this resistance will vary ±20 percent, so a 0.001 μF capacitor is usually sufficient.

PULSE COUNTING (DIGITAL) FM DETECTORS

Figure 12-13 shows a "coil-less" FM detector which, until recently, was pretty much restricted to use in FM telemetry, analog FM data recorders, etc. At least one manufacturer, and

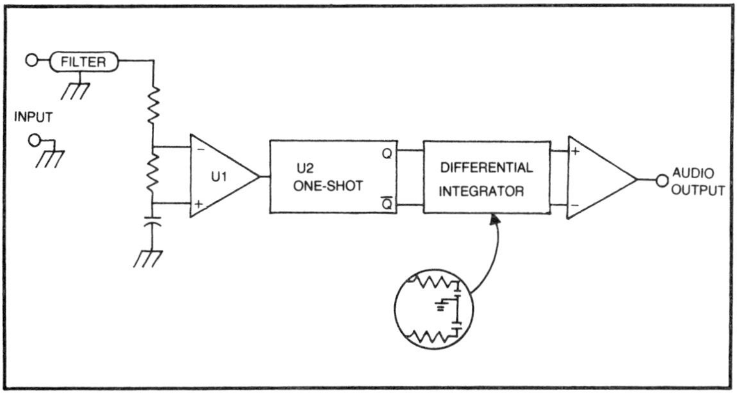

Fig. 12-13. Digital, or pulse-counting, detector.

several articles in the Amateur press, have used this detector in FM receivers. In fact, the Heath Company used the design in their first high-dollar "digital" FM tuner.

FL1 is a bandpass filter which might be either a piezoelectric (ceramic) type, or an *LC* tank circuit. U1 contains a gain stage and limiter amplifiers. IC U2, however, is a special TTL digital-logic element called a "retriggerable monostable multivibrator" (one-shot). This chip will produce a single pulse every time it is triggered by an input pulse from U1. These pulses will have constant amplitudes and durations, and only their repetition rate will vary with signal deviation. There are two complementary outputs from U2, designated Q and not-Q. Each of these opposite outputs is fed to an *RC* integrator which averages the signals to obtain an audio signal. These audio signals can, in turn, be fed to the differential inputs of an IC operational, or Norton, amplifier.

Chapter 13

Troubleshooting Distortion Problems

Distortion is any change of a signal as it passes through circuitry. All circuits possess some distortion; none will actually pass a signal totally unmodified except for amplitude. In the high-fidelity world, of course, distortion is a big bugaboo, and much money is spent in an effort to eliminate as much as possible. Two decades ago, hi-fi enthusiasts would boast of their wonderful new (usually monaural) amplifier that produces only 1.5 percent total harmonic distortion. But, today, we find that amplifiers are able to boast of less than 0.05 percent THD, which is lower than the most popular professional THD analyzer of ten years ago could measure! Would it surprise you to know that the output signal from many Amateur Radio receivers has a total harmonic distortion in the 5- to 10-percent region? Of course, this is not hi-fi, and it has been found that a little distortion does not hurt the intelligibility of the communications. In fact, at least one person has claimed that a small amount of THD actually improves intelligibility. I can't vouch for that claim, but I am certain that distortion does not necessarily deteriorate the performance of receivers.

INTERMODULATION

What is "distortion" and why are we so concerned about it? There are actually two basic types of distortion, three if you read hi-fi magazines. Intermodulation distortion (IM) is the mixing together, i.e., heterodyning, of two signals. You may already be familiar with this process in another context; it is also called "hetrodyning." If two signals are mixed together in a nonlinear circuit, then they will hetrodyne together to produce additional frequencies that did not exist before. For example, if we have two frequencies, A and B, mixed together in a nonlinear circuit, the output will contain A, B, A+B, and A–B, as well as certain additional products of their respective harmonics. In AM transmitters we seek this process, and call it "modulation." In that case, the two frequencies are the carrier and the audio modulating signal, while the product is the modulated rf carrier applied to the antenna. In receivers, and certain types of transmitters, we hetrodyne together two rf signals to produce a third signal. In these cases, the circuit must be nonlinear for the action to occur, and the result is highly desirable. But, in an amplifier, we do not want any nonlinearity. We can measure IM by a technique that will be covered in our discussion on measurements on untuned amplifiers (Chapter 21).

TOTAL HARMONIC DISTORTION

The other type of distortion is called total harmonic distortion, or THD. We know that the sinewave is the only "pure" waveform. It contains no frequency components other than the fundamental frequency. All nonsinusoidal waveforms contain harmonics. The exact harmonics present, and their relative amplitude and phase relationships, determine the actual shape of the waveform. If an amplifier is nonlinear, then it will produce harmonics from an otherwise pure waveform. This, incidentally, is one mechanism by which an overloaded TV receiver can seem to be suffering from harmonic radiation, even though your transmitter is clean. A nonlinear front end in the TV, whether due to overload from your nearby transmitter signal, or due to a malfunction in the rf amplifier of the TV, will generate 2nd and 3rd harmonics where none existed on the signal coming in. We measure these harmonics by comparing the amplitude of the output signal with and without the fundamental present. This is the job of THD analyzers.

In this chapter, we will not dwell too much on the theoretical whys and wherefores of distortion, but merely point out some of the more common causes, and how to fix them. Keep in mind, however, that distortion can often be caused by subtle problems that are difficult to find, regardless of the troubleshooting technique, or the ability of the troubleshooter. In many cases, either dumb luck or "shotgunning" is the only way problems are located.

AUDIO CIRCUIT FAULTS

The primary seats of ordinary distortion are the audio

amplifiers. But we must also take into account certain non-audio causes, such as oscillation, overload, and agc-circuit problems.

The audio stages of most receivers are now direct-coupled, cascade circuits using one or two preamplifiers and a power amplifier. Figure 13-1 shows a typical audio output stage from a mobile receiver. The output transistor is a PNP germanium type. These have been replaced in more recent designs, but were quite common up until a couple of years ago. The power transistor, Q3, derives its operating bias from the voltage drop across the 47-ohm resistor (R4) in the base circuit. This voltage drop is caused by the flow of collector current in NPN transistor Q2. This means that the collector current of Q3 will increase as the collector current of Q2 increases. Q2, in turn, is biased by Q1, and the voltage drop across resistor R2 (3.3 kohms). This network functions similarly to a voltage divider, with Q1 acting as a variable resistor. The voltage appearing between the base of Q2 and ground will decrease as more current is drawn by the collector of Q1, because more of the available power-supply voltage will be dropped across R2 by the heavier current flow. Since Q2 is an NPN transistor, this will tend to reduce the amount of current in its collector circuit, which in turn, will reduce its gain.

The 600-ohm variable resistor in the emitter circuit of Q1 (R1 in Fig. 13-1) sets the operating conditions for all three audio stages by controlling the amount of current drawn by Q1. Since we have one preamplifier transistor turning on, while the other is turning off, some people have dubbed this a "see-saw" amplifier.

Transistor defects are the major source of trouble in this type of circuit. Germanium output transistors are especially prone to

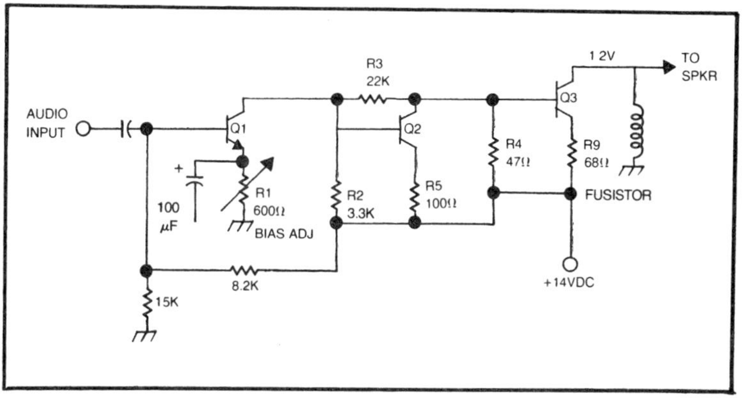

Fig. 13-1. Single-ended class-A audio output stage (PNP output).

becoming leaky, or developing collector-to-emitter short circuits. This often causes excess current to flow, resulting in blowing fuse resistor R9. The symptom often reported is "weak and distorted" audio.

A lot of different problems can cause weak and distorted audio. But there are also other tell-tale signs that point it to the audio transistor. One is that there is little or no "thump" as you turn the receiver on and off. Secondly, you might notice that either the body of the fuse resistor, or the printed circuit board underneath it (the resistor gets very hot before it pops) is charred, or completely burned. I have seen PC boards turned into charcoal because the user kept on operating the equipment after the defect was noted... on the grounds that it wasn't totally dead, so why not? Thirdly, you might make a quick voltmeter check of the collector potential, and note that it is missing. This is a negative-ground circuit because it was intended for use in a mobile receiver, so the collector of a PNP transistor used in the circuit will be slightly positive with respect to ground. In most cases, this means some value between 0.5 and 2.0 volts.

An alternate fault, due to the same problem, is that the receiver might draw too much current from the power supply. In this case, the fuse resistor has not yet blown, so the excess collector current in the power transistor is just serving to burn the rest of the circuitry. The collector voltage, if the circuit is of the type shown in Fig. 13-1, will increase from its normal value under 2.0 volts to 2.5 to 5 volts.

Troubleshooting these circuits can be time consuming unless proper procedure is followed, because all three stages can cause the same output indication. It doesn't help too much to try to "overisolate" the fault too early in the game. These transistors all interact with each other. Because of this, we must treat the entire three-stage chain as a whole.

A small collection of miniature alligator clip leads is a real help in this case. But, a word of caution; in fact, two words. One, do not try using the regular-size alligator clips. They are too large to get into many areas. With the reduction of equipment size, or increased component density, large clips will be a hindrance more than a help. Use, instead, the small type, or the clip-on Klep-Klips by Pomona Electronics. They are a lot safer in tight spots. In any event, use only high-grade alligator clips. They cost the same as the cheapies, but perform so much better. Second, do not even bother buying one of those multi-colored assortments of short alligator clip leads.

These are usually low-grade alligator clips crimped over a poor grade of wire. They will inevitably come loose, or oxidize, and cause problems. I have seen collections of these clips where every one was an open circuit within a week after purchase! We also see many cases where the alligator clip lead develops a resistance (1 or 2 ohms) that could be critical in some circuits. In fact, in an impromptu session at the work bench, I once dropped 12 volts, from a high-current filament transformer, to 10 the volts required by the heater of a transmitting tube by connecting three alligator clip leads from a $2.98 assortment in series!

Clip leads are especially useful in determining the cause of excessive current drain in audio power-amplifier circuits, such as Fig. 13-1. Use one clip lead to short together the emitter and base terminals of the power transistor. If the circuit still draws excessive current, then investigate the power transistor; it is probably shorted. But, if the current drain drops to only a few milliamperes, then you know that the base of the power transistor is still capable of controlling the collector current. The problem is most likely to be some place else in the circuit. In circuits such as Fig. 13-1, the problem would probably be found in one of the NPN preamplifier stages.

The next test point will be the base of driver transistor, Q2. Remove the alligator clip lead used to short the b-e junction Q3, and use it to short the b-e junction of Q2. If this causes the output current to cut off, then you are reasonably sure that Q2 and its associated circuitry are okay.

But, let me issue a word of advice. Analyze the circuit before connecting jumpers! In the case of the output transistor, it is usually okay to just go right ahead. But, in earlier stages, we must know just what effect cutting off the stage will have on later stages. This can prove fatal in direct-coupled circuits. Cutting off Q1, for example, will increase current flow in Q3! This example about cutting off Q2 applies *only* in circuits such as Fig. 13-1! In other circuits, you are on your own.

For reasons just given above, you may not be able to use the clip lead to tell anything about transistor Q1. Here, you must refer to voltmeters for information. The "see-saw effect" requires more careful testing in this stage than in the others.

We all develop certain ideas about which components are most likely to fail in any given circuit. In fact, many experienced troubleshooters will, when faced with a large number of parts to check, inspect first those "most likely to fail." This often yields results, even though it is not intellectually elegant. We all know, for example, that PNP germanium power transistors usually fail by developing collector-emitter leakage or shorts. Plastic-case power transistors, whether PNP or NPN, usually fail by developing a base-emitter open circuit. Similarly, small plastic transistors fail more frequently than those in TO-5 or TO-99 metal cases. Also, we know that potentiometers are a frequent source of problems.

COMPONENT PROBLEMS

In the circuit of Fig. 13-1, the 600-ohm potentiometer in the emitter circuit of the preamplifier transistor can develop an open-wiper condition, not uncommon where DC is allowed to pass through the potentiometer. These problems are often intermittent in nature, and so are the very dickens to troubleshoot. In fact, should you see a case where the Q3 current-drain increases intermittently, then suspect the current-setting potentiometer. Replacement is the best cure, but cleaning might be helpful if no replacement is available. This was a problem so great in some early 1960s car radios that the manufacturer redesigned the circuit to eliminate the potentiometer. I have seen similar circuits in Amateur and CB gear, so you are forewarned.

You can often find bad potentiometers through the simple technique of tapping the suspect pot with a pencil eraser (lightly!). If the current flowing in the Q3 collector circuit increase (or decreases), or if the distortion changes, then the potentiometer is bad. This part is suspect in most "no-apparent-reason" cases where the fusistor (fuse-resistor) is found blown.

Figure 13-2 shows a slightly different circuit, in which an NPN plastic-cased power transistor is used in the output stage. This type of circuit has replaced the PNP germanium types of earlier years. Note that the power transistor is plastic. These transistors cannot tolerate the abuse that can be given to TO-3 devices. If you bend the leads side to side, for example, they will snap the connection inside of the case, resulting in an open transistor! Bend the leads fore and aft, or at the extremities. The particular bending that causes the problem is when the leads are bent side to side, right at the point where they disappear inside of the TO-66 plastic package. Be careful.

In the circuit of Fig. 13-2, an NPN silicon transistor is used as the power amplifier. In the negative-ground system that is so common, this means that the collector will be at, or near, the power-supply potential, while the base and emitter terminals will be

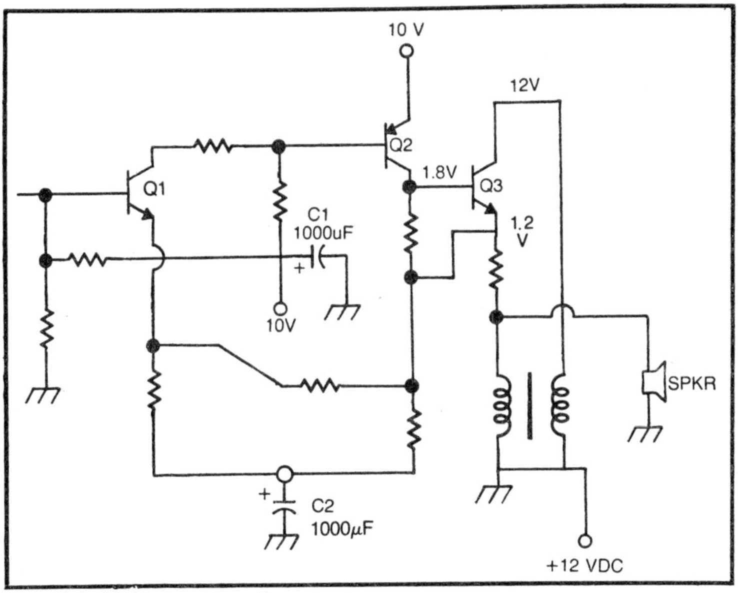

Fig. 13-2. NPN class-A audio output.

at a low-positive to ground potential. In the most common fault found on the plastic power transistor, the b-e junction opens. This will cause the emitter potential to drop to zero, and the base potential to rise to a point close to, but generally less than, the power-supply potential applied to the collector.

The circuit of Fig. 13-2 is predominantly direct-coupled, so most faults that cause distortion will be defective transistors. Only in a few cases will you find resistors at fault. In AC-coupled circuits, however, you will find leaky and shorted coupling capacitors causing bias conditions that result in decreased audio output, and a tremendous increase in distortion.

The capacitors in Fig. 13-2 are bypass capacitors, and are used for decoupling. If either capacitor opens up, a common occurence, then the result will be either motorboating (a low frequency—less than 10 Hz—oscillation) or a loss of gain. In the latter case, the distortion is not markedly increased, but the audio is weak; too weak to drive a loudspeaker.

AM DETECTOR DEFECTS

Most AM receivers use a simple envelope detector to demodulate the signal from the I-F amplifier. A germanium or silicon detector diode is used (Fig. 13-3) to rectify the signal. If the diode is overloaded by a strong signal, as happens sometimes—like when your neighbor fires up a kilowatt power amplifier with the beam aimed right at you—then the detector diode might short out. Normally, one would expect this to cause the receiver to go entirely dead. But, if the input signal is strong enough, it can feed sufficient signal through the shorted diode, the 5-kHz "tweet filter" following the diode, and the volume control, to the first audio preamplifier, where it is rectified. The rectified signal will most likely overdrive the preamplifier, causing the output to be distorted. If the symptom is distortion on strong stations, and practically zero output on weaker stations, then suspect the detector diode.

This same type of fault can exist on FM detectors, especially those that use diodes (discriminators and ratio detectors). In the FM detector, there is also the added problem of an open or shorted diode upsetting the balance of the circuit, thereby creating distortion.

FM DETECTOR PROBLEMS

FM detectors of all types, except the non-tuned types (phase locked loops and "pulse-counting," or "digital" types), will produce a distorted output if the demodulator circuit balance is upset. In a new set, or one where the detector transformer has been recently replaced, the alignment of the detector transformer secondary is suspect. The alignment balances the circuit, and if incorrect, will produce a non-symmetrical response curve, hence a distorted output.

Note well, however, that distortion that appears suddenly after the set has been working for awhile is probably due to something besides alignment. Do not attempt to align, or troubleshoot by alignment, until you have exhausted all other possibilities. The "galloping diddle stick" (alignment tool) is the surest sign of the troubleshooting novice. If the transformer is faulty, then it should be repaired or replaced, but alignment will not solve that type of problem.

SSB DETECTOR DEFECTS

Single sideband (SSB) signals are demodulated in a circuit called a product detector. These are hetrodyne mixers in which the SSB signal from the I-F amplifier is beat against a local signal derived from a beat-frequency-oscillator. If the distortion is traced to the product detector, then look for the same types of faults as in the simple AM detector and the FM detector. Additionally, it might also be that the BFO frequency is too far off, and the signal is never properly demodulated.

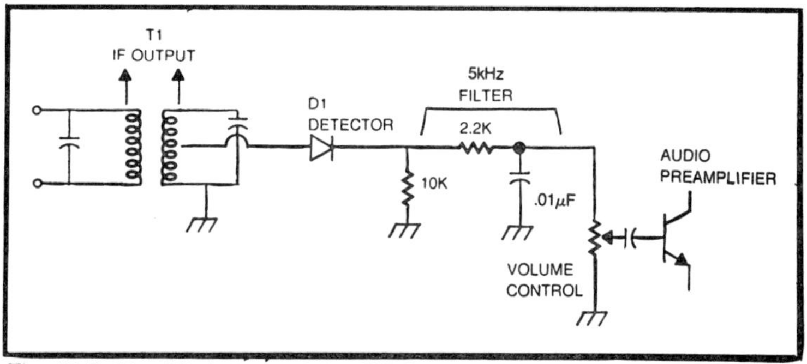

Fig. 13-3. Simplified circuit of the AM detector, filter and volume control.

In sets that use an IC demodulator, whether AM, SSB, FM, or all three, look for the IC to be bad. Of course, one should check all of the components related to the detector (separately if need be) before replacing an IC detector. Keep in mind that these ICs are very-special-purpose devices. Most equipment manufacturers do not, repeat *do not* make their own ICs. They buy them from companies like Motorola, RCA, Fairchild, National Semiconductor. Most of these ICs are catalogue items, and can be bought for a lot less than the equipment manufacturer charges. Many years ago, I was in the car-radio business as a service technician. Two auto-radio makers went to an IC quadrature detector for their FM car radios. It turned out that both were using the Motorola MC1357P or the ULN2111 devices. These ICs could be purchased from industrial distributors for less than $5 each, and from hobby-market distributors (who often charge less than industrial distributors in the onesey-twosey market) for around $3.50. The car-radio makers were charging $9 to $12 for the same IC under their "house brand" part numbers. A little judicious probing of IC-manufacturers data books and applications notes might turn up, not only the right chip, but the same, exact, circuit used in your rig!

AGC-RELATED DISTORTION

An automatic-gain control is a feedback control system that varies the gain of a receiver to keep the output level relatively constant as you tune across the band, encountering signals of widely varying strengths. Most of these circuits operate by sampling the signal in the last, or next to last, I-F amplifier stage, and then using it to set the gain of the rf amplifier, or one or more of the I-F amplifiers. Figure 13-4 shows a typical, if simplified, agc system for a receiver. A small-value capacitor takes a sample of the signal from the collector of one of the amplifier transistors. This signal is rectified by diode D1, and then smoothed in a filter circuit. The resultant DC bias is used to add to, or subtract from, the bias on the rf-amplifier transistor. If the signal is strong, then the agc bias bucks the regular bias, reducing the gain of the rf amplifier. Similarly, if the signal at the collector of the amplifier is weak, then the bias on the rf transistor is appropriate for maximum gain, thereby increasing signal strength. In theory, the idea is to keep the signal at the collector of the I-F amplifier (hence the output) constant for a wide range of input signals.

The distortion caused by the agc circuit, especially in AM and FM sets, sounds very similar to audio distortion, and may be incorrectly identified as audio distortion. There are two indications, however, that the agc is the problem. One is that the distortion is often accompanied by squeals and "birdies" because the feedback line is now essentially undecoupled. This is easy to spot on an AM broadcast radio, difficult on most ham bands, and impossible on CB (where howls and squeals are normal due to too many stations on too few channels). The other sign is that the distortion seems worse on strong stations, and may disappear altogether on weak stations! If this is noted, do not waste time in the audio amplifier until you have cleared the agc system and the rf amplifier. Check both the diode and the bypass capacitors in the agc system.

RF AMPLIFIER DEFECTS

The rf amplifier can produce symptoms very much like the agc distortion discussed above. In transistor sets, especially those older designs that used PNP germanium transistors in the rf amplifier, the most common cause of this problem is collector-to-base leakage. This will produce distortion on the stronger stations, but not on the weaker. It will also show up as incorrect bias or conduction voltages in the rf amplifier. Measure first the agc voltage to see that it is changing properly as you tune the receiver across the dial. Next, measure the conduction voltage in the rf amplifier to make sure that it also changes as you tune across the dial. The conduction voltage will be the voltage across the emitter resistor, or in the case of PNP transistors operated in negative-ground sets (very common), the collector-to-ground voltage. If this voltage remains fairly constant with changes of input-signal strength, then suspect the transistor. Replacement transistors, incidentally, are relatively easy to obtain these days, so substitute any good rf transistor (especially those service-grade replacements marked for

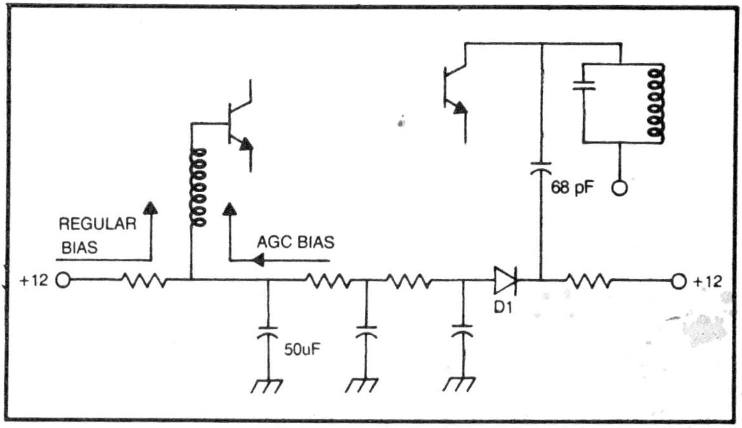

Fig. 13-4. Agc circuit.

FM or VHF use). If it solves the problem, then either order the correct transistor, or leave the replacement in.

Vacuum-tube rf amplifiers also suffer from this type of problem. In this case the fault is an interelectrode short circuit. If the signal level is low enough, the deterioration will not affect the output that much, but on strong stations the problem surfaces radically. Note that shorted vacuum-tube rf amplifiers often take with them either a cathode- or plate-load resistor. In many cases, the designers placed a 1000-ohm (or less) resistor in series with the B+ lead to the rf-plate transformer in an effort to secure a little more gain and better decoupling. If the rf-amplifier tube shorts, then this resistor (or the cathode resistor) may burn up. In the usual failure mode, the resistor does not burn open, it merely increases in value due to the overheating. This lowers the plate voltage on the rf amplifier (reducing gain, incidentally). A weak signal will not be affected by the lowered plate voltage, but a strong signal could easily drive the rf amplifier into saturation.

Note that most of these problems in the rf amplifier affect only A-M and SSB receivers, whose output waveforms are sensitive to distortion of the rf signal. But, fm sets use the frequency of the rf signal to contain the information; amplitude distortion of the type mentioned will not affect these sets.

I-F OSCILLATION

Another non-audio source of distortion is I-F amplifier (or other stage) oscillation. If the amplifier begins to oscillate, the agc system sees a tremendous signal, and thereby reduces the gain of the rf amplifier. This is likely to reduce signal strength and cause distortion. But, since the bulk of a receiver's gain is in stages other than the rf amplifier, it will be the distortion that draws attention to the problem. Unfortunately, I-F stage oscillations are not always audible. They will produce beat notes with the signal that are often inaudible. In this case, an oscilloscope is very valuable in locating the problem! If you lack an oscilloscope, however, there are certain DC indications that tend to support the idea of an oscillation. For one thing, is the bias on the amplifier normal? Second, if the set has an S-meter, does it read over S-9 all of the time, even when the antenna is disconnected? Similarly, does the agc voltage, or the rf-amplifier conduction voltage, remain nearly constant as you tune across the band (assuming, of course, that the rf transistor has been checked)? If these are true, then think about the possibility of an oscillation. Build an rf probe (Chapter 8) for your voltmeter. This will register zero on a DC voltmeter if there is no signal present in the I-F amplifier, and some DC level if there is a signal present. Disconnect the antenna and short out the antenna terminals. Then measure the rf voltage present at the output of the I-F amplifier, or on the collector of the last transistor.

The use of the rf voltmeter probe is almost mandatory in equipment using IC I-F amplifiers. In the stereo-receiver business, a few years back, all of the major manufacturers went to the μA703 amplifier for the I-F stages. These had to be checked with a wideband oscilloscope or an rf voltmeter probe attached to a DC voltmeter. The same tactic will also work on the IC amplifier devices used in present day Amateur and CB equipment.

POWER SUPPLY DEFECTS

Defects in the DC power supply of most equipment accounts for the majority of problems encountered in troubleshooting. In fact, it is reasonably good advice to the newcomer in the troubleshooting business to always check out the DC power supplies first, regardless of the symptoms or customer complaints. This will often save a lot of hassles. . . power supply faults often imitate other faults. In most cases, you are looking for a voltage that is too low, or too high. A too-low voltage could be caused by any number of defects (see Chapter 10), but the too-high voltage is usually caused by a defective regulator.

Additionally, look for the ripple on the DC power supply. This requires an oscilloscope, but since it is at a frequency of 60 or 120 Hz, almost any oscilloscope can be pressed into service!

Chapter 14

Troubleshooting the Dead Oscillator

Let's ask a rhetorical question: "How do you troubleshoot an oscillator circuit to tell whether it is, a) oscillating, b) on the correct frequency?" Answering this question is actually somewhat more complicated than it sounds because so many of the normal troubleshooting methods can shift the frequency of oscillation enough to put the circuit out of commission. Much depends upon the type of oscillator circuit and the frequency of operation. As a general rule, the higher the oscillation frequency, the harder the job of troubleshooting the circuit.

USING THE VOLTMETER

A partial schematic of a vacuum-tube oscillator circuit is shown in Fig. 14-1. This type of circuit may be a fixed-frequency crystal oscillator, or a variable-frequency oscillator (VFO) tuned to resonance by an *LC* tank circuit. Bias for the tube is supplied partially by the grid-leak action of resistor *R1* and capacitor C1. As long as a signal, supplied by the oscillator action of the tube, passes through capacitor C1, there will be a small negative voltage on the grid of vacuum tube V1. Furthermore, this voltage will vary as the resonant frequency of the tank is varied. You can observe this change with a DC voltmeter on the grid as you tune the VFO through its entire range. In crystal circuits, the voltage will drop to almost zero when the crystal is either removed from the circuit or disabled.

Transistor Oscillator Circuits

A situation similar to the tube-type oscillators occurs in transistor oscillator circuits. The two transistor circuits shown in Fig. 14-2

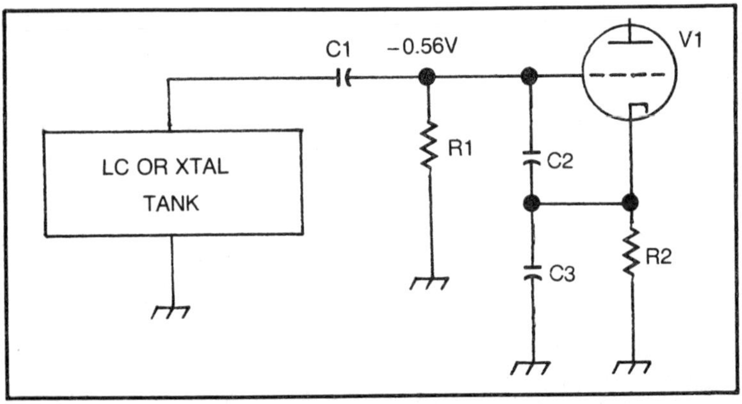

Fig. 14-1. Voltage on grid of oscillator is a clue to its operation.

represent the emitter circuits in most oscillators. Figure 14-2A is for PNP circuits, while Fig. 14-2B is for NPN.

To determine whether the circuit is oscillating, try the standard troubleshooting technique used in almost all solid-state circuits—that of measuring the voltage drop across the emitter resistor. Just the existence of the votage drop is generally a good indication that the stage is conducting current, and if the circuit is oscillating, the emitter-voltage-drop should change as you tune the oscillator through its range.

Start at the low end of the range, and tune to the high end, while watching the voltmeter connected across the emitter resistor. You should notice a smooth, continuous, change in the reading as you tune the oscillator. If the reading does not change, then the circuit is not oscillating, and the emitter current may be due to either a short or, simply, to the static bias on the stage.

In crystal oscillators, there will be an abrupt change in the voltage drop as the crystal is either removed from the circuit or disabled. Be careful when shorting out crystals, incidentally, because in some circuits (such as the Pierce oscillator) a short across the crystal will place the collector voltage on the base, resulting in a possible burn-out of the transistor. Use a high-value capacitor across the crystal to do this job. 1000 pF seems about right in most cases.

When measuring the voltage drop in medium-wave and lower high-frequency oscillators, the capacitance of the probe will detune the oscillator to an incorrect frequency, but will probably not kill it all together. In fact, if the oscillator does die, then this could indicate a problem.

When working in oscillator circuits use either a VTVM, a VOM with a very high-sensitivity specification, or an electronic voltmeter. Cheap, low-sensitivity VOMs tend to load the circuit too much, and could kill the oscillation. Be careful, though, high rf fields, such as present in transmitters, can often get into VTVM, FETVM, TVM or DVM type instruments (almost all electronic voltmeters, in fact), and bias their active input elements into nonlinearity, or into saturation. The standard advice, in these cases, is to use a shielded rf choke (Fig. 14-3A), or limit your selection or voltmeter to the VOM type of instrument.

Another way to use the voltmeter to check a stage for oscillation is to equip it with a demodulator, or "rf", probe. This type of probe (see Fig. 14-3B) rectifies and filters the rf signal to produce a DC level. You can either build your own probe, using a circuit such as Fig. 14-3B, or you can buy one for a minimal cost. If the demodulator probe produces a DC output, then it is probable that the oscillator is working.

Up to this point, I have outlined methods for determining whether or not an oscillator is working. But, none of these techniques will help determine whether or not the oscillator is on, or even near, the correct frequency.

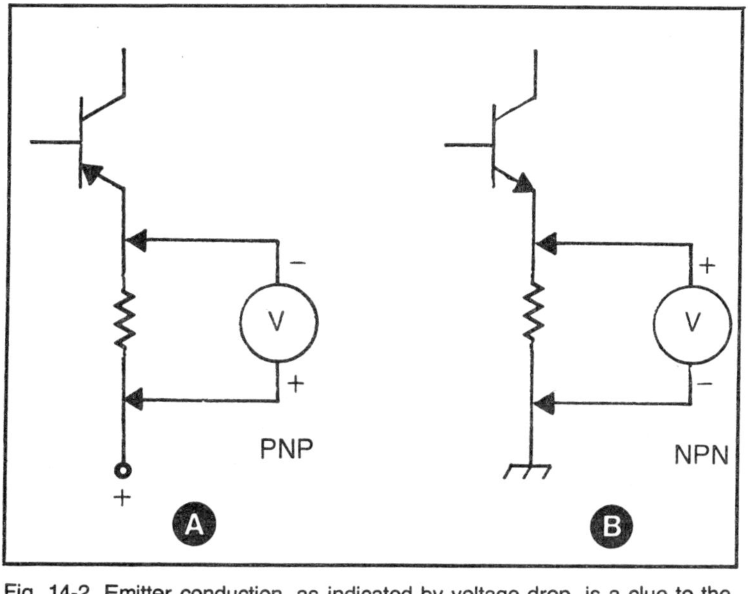

Fig. 14-2. Emitter conduction, as indicated by voltage drop, is a clue to the operation of a transistor oscillator.

THE OSCILLOSCOPE

The wideband, high-frequency, oscilloscope has become part of the standard test-equipment inventory of professional electronic service shops. It can be used to develop information as to the operation of oscillator circuits, but (unless very precise calibration is possible) will not yield much more information about whether the oscillator is on the correct frequency than will the voltmeter. Some may argue this point, but they should consider that, in narrow-band communications receivers, the error that causes the set to appear "dead" is well within the specification error of most oscilloscopes.

An oscilloscope can be used to examine the output waveform of an oscillator, up to the bandwidth limitations of the oscilloscope being used. This feature allows us to tell whether or not the oscillator signal is present, but does not yield more than "ballpark" measurements of the oscillator frequency.

At any frequency above what could be termed "moderately high ultrasonic," a low-capacitance probe (Fig. 14-3C) must be used to avoid a change in oscillator frequency due to capacitive loading. In addition to capacitive-loading problems, there also is a signal-attenuating voltage-divider action, consisting of the scope-input resistance, the circuit impedance, and the capacitive reactance of both the probe and the scope input capacitance. Also remember that, although sinewave signals can be displayed up to the practical bandwidth limit of the oscilloscope, more complex waveforms will be deteriorated if their fundamental frequency is greater than a relatively small fraction of the practical bandwidth of the scope. For example, a 1-MHz time-base oscillator in a frequency counter will appear to be generating sinewaves instead of squarewaves if a 4.5 MHz "TV-service" oscilloscope is used.

It is worth noting here that a DC-coupled oscilloscope is also a DC voltmeter, with the amount of deflection proportional to the input voltage. Thus, any technique using the voltmeter can also be applied to the dc-coupled oscilloscope.

The problems of oscillator troubleshooting become even more difficult when they involve an FM or other VHF local oscillator. Even if you are lucky enough to own one of the newer Tektronix or Helwlett-Packard oscilloscopes, you will find many circuits in which even the specially designed VHF probes will cause excessive loading of the signal. The answer in these cases is to use an instrument that is not commonly used for this type of troubleshooting, the *dip oscillator* (formerly known as the *grid-dip meter* when vacuum-tube technology was used).

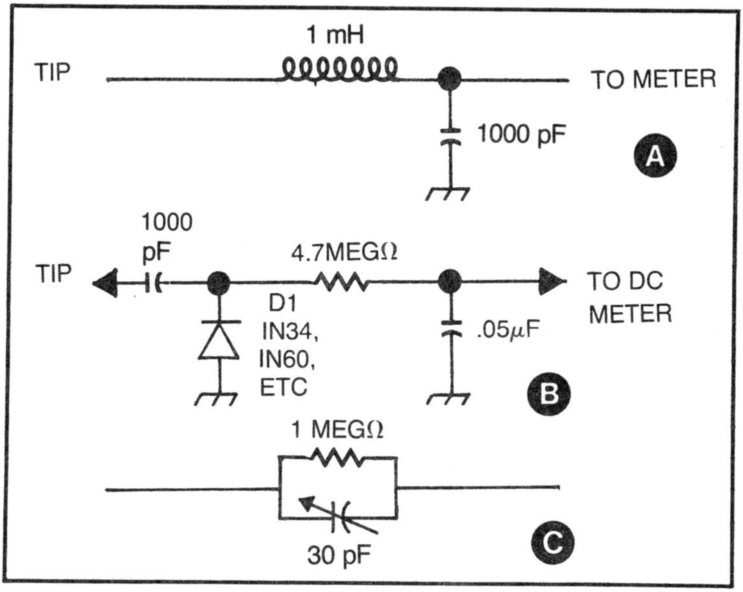

Fig. 14-3. (A) Rf-choke probe; (B) passive demodulator probe; (C) low-capacitance probe.

USING THE DIP METER

The dip meter operates as an oscillator in which the inductor portion of the resonant tank circuit is mounted external to the instrument case (see Fig. 14-4 for a generic dip-meter circuit). The dip-meter oscillator is a VFO and it has a dial calibrated as to frequency. In normal operation, the dip-meter coil is coupled to the circuit being tested so that some energy is transferred. When the dip-meter VFO is tuned to the resonant frequency of the circuit under examination, a large increase in energy transfer occurs, and that causes the meter to deflect sharply downwards (in other words, it "dips"). The dip meter, then, can be used to find the resonant frequency of a tank circuit, and it will also tell us if an oscillator is oscillating.

There are two ways you may use the dip oscillator to determine if the oscillator is running: as a substitute oscillator or as an oscillating detector.

Substitute Oscillator. If your are checking the local oscillator (LO) of any HF and VHF/UHF receiver, loosely couple the oscillator to the inductor of the dip meter (Fig. 14-5) and note what happens as you tune the dip meter across the appropriate band. You will hear stations (if the band is active!) if the receiver LO is dead, or

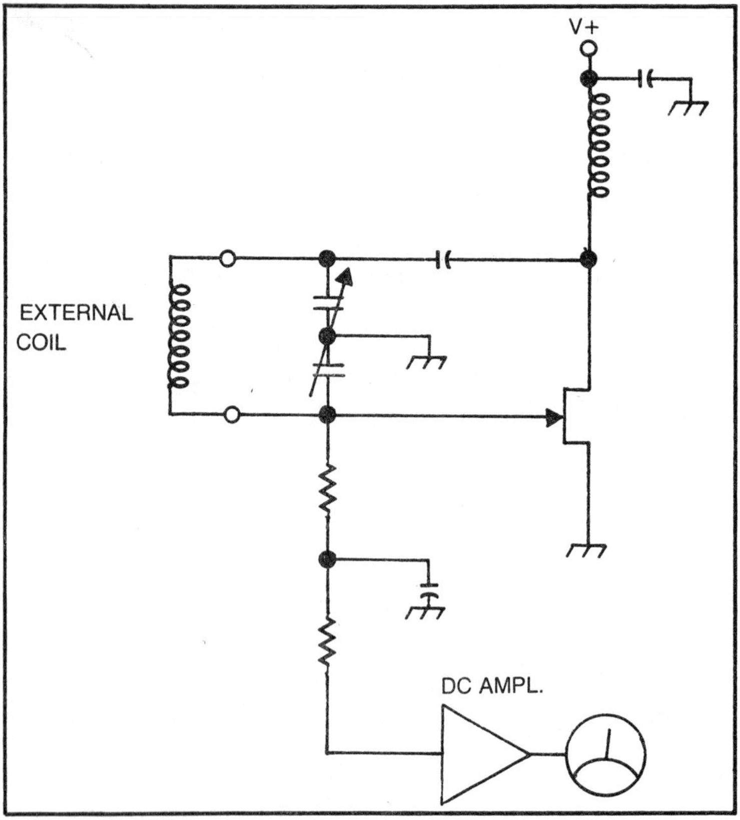

Fig. 14-4. Simplified dip-meter schematic.

will see "information" flash across the screen if working on a television receiver.

Tune the receiver to approximately the middle of the band (or where there are a lot of stations). Adjust the VFO of the dip oscillator, while the coil is loosely coupled to the receiver LO circuit, tuning through the range of expected LO frequencies (rf plus or minus i-f). If the receiver LO is working, then you will hear birdies and whistles. But, if the LO is dead, then you will hear stations. Note that, while it is possible to tune rather sloppily when using this technique on FM-broadcast receivers, the narrow-band characteristics of FM-communications receivers (and most other Amateur receivers) makes it imperative that you tune the VFO of the dip oscillator slowly and carefully.

Oscillating Detector. Many dip meters have a provision for replacing the meter with a pair of earphones. You will normally hear only the hiss of thermal noise in the earphones. If the receiver's local oscillator is operating, however, the oscillator will couple energy into the dip-meter coil so that when the VFO of the dip meter is tuned near the frequency of the LO, hetrodyning will occur and a beat note is established. Once the beat note is heard in the earphones, turn off the receiver to verify that the hetrodyning is due to the LO signal, and not some other nearby source. At some frequencies, the dip meter will pick up oscillation from a distance of several feet. Also, consider the frequency at which the beat note occurs; there are usually several signal sources in a typical receiver, any of which might produce a beat note by interfering with the fundamental or harmonic frequency of the dip-meter VFO. Consequently, it is sometimes wiser to disable the LO, rather than the entire receiver, when checking the validity of the beat note.

FINDING THE FREQUENCY

As I have indicated, none of these methods (voltmeter, oscilloscope, tracer, or dip oscillator) really gives any precise, definite, information about the frequency of operation of an oscillator circuit. A dip meter tries, but the calibration of these instruments is best described (charitably, at that!) as terrible. For a precise determination of frequency, a frequency meter or frequency counter is needed.

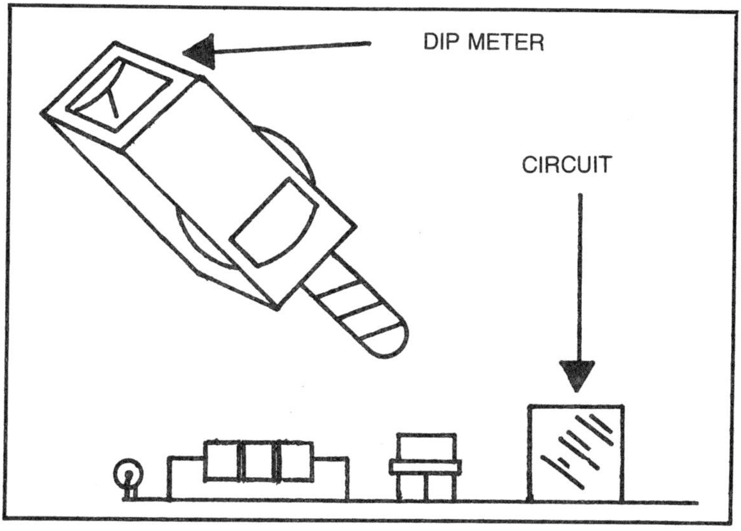

Fig. 14-5. Close-couple the dipper to the circuit to use it as a substitute oscillator.

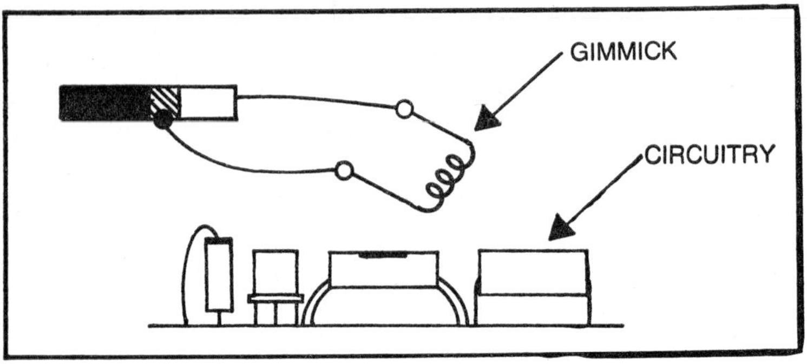

Fig. 14-6. Couple the signal source to a circuit through coax cable and a "gimmick."

A *heterodyne frequency meter* tells us the oscillator frequency by beating against an unknown frequency, and then comparing the point at which zero beat occurs with the known dial calibration.

Today, however, many servicers prefer to use the *digital frequency counter* many of which are available at low cost (do not, however, disdain the older heterodyne units. They can be bought for a song, and can still make valid measurements). A disadvantage to the counter, in most cases, is the necessity to physically connect the counter to the oscillator under test. This tends to load the circuit, in many instances, thereby shifting the frequency somewhat.

The loading effects of the counter can be reduced somewhat, thus improving accuracy, through the use of a "gimmick" (see Fig. 14-6). If the oscillator is really low level, or if a lesser coupling is required to eliminate frequency-pulling effects, you may want to use an amplifier between the gimmick and the input of the counter. A transistorized broadband amplifier can be constructed for this purpose, although it is often simpler to buy one of the International Crystal kits for this purpose. One of the International Crystal kits is a broadband 1-150-MHz amplifier built on a 1.5 × 1.5-inch-square printed circuit board.

Chapter 15
Solid-State Power Amplifiers

Many of the newer Amateur transmitters, and all of the current crop of CB transmitters, use solid-state devices in the power-amplifier section. To many Amateurs who were right at home troubleshooting a vacuum tube power amp, this is something a little unnerving. But, solid-state power amps are not a whole lot different from any other solid-state amplifier, and the same tactics can be used.

Figure 15-1 shows a composite power amplifier, with several of the different techniques used in these circuits. The bias may be from a resistor network, as in transistor Q1, or may be self-bias as in Q2 and Q3. Transistor amplifiers operated class-C often use a single resistor (such as R4 and R6 for Q2 and Q3, respectively) to develop the bias voltage. The existence of a DC voltage drop across these resistors is proof that the transistor is being driven by the previous stage.

But, where the DC voltage is developed from a power supply and resistor network, as in the case of Q1, the voltage drop across the resistor (R3) may not be a meaningful indication of the existence of a drive signal. It will, however, tell us whether or not the DC operation of the transistor is normal, as we described in Chapter 3.

There are also different ways to handle the emitter circuit of the power amplifier stages. In Q1, for example, an emitter resistor is used, and it is bypassed to ground for rf. This keeps the emitter at a slightly-higher-than-ground DC potential, but the signal sees a short to ground. The capacitance of C1 is selected to be less than one-tenth the resistance of R1 at the lowest frequency of operation.

In stage Q2, a slightly different approach is used. Here the emitter resistor is *not* bypassed. This type of circuit is an attempt at obtaining a little bit of negative feedback to stabilize the stage.

The transistor in stage Q3 is operated with the emitter directly grounded. This is usually the case in the higher power stages.

For the most part, tuning is by slug-tuned coil in the drivers and preamplifiers, and by capacitor and coil in the final-amplifier output stage. A few transmitters use variable capacitors, but most use variable inductors to adjust the tuning of these stages. This is especially true where the transmitter is tuned only once, or when a channel change is made.

Some manufacturers specify an order in which the tuning adjustments must be made, especially in power amplifiers. This instruction must be obeyed. . .try to ad lib and you will lose a power transistor. The problem seems to be a low-frequency oscillation that is created when the adjustments are made out of sequence. If allowed to continue, it will destroy the power-amplifier transistor.

In most cases, a transmitter that is bad will have the complaint "no transmit." But this could mean two things: no rf production or no modulation. In many instances, you will not know which is true unless you make some checks. All that you know for sure is that "you ain't gettin' out." In some cases, especially local VHF repeater operations, other stations will tell you that you are quieting the repeater, but are not producing any audio. This is a good indication that the rf circuits are okay, and that the problem is in the modulator circuit or the microphone.

One prime piece of advice. . .*do not overlook the obvious*! I have serviced many transmitters where the only fault was a bad microphone. The fault could be a bad element inside of the mike, but most of the time it is a broken wire inside of the connector at the transmitter end of the cable (and less often inside of the microphone housing). So, before you tear into the transmitter, make sure that, a) the microphone is working, and , b) the push-to-talk switch is closing properly.

Once it is determined that the problem is no rf output from the transmitter, then you must find a way to determine which stage is bad. Rf power transistors are too costly to try the old troubleshoot-by-substitution, or "shotgun" method!

If the stages are biased like Q2 or Q3 in Fig. 15-1, then a simple DC voltmeter will tell you which is the bad stage. But keep in mind that electronic voltmeters (VTVM, FETVMs, DVMs, etc) are sensitive to rf fields, and will read erroneously, if at all, when used

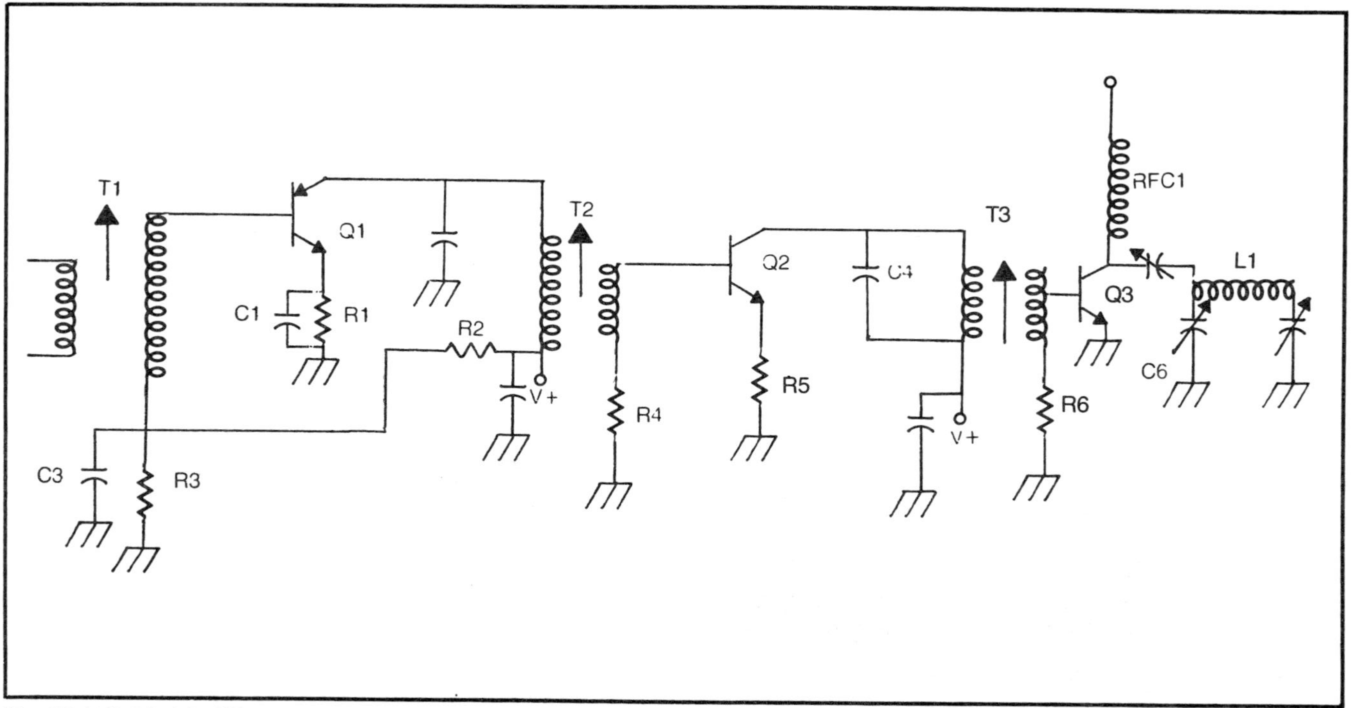

Fig. 15-1. Solid-state RF power amplifier.

on transmitter circuits. This is a powerful argument for keeping, in addition to the fancy electronic VM, and old-fashioned VOM. In recent years, both the Japanese imports and the instruments made in the USA have come down in price quite a bit. There is no reason for an Amateur not to have one as part of the station test equipment. I bought a 30,000 ohms/volt VOM with a 0.5 volt low scale (DC) for less than $35 just a few weeks before this was written. . . and from a dealer not known for his charity.

In stages such as Q1 in Fig. 15-1, you can signal trace with either an rf voltmeter capable of measuring down to about 0.5 volts peak rf at the frequency of operation, or an oscilloscope with a vertical bandwidth capable of seeing a signal at the frequency of operation. In this latter respect, note that it is not necessary to have a 30 MHz scope in order to see a 28-MHz signal! If the −3 dB bandwidth is, say, 15 MHz, then the scope may be capable of displaying (at greatly reduced amplitude) signals as high as 30 to 40 MHz. The signal in the transmitter will be in the 0.5-volt range, but it may easily be viewable on many 10 - 15 MHz scopes if the vertical sensitivity is less than, maybe, 100 mV. The idea is not to *measure* the signal, but to find out if it *exists*.

Before replacing an expensive rf-power transistor, you might want to try using the ohmmeter to examine some of the other components in the circuit. Make sure that the capacitors are not shorted, and that the coils have continuity. Look at the resistors to determine if they have the correct value. Then look at the voltages on the transistor, using an appropriate rf choke in series with the VOM probe, to see if they are as expected (see the service manual). Review Chapter 3 for information on troubleshooting transistor circuits in general.

If these tests do not turn up another component as the fault, then try replacing the power transistor. Note that the simple power transistor tests of Chapter 19 do not always work for rf power transistors, especially VHF and UHF types, and that replacement may be the only way to find out for sure. Of course, if the ohmmeter test shows the darn thing to be dead shorted, then believe it. . . it is dead shorted. But if it fails to pass the forward and reverse resistance tests on the junctions, then believe it with a big "maybe."

Replace the transistor with a direct replacement. Guessing at a replacement does not always work in rf power amplifiers. I have gotten away with it, especially in CB sets, but have been burned more often than I have been successful. If the manufacturer designed the circuit around a 2NXXXX, then it may well fail to operate if a 2NXXXY is used, no matter how similar the two devices are.

Chapter 16
Troubleshooting Internal Noise Problems

Spurious noises in receivers, amplifiers, and modulators are signals that should not exist, but do, and originate within the device itself. We are not referring to a static crash from a bolt of lightning, or electrical-arc interference from a motor, but to noises, static, and oscillations that originate from the circuits of the receiver itself.

OSCILLATIONS

If an amplifier circuit becomes unstable, it may begin to oscillate. If this happens, it generally means that some component has failed. The exception, of course, is the newly constructed project or circuit, in which case layout might be a problem.

There are two basic types of oscillation in receivers: tunable and nontunable. The nontunable oscillation generally occurs in an audio-amplifier stages. Tuning the receiver across the band will not affect the *pitch* of the oscillation, although it might affect the apparent amplitude due to signal-to-noise variations. A tunable oscillation usually indicates that the problem is in the rf amplifier, I-F amplifier, or the oscillator/mixer (or converter, if such a stage is used).

In vacuum-tube circuits, it is very common to find oscillations due to an open suppressor grid. The cure, of course, is to replace the vacuum tube. Incidentally, there are also defects in solid-state devices that can cause oscillation, but these are rare enough to make the method of testing by substitution more of a problem than it is worth.

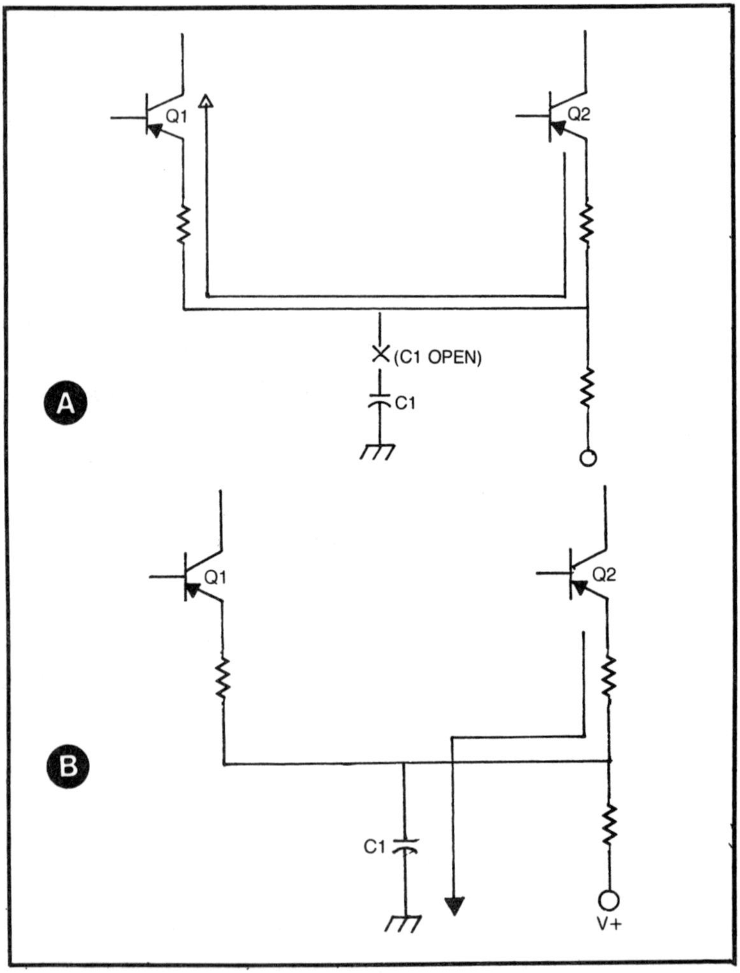

Fig. 16-1. (A) Normal path for ac signals; (B) path when capacitor C1 is open.

The major causes of oscillations are open capacitors. One function of bypass capacitors, in both solid-state and vacuum-tube circuits, is to *decouple* the power supply line between two stages so that a feedback path does not exist. Figure 16-1 shows two transistor stages fed from the same power supply line. In Fig. 16-1A capacitor C1 causes the V+ line to have a low AC impedance to ground, while remaining at a high DC resistance. Any voltage variations in the supply due to loading by either Q1 or Q2 will be by passed to ground by C1. But, in Fig. 16-1B we see the situation if capacitor C1 become open. The V+ line is no longer at a low AC impedance, but has roughly the same impedance as DC. Any voltage variations reflected in V+, due to the loading of one of the stages, forms a feedback signal to the other stage. Consider the case where signal from Q2 gets onto the V+ line, and there is no decoupling. This signal is applied to Q1, and sees Q1 as a grounded-base amplifier. Since this signal is from the emitter of Q2, and is fed through Q1 as a grounded-base amplifier, it is applied in-phase back to the base of Q2 (via the Q1 collector). The requirements for oscillation in any electronic circuit are a) loop-gain greater than unity, and, b) in-phase feedback. The circuit in Fig. 16-1B meets both of these criteria, so it will oscillate.

Another possible cause of oscillation is the screen bypass capacitor in vacuum-tube circuits. Figure 16-2 shows a typical tube stage. Resistor R1 is the plate load, R2 is the screen dropping resistor, and C1 is the screen bypass capacitor. Ordinarily, the screen is decoupled to ground through C1 (the screen sees a low AC impedance to ground). If C1 becomes open, then the circuit will oscillate.

In vacuum-tube receivers, the open screen bypass capacitor is a frequent cause of tunable oscillations. Unfortunately, these circuits usually have a feedback path in the form of the automatic gain control (agc) circuit so the troubleshooting method is sometimes

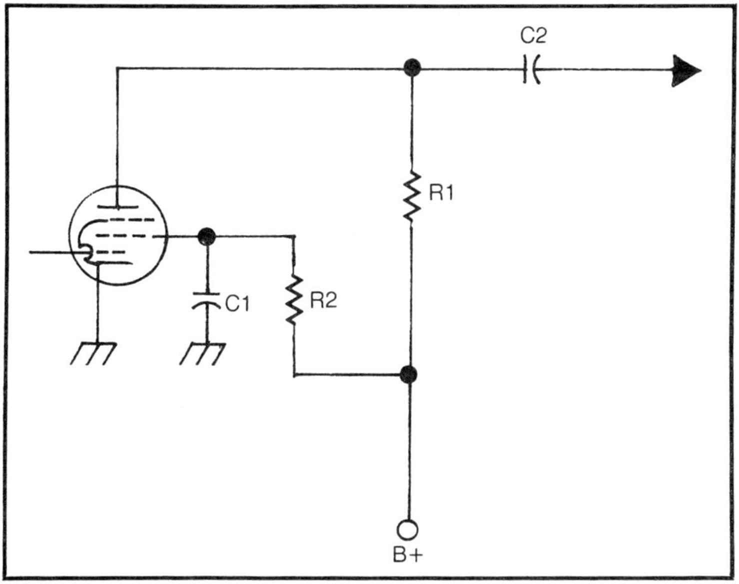

Fig. 16-2. Capacitor-coupled pentode stage, with screen bypass (C1).

less than clear. But, fortunately, there are only a few capacitors in the receiver that can cause this problem. To troubleshoot the problem of tunable oscillations, just bridge a known-good capacitor across each one in the circuit, in succession, until the open capacitor is found.

Some technicians troubleshoot multi-stage cascade chains of amplifiers (as in receivers) for oscillation by using a bypass capacitor to ground the plate or collector of each stage. If the oscillation continues in the next stage, then it is the next stage that is the problem. I personally have not found this technique to be always reliable, but you can try it.

Figure 16-3 shows the block diagram of a receiver with an agc circuit. A sample of the I-F signal is taken (usually from the output I-F stage), and is then rectified and filtered to form a DC bias that controls the gain of the rf amplifier (and sometimes the I-F amplifier as well). Capacitor C1 acts in the same manner as a smoothing filter in DC power supplies, but also serves to decouple the agc feedback line. If this capacitor becomes open, then the result will be an in-phase feedback path for the ripple resulting from the rectification process.

Again, the troubleshooting procedure is to bypass the agc filter capacitor (C1 in Fig. 16-3) with a known-good capacitor of similar value.

There are two oscillation problems that are caused by plain old dirt. In one, the wiper contact that serves as the electrical connection between the rotor plates of a variable capacitor and its frame becomes loose, or corrosion builds up underneath it. This problem results in a tunable oscillation that is often accompanied by static crashes, or will appear only in certain portions or the band. In communications receivers, the oscillation may appear only on one band, usually the highest band on the receiver. I once had a Hammarlund HQ-145 receiver. It was purchased originally for a friend of mine, as a present for Chrsitmas, 1960. Fifteen years later I bought it from him in non-working condition. The major problem, it seems, was a tunable oscillation on the 12-30 MHz band. As an old hand at service technicianing, I knew that this type of problem could be one of the worst to find. Fortunately, part of my own clean-up protocol for any old equipment is to clean the bearing and contact spring in all the variable capacitors. As soon as this was done, the receiver operated normally.

The other dirt-caused type of oscillation involves the normal film that always seems to build up on the chassis of electronic

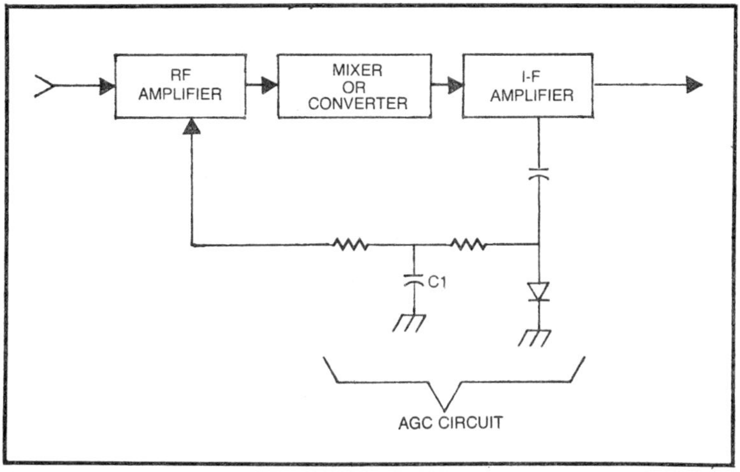

Fig. 16-3. Agc loop.

equipment. If this film spreads far enough, then it can form feedback paths between stages. If there seems to be no obvious solution (i.e., bypass capacitors, open suppressor grids, etc), then look to the chassis film. I personally prefer buying one of the standard electronic degreaser/cleansers, or some Freon TF solvent. These seem to work nicely on electronic equipment, with little danger of leaving a conductive residue or of damaging plastic or other non-metallic components. Some people use ordinary soap and water for this purpose. In fact, a popular variation on the theme a few years ago was to mix a pinch (a very small amount!) of acetone into a solution of one quarter *Lestoil* (or some similar liquid cleanser) and tap water. The mess was then used to clean the chassis, followed by a rinsing with clear water. I have used this technique with good results, but still prefer the nice, easy-to-use, aerosol cans of electronics cleaner to the goop mentioned above.

Hint. Very rarely will alignment, or deterioration of same, cause oscillations! Invariably, though, neophyte servicers tend to reach for the diddle stick and go to it . . . ruining the already perfect alignment job! I personally feel that "they" (the great, mysterious "they" in the sky) ought to lock up all alignment tools, and issue them only once you have convinced the great servicer in the sky that alignment is needed or justified.

One oscillation problem that I saw some years ago (when people still brought in lots of vacuum-tube equipment to be fixed) was due to a pilot lamp . . . and an open bypass capacitor. Figure 16-4 shows the circuit. The designer of the equipment probably

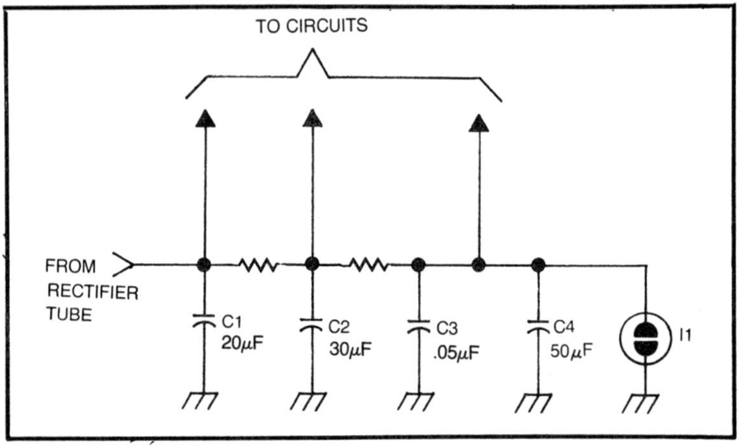

Fig. 16-4. Power-supply circuit.

wanted to use the neon (NE-51) pilot lamp as a low-voltage regulator (not an uncommon practice in those days), because the normal series resistor was deleted. Ordinarily, there would be a resistor in series with lamp I1 to limit the current flow through it.

Capacitors C1, C2, C3, and C4 serve as decouplers, while C3 also serve as an rf/i-f decoupler. The other capacitors are electrolytic power-supply filters, so will not work well at higher frequencies.

The techniques of using a small value disc or mica capacitor shunted across an electrolytic (of much higher value) is often used in circuits where high frequencies are present. This becomes necessary because the high-value capacitors needed to smooth the pulsating DC ripple from the power supply are usually ineffective in decoupling high fequencies. Part of the problem is the high series inductance of electrolytics, but most of it is due more to the nature of electrolytic capacitors (the inductance not withstanding).

In the circuit of Fig. 16-4, the oscillation occurred when capacitor C4 became open, allowing the C3/R2/I1 combination to operate as an audio-frequency-range relaxation oscillator. The frequency of oscillation is set by the time constant R2-C3, the applied voltage, and the firing voltage of I1.

Motorboating is a low frequency oscillation (usually in the 0.25- to 5-Hz range) that makes the output of a receiver sound like an outboard motor boat in need of an engine tune-up . . . badly. This problem is most likely to occur in the audio stages or in the AGC loop. It is more common in transistor circuits than in vacuum tubes (as I seem to recall), especially where the agc loop is involved.

Motorboating is almost invariably caused by open bypass or

decoupling capacitors, usually those with a high value (over 100 μF). There have been some cases, however, where collector-emitter or collector-base leakage in transistor amplifiers have caused motorboating.

Flutter is motorboating that occurs in the 5- to 20-Hz range (my own definition). Both flutter and motorboating are due to the same causes, which need not be repeated here (see above).

These oscillations tend to cut the stage(s) on and off, but (because of their low frequency) might otherwise go unnoticed. However, when the amplifier or receiver cuts on and off at a 5-Hz rate, you know there is a problem!

Other Noise Problems

In solid-state circuits, the noise may come from a number of different sources. The reverse biased PN junction, for example, can create a hiss-like "white noise." In the test circuit of Fig. 16-5, we see a semiconductor diode reverse biased by a battery, or other voltage source, and a series resistor to protect the diode should its Zener point be reached. If an oscilloscope is connected across points A-B (provided it has high enough gain and bandwidth), we should see a lot of "grass" on the CRT screen.

It sometimes happens that faults in capacitors, resistors, and other semiconductors may cause a certain PN junction to become reverse biased.

Consider the circuit of Fig. 16-6. This is part of an audio power-amplifier circuit, and such a stage might easily be found in a solid-state receiver or transceiver (most stereo or hi-fi power amplifiers are direct-coupled, instead of AC-coupled as is Fig. 16-6).

This circuit operates from a dual-polarity power supply in which Vcc is positive with respect to ground and Vee is negative

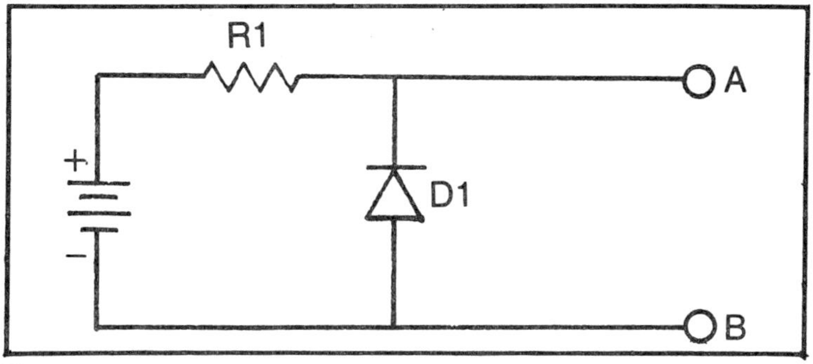

Fig. 16-5. "Noise generator".

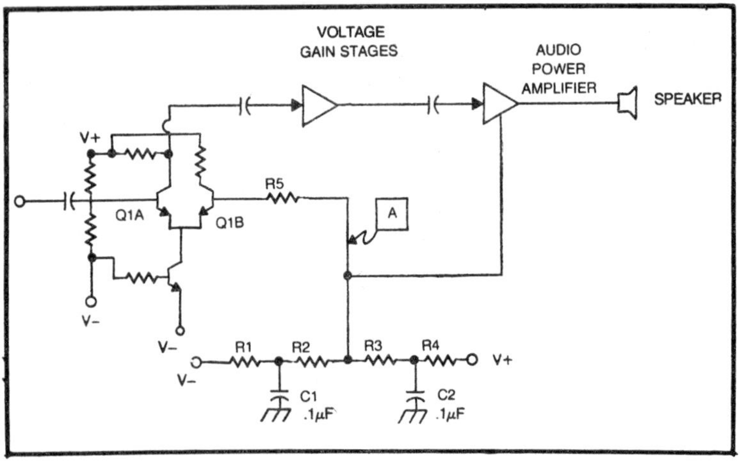

Fig. 16-6. How it can occur in circuits!

with respect to ground. The input stage, Q1A and Q1B, is a differential amplifier in which one input (the base of Q1B) is held to a constant DC potential. Bias to transistor Q1B is held constant through the action of resistor network R1 through R5.

In a problem involving this circuit, capacitor C1 became leaky, and that substantially reduced the contribution of the Vcc (+) supply to voltage E_A. This caused the base-emitter junction of Q1B to become slightly reverse biased. Of course, this turned Q1B into a noise generator, with its output coupled to the rest of the circuit through Q1A.

This problem might not have occurred in a higher level stage. It became acute only because the defect occurred in the first amplifier stage in the cascade. After the defect was apparent, some DC measurements were found to be incorrect, and a sinewave on our oscilloscope was noticed to be slightly distorted.

Time and time again we hear "old hand" servicers giving advice to newcomers about using the DC voltmeter *and* an oscilloscope for troubleshooting noise problems. Perhaps it is wise for us "old hands" to listen to our own burblings and apply the advice ourselves! This problem would have been a lot easier to spot if I had used the meter and oscilloscope to check DC levels and distortion.

NOISY TRANSISTORS

There is always the possibility that transistors will become defective in such a way as to produce large amounts of noise. If the transistor is in a circuit such as Q1A, or any other circuit position far enough back toward the input in a cascade chain, then only a slight amount of noise from a transistor will produce a large noise in the output. This is because the circuit amplification will increase the noise amplitude, just as it would with any other signal.

Because of acute noise problems in modern high-fidelity amplifiers, designers often specify low-noise, high-reliability transistors and resistors for the early stages. In fact, the use of "off-the-wall" substitutes has been responsible for the failure of many projects built from magazine articles, especially in the noise department.

If you plow through an engineering textbook on the subject, you will find equations that give rms value of noise-voltage signals. At least one of these, the one most commonly encountered, will tell you that the noise amplitude is proportional to the ambient temperature, circuit resistance, and bandwidth. Of these, circuit resistance and bandwidth are set by the designer, and are either difficult or irrelevent to troubleshooters seeking the cause of a problem.

Troubleshooting with Temperature

The ambient temperature of electronic circuits and components can be a worthwhile clue in troubleshooting. However, there

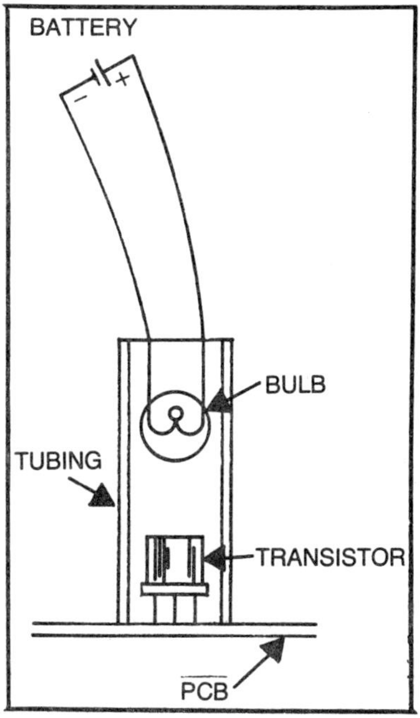

Fig. 16-7. Heating a component to see if it becomes noisy, or ceases operating when hot.

is a problem to overcome when you attempt to raise the temperature in a circuit for purposes of troubleshooting. It seems that lots of possible techniques spread the new temperature over too large an area.

For example, take the common use of Freon areosol "circuit cooler." Even with a long, thin, nozzle, you'll find that without certain precautions, the spray covers more components than is desired. The same holds true for heat lamps and other heat sources. The cold and heat covers more than just the component being checked, and you often wind up either not knowing which it is, or with false results.

What is needed is a means for localizing the cold or heat. One solution I've used on many occasions is shown in Fig. 16-7. Use one of those oversize pieces of insulated sleeving that are usually called "spaghetti." Simply cut one -to three-inch lengths of the tubing to fit snugly over resistors and transistors under suspicion. If cooling is your goal, the tubing allows you to concentrate the spray over only the component being tested. For the heat treatment, simply place a small incandescent pilot lamp in the open end of the tubing and connect is to a battery or DC power supply, as shown in Fig. 16-7. It has been my experience that, in most cases, any problem truly heat-caused will show up in not more than five minutes, and usually a lot sooner. Many problems will succumb to the heat in less than sixty seconds!

Resistors as Noise Sources. Do not overlook the possibility that resistors can cause noises. A defective resistor often produces noises that are a combination of white noise, frying-egg sounds, and popcorn popping. Thermal troubleshooting methods will generally reveal the problem.

Very lightly (don't give it a resounding thump!) tap the suspect resistor while monitoring the output of the circuit. Look for either a change, up or down, in noise level, or a single "pop-fry" burst of noise as the resistor is tapped. The "pop-fry" burst may not be repeatable, so don't just assume that the noise trouble has simply disappeared if the burst does not repeat itself. You may rest assured that the noise *will* reappear when you put the equipment back together and try to use it.

Remember the well-known corrollary to Murphy's Law: Anything that will go bad, will do so at the least convenient time.

Resistors are cheap . . . consider it cheap call-back insurance to replace the suspect resistor while you have the cabinet unbuttoned and the rig is conveniently on the bench.

Internal Component Arcing. Some noises show up in the loudspeaker as a result of internal electrical arcs inside of components. These can be among the most miserable problems to locate—short of "shotgunning" the whole circuit.

One very effective way to locate arcing inside of components or on printed circuit boards is the use of a long piece of rubber or neoprene tubing, or even a length of radio-tv spaghetti. Hold one end of the tubing gently in your ear and use the other end as a probe to find the component that is arcing. The tubing will transmit what is normally a barely audible "click" in full amplitude to your ear.

Actually, I prefer to use a modified medical-style stethoscope for this purpose. Although the professional grade of stethoscope used by physicians runs anywhere from 25 to $100 or more, cheaper types are available from local drug stores or mail-order outlets as part of blood-pressure checking kits. A mail-order source

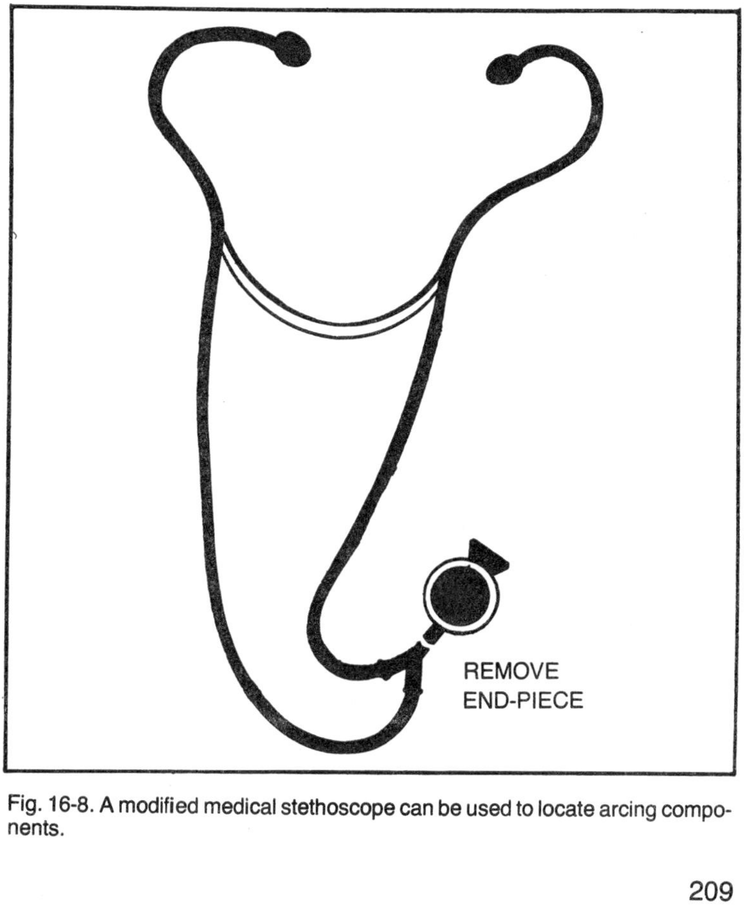

Fig. 16-8. A modified medical stethoscope can be used to locate arcing components.

is Edmund Scientific of Barrington, NJ. You can buy good models for less the $15, but even a $2 model from a child's "play doctor" toy kit is usable.

When using the stethoscope method for locating the clicking of an electrical arc, remove the metal end-piece and use only the tubing as a probe (Fig. 16-8). This will serve both to localize the source of the arcing, and will prevent accidental shorts, and possible electrical shock (to you), from the metal bell or diaphragm if it touches any current-carrying points.

It is necessary to scan the whole component in many cases because the small opening of the tubing covers only a small area at a time. In fact, the resolution is so good that you should be able to tell which end of a mylar capacitor contains the arcing!

Noisy I-F Transformers and Plate Loads

In many cases the plate or collector load of a stage may be a resistor, I-F transformer primary, or an rf tuning coil. If these become noisy, then it may prove difficult to discover which component in the circuit is at fault. Arcing trimmer capacitors sound almost the same as arcing resistors and coils!

Figure 16-9 shows one method for troubleshooting these problems. Break the lead to the transistor collector or the plate of the

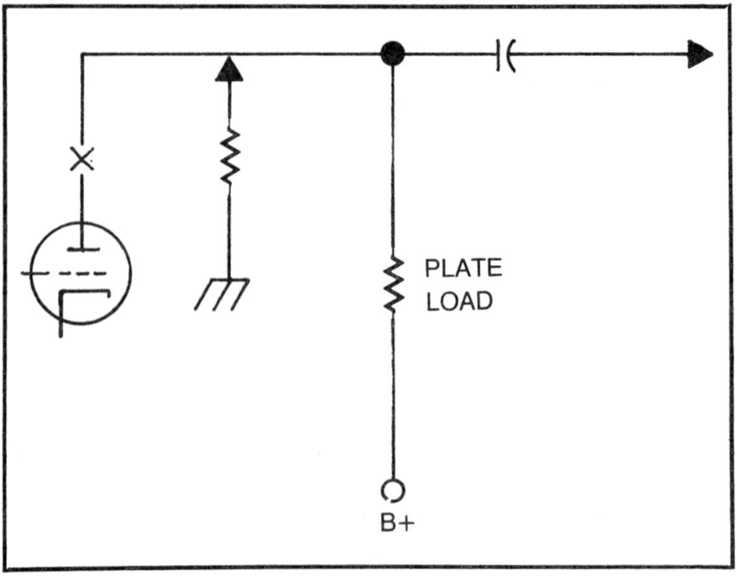

Fig. 16-9. Locating a noisy plate load.

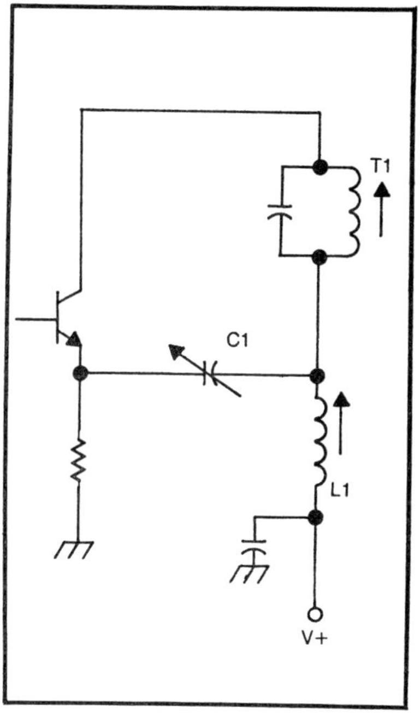

Fig. 16-10. The case of the noisy trimmer capacitor (C1).

tube, and then connect a resistor between the "hot" end of the break and ground. The value of the resistor should approximate the plate or collector resistance of the device disconnected. The idea is to cause the same amount of current to flow in the plate/collector load, yet eliminate the active device as a source of noise. If the plate/collector load component is at fault, then the noise will continue essentially unabated. But, if the active element is at fault, then the noise will disappear.

Noisy Trimmer Capacitors. Figure 16-10 shows a part of a circuit used as a converter (mixer and LO in one circuit) in some receivers. T1 is the i-f transformer, while L1 is the local oscillator coil. Trimmer capacitor C1 is used to resonate coil L1. Mica compression trimmers can develop arcs across the insulator which causes a sound in the output much like static from nearby lightning. This problem is worse in humid areas than in dry, desert areas.

Unfortunately, trimmers are difficult to check by substitution because they, a) affect the alignment, and, b) are usually not easily obtainable. We can, however, *check* the trimmer easily. Disconnect one end, and then place in series with it a 0.01 to 0.1 μF disc ceramic or mica capacitor. If the noise disappears, then buy a new trimmer. This is not a fix ... only a check!

Chapter 17

Mobile Noise Problems

The automobile is one of the most hostile environments in which a radio receiver must perform. There are, in addition to vibration and extremes of temperature, many different sources of interference. In general, any electrical device on the car is capable of creating noise interference to the mobile radio receiver. Even the much touted "noise free" operation of fm receivers is conditional on strong signals and certain other not-always obtainable factors. To make matters worse, troubleshooting mobile noise problems is one of the most difficult jobs going! There are simply too many sources of noise that sound similar, and these sources are not always apparent.

The first step in troubleshooting mobile noise is to attempt to identify the *source* and the *type(s)* of noise present. Is it ignition noise? Is it alternator/generator noise? Or is it the ticking of gasoline gauge noise? Do not make the mistake of assuming that only one type of noise is present; many noise sources can create more than one type of noise in the radio, while in other cases the same noise may be due to different sources!

The most common types of noise, and their *most probable* sources, along with typical recommended cures, are shown in Table 17-I. Be aware, though, that many car manufacturers may have their own "pet fixes" or recommendations. A call to the local dealership, or auto manufacturer's *radio* service representative, can often prove helpful in this matter. Also, let me point out right

Table 17-1. Troubleshooting Chart For Mobile Noise.

Sound Made	Cause	Cure	Comments
Popping static at regular rate, varies with engine speed	Ignition system	Check condition of Radio-TV high voltage wire. Bypass capacitor (0.5-μF) from ignition coil's battery terminal to ground	Also check antenna and radio chassis grounding
High pitched, raucous howl which varies with engine speed	Generator	Bypass armature terminal to ground with 0.5-μF. DO NOT bypass field wire. The correct wire is the heavy cable	Only found in older American-made made and some current foreign cars
High pitched whistle, varies with engine speed. More "pure" than generator noise	Alternator	Check manufacturer's service manuals and bulletins. Install L-section filter in radio power lead	
Foutter at low engien RPM and when receiver set to low volume	Breaker points		

now that the bypass capacitor that works fine in vehicles equipped with the traditional coil/breaker-point Kettering ignition system may not work at all on modern solid-state ignition systems.

Note, also, that the ignition system is not the only cause of spark-like noises. Figures 17-1A and 17-1B shows sources such as the ignition system and the numerous small DC motors used on modern automobiles. Often overlooked, incidentally, are the blower motors in heating/air-conditioning systems.

DEFECTS IN THE RECEIVER

It is wise to realize that certain radio-receiver defects can also create noise problems. Typical internal causes of noise problems in radio receivers include defective components in the power supply input circuit (a shorted L1, or open C1 or C2 in Fig. 17-2) and incorrect positioning, or "dress," of the power wiring and antenna-input circuitry. The main power lead for a mobile radio is called the "A-lead." If the A-lead is positioned parallel to the antenna input circuit, or the volume/gain-control circuitry, then it is possible that noise pulses appearing on the A-lead will be coupled to the radio circuitry at particularly vulnerable points.

To find out if the defect is in the radio, it is best to try substitution of another rig for the noisy one. It is best to use the same make and model, but any similar rig may be pressed into service in a pinch. If you do not have a second rig (few of us do), then borrow a rig from a friend. Connect the new rig in place of the old, and check to see if noise is still present. Remember, though, because the substitution radio will probably not mount in the same manner as the original, there might be *some* noise present. So look, instead, for a substantial *reduction* in the noise level with the new rig.

ANTENNA AND RADIO GROUNDING

Do not overlook the grounding connections of *both* the antenna and the radio receiver/transceiver. These are frequent causes of problems.

The antenna ground may be checked with an ohmmeter, but realize that a good DC ground does not necessarily imply a good rf ground. If there is any doubt, then check the antenna by replacement. This advvadvvadvic iss is especially true if there seems to be no other source, and the antenna is one of the shortened HF antennas, one of the VHF antennas, or any antenna that uses a base-coil mounting assembly, or press-fit (instead of solder) connectors. In the latter case, the ground connection is made by press fitting the rf connector against the coaxial-cable outer shield. As vibration and corrosion take their toll, however, the grounding becomes less and less effective. There will come a point where the ground appears "good" to a DC ohmmeter, but is actually a high impedance at rf.

If a substitution antenna is used, it is not always necessary to actually mount the antenna on the vehicle. But, be sure to ground the test antenna against some metal fitting on the car body that is directly grounded. Some of the chrome trim and bolts around the front door usually work well for this purpose, if you hold the antenna base hard against them. These points can usually be reached easily from inside of the car, where you will most likely be working.

Before moving on, let me have another brief word about the use of a simple DC ohmmeter to check proper grounds in a mobile installation. This instrument can be highly misleading. Like the standard advice as to vacuum-tube testers, if it reads "bad," then believe it; but if it reads "good," then accept the fact as only tentative. . . because it may still be bad. In the case of a mobile antenna, the ohmmeter may tell you that the outer shield is open. In that case, believe it. . . the antenna lead *is* bad. But, if the ohmmeter reads a relatively low resistance, then believe it only after you have exhausted all other causes and made the substitution test!

Similarly, to the DC power supply in a receiver, a radio chassis may appear grounded. But, to rf, it actually might be a relatively high impedance. Fortunately, in this case, amateurs may have the advantage over car radio and CB servicers; their transmitters draw many amperes in the transmit mode. If there is a substantial voltage drop on the A-lead when you transmit, then suspect ground or battery connections. In fact, some people have obtained good results by troubleshooting with a DC voltmeter. A bad ground or battery connection should show only a few hundred millivolts of

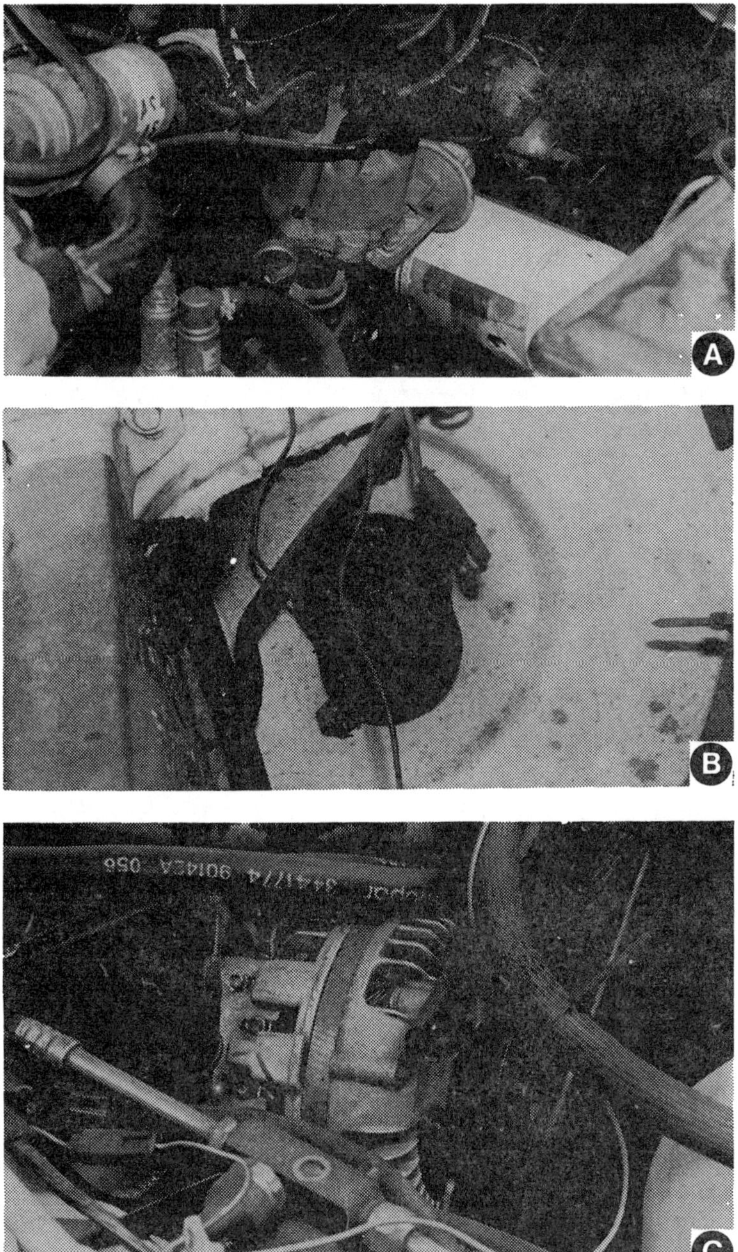

Fig. 17-1. Common auto noise causers: (A) ignition; (B) small DC motors (such as fan or blower); (C) alternator or generator.

drop when transmitting. If the drop approaches 1 volt (or more), then clean the connection, or make it tighter!

Anti-rattle compound is a noise-troubleshooter's nightmare. Modern automobiles use a substance like children's modeling clay as an anti-rattle compound between the dashboard assembly and the car's firewall. If the dashboard mounting struts or brackets make up part of the ground return path for the radio, then this compound may cause both noise and a DC voltage-drop in the transmit mode. Similarly, the compound makes the dashboard appear to be "floating" to rf, which may reradiate noise into the radio, even when the radio chassis is grounded directly to the firewall. In fact, problems involving anti-rattle compound are some of the most difficult to trace; the noise seems to come from everywhere.

The solution is to connect a *heavy* conductor between the radio chassis and the firewall. Use battery braid, ground braid, or the outer conductor of a coaxial cable such as RG-8/U or RG-11/U.

The problem involving noise created by poor grounding of the receiver has been compounded in recent years by the use of lock-mounts as mounting brackets. Theft of mobile rigs has become a fact of life in many areas of the country. So much so, in fact, that some insurance companies no longer cover a radio transceiver (CB or 2-meter rig) on the standard comprehensive policy without an additional premium. Many installers, and that includes some professional installers, will use the outer shield of the coaxial cable as the power-supply DC return line. This is, from the noise point of view, is always bad practice. If the radio has a ground wire, then use it. If the radio does not have a ground wire, then make one by connecting a heavy length of flexible wire from a grounded point on the chassis to the firewall of the vehicle.

I have found that, in low-power rigs, such as 2-meter FM and CB models, that the copper spring clips used on lock-mount brackets cause few problems when they are in good repair and clean. But, when they lose some of their tension, or when the copper surfaces become corroded, then troubles (in the form of reduced output power and noise in the receiver) will develop.

When using a lock-mount, or transmission hump "anti-theft" console, make sure that the grounding wire is heavy, and that it goes directly to the firewall. The importance of a good radio-chassis ground in mobile installations cannot be overemphasized.

Noise Via Antenna and Power Lines

Once the antenna, radio, and chassis-grounding connections have been eliminated as possible sources of noise, the next consid-

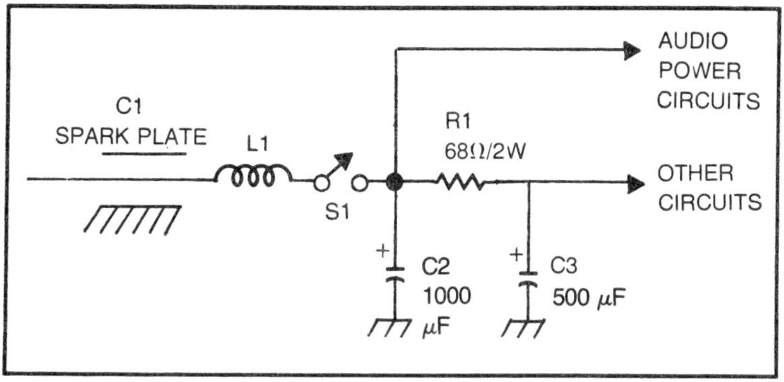

Fig. 17-2. Typical mobile receiver input circuitry.

eration is whether or not the noise level is reduced. Some servicers erroneously believe that the type of noise will tell you where it enters the receiver. This is not always true. Ignition noise, for example, can enter the receiver through either the power lead or the antenna lead. The corrective measures appropriate for each are different.

Antenna-borne noise, for example, is most likely caused by inadequate shielding action of the hood and body panels. Or, it might be caused by reradiation of rf noise from parts such as the tailpipe.

Noise that enters via the power lead, on the other hand, may be caused by improper dress of the power lead itself, or by induction from another cable. Once you realize this fact, and just what it can mean, you will understand the old timers who wisely tell us that once you have eliminated the common and the obvious causes of noise, finding the remaining noise sources is a lot like trying to skin an amoeba! The job simply is not very easy after that (in fact, it is never really easy; it is always harder work than "benchwork").

Also, it may be that the noise occurred right after the car was serviced, or was in an accident. The extensive use of fiberglass and plastic body components to repair the damage, and disconnected antenna grounds, can raise the dickens with noise interference.

I recall a Chevrolet *Stingray* (with an all-fiberglass body) that had recently been to a body shop for extensive rear right quarter-panel repairs following an accident. The body repairmen had left the ground braid used on the *Stingray* AM/FM broadcast antenna disconnected (it normally connected to the tailpipe). The result was tremendous ignition noise. In this instance, we are also given a lesson for mobile installers who are confronted with an all-fiberglass car. Be wary of mounting locations that are not near a metal chassis piece, or some other electrical ground point. Even a short run of ground braid can form a high impedance at VHF.

Another problem involves noise that shows up immediately following a tune-up of the vehicle's engine. If the mere mention of the word "tune-up" prompts you to recall that the problem was noticed immediately after the car had been to the mechanic for a tune-up, then suspect a problem in that area.

In most cases, you will probably discover that the mechanic has replaced the carbon filament "radio-TV" ignition wires with plain copper wire. The carbon wire has a high resistance (10-kohms/foot), and is specified by the car manufacturer to eliminate (or reduce) ignition noise in the car radio, and the radios of nearby cars. Some mechanics, out of honest conviction that the wires deteriorate auto performance, or from greed (the straight wire often sells for less wholesale, but can be priced to the customer at carbon-wire rates!), will use the straight wire.

It is probably true that old carbon-ignition wires will cause poorer car performance. But, this is usually rectified by the use of new wires. . . renew them periodically. If you are one of these car hobbyists who demand the alleged advantages of straight wires, then you must accept noise in the receiver as a consequence.

Some people use the in-line, add-on carbon resistors called *noise suppressors* (Fig. 17-3) in place of the carbon wires. These reduce performance almost as much as the wires (maybe more!), and have the added disadvantage that they tend to fall off when old! This will leave you stranded without an ignition system because these resistors are usually installed in the high tension lead between the distributor and the ignition coil!

Resistor spark plugs, when available, are almost as effective as carbon wires in reducing interference, and are considerably more reliable than the in-line resistors.

Some companies manufacture ignition-line shielding for really tough, or sensitive, cases. Although this is a real hassle, and is expensive, the results are impressive.

If you suspect that the carbon radio-TV wires have been replaced with regular wire, and the fact isn't reflected in the labeling on wire, then use an ohmmeter to make the determination. Regular ignition wires will show show a short circuit from end to end, instead of several kohms of resistance.

It is equally important to determine whether the car has not been tuned up recently, and should have been! If the car has not been tuned up in the past 25,000 - 30,000 miles, then it is a good bet

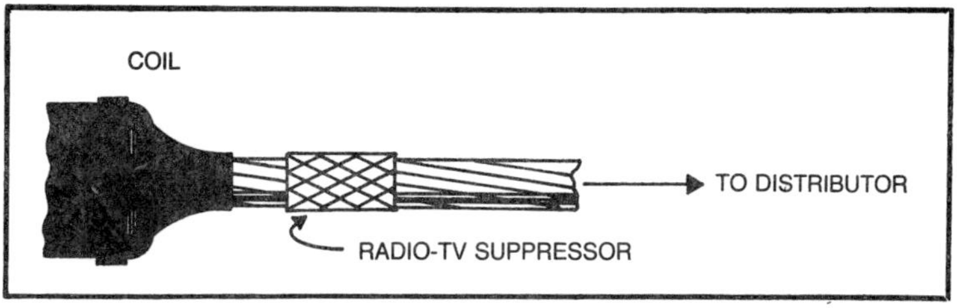

Fig. 17-3. In-line ignition-noise suppressor between coil and distributor.

that the noise is a direct result of normal deterioration, especially if the car seems to be running poorly. The carbon filament used as the "wire" in radio-TV ignition cable will develop cracks. Also, distributor points, breaker points, and electrical connectors will all wear out in time. Such components are usually replaced by the mechanic who tunes up the engine. Once the deterioration passes a certain point, then RFI/EMI in the radio will result.

Besides rough idling, one telltale sign of a bad ignition cable is a "pop-pop-pop" sound as the engine idles. This is due to one ignition cable being open, or off the spark-plug terminal. If you have ever seen the firing parade of a multi-cylinder engine on a mechanics tune-up oscilloscope, then you will recognize the missing or open wire as the pulse with an extremely high amplitude, off the screen, in fact.

Many professional servicers will refuse to attempt to troubleshoot noise problems on vehicles they deem to have been too long without a tune-up. The effort is simply too much, with too little chance of success.

AUTO ELECTRICAL SYSTEM AS A NOISE SOURCE

In general, in mobile installations, if it is electrically operated, then it is a suspect as a noise source!

One major source of noise problems is the alternator/generator/voltage regulator. The whine, and other noises, produced by these systems, and noises produced in the ignition system, can often be eliminated by the installation of appropriate bypass capacitors, or an L-section lowpass filter. In Fig. 17-4, a bypass capacitor has been installed on the alternator, and in Fig. 17-5, on a generator. In the case of an alternator, it is wise to follow the manufacturer's recommendations as to the capacitor value to be used. With the wrong value, it is possible to resonate the circuit at some engine speed resulting in poor performance, or burn-out of the alternator coils.

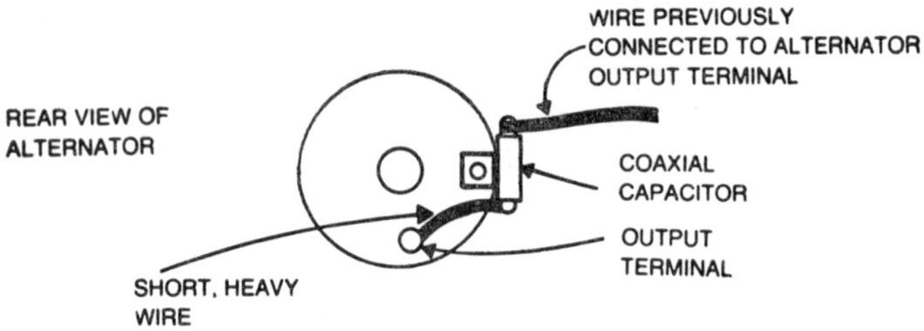

Fig. 17-4. Connecting a capacitor to the alternator.

In Fig. 17-6, a capacitor has been installed on a voltage regulator to filter out noise. On older regulators, using two or three relays, it has been customary to use 0.1 μF on each of the high-current terminals (battery, and either generator or alternator), and the *RC* network shown for the field terminal.

In Fig. 17-7 we see the use of an L-section filter to eliminate alternator whine. Bear in mind that this technique works best on installations that draw little current, and does not work well on high-current installations. In some cases, success has been achieved by separating the receiver and transmitter A-leads inside of the case, then installing a circuit such as Fig. 17-7 in the receiver line (internal to the case). Otherwise, the circuit can be installed external to the transceiver or receiver. I have not tried high-current chokes, but it appears possible to make such a choke, using heavy gauge wire and one of those large toroid cores used to make high-frequency antenna-balun coils.

For low-current-drain receivers, the choke in Fig. 17-7 may be almost any car-radio input choke in the 0.5- to 5-Henry range, provided it has low DC resistance.

INDUCTIVE AND CAPACITIVE PICK UP

At this point, we can claim to have covered all of the major problems in mobile-noise elimination; at least those problems usually discussed in book and magazine-article treatments of the subject. Those problems that remain, however, can be devilishly difficult to service. To the professional, such problems invariably prove unprofitable.

Here's an example. In the early days of CB (*circa* 1960) I did an installation of a CB transceiver in a 1956 Ford *Crown Victoria*, and it had a tremendous amount of motor noise. The boss and I had spent

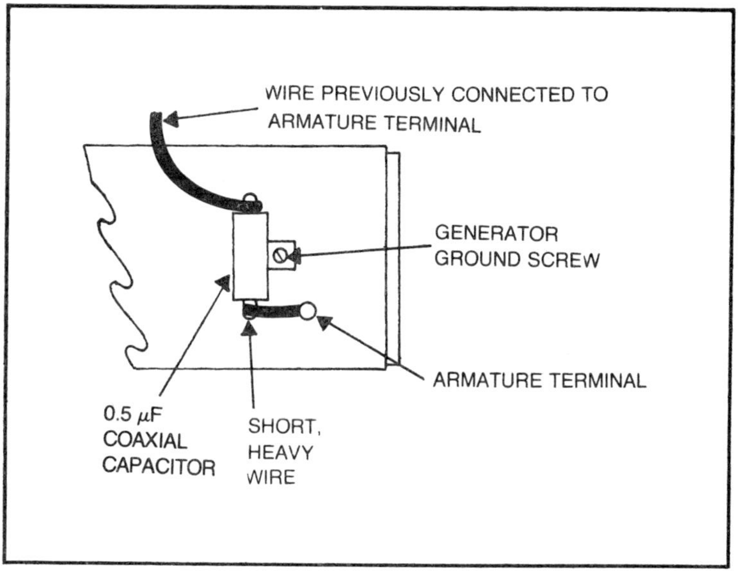

Fig. 17-5. Connecting a capacitor to the generator.

most of the morning and afternoon trying to pin down the problem. We had all but admitted defeat. To ease the pain in my lower back, caused by being too fat and bending over an automobile engine too long, I leaned against the roof-line and side of the car. . . and the noise stopped!

Inspection revealed that a decorative crome strip around the roof line was loose, and the few remaining fasteners were corroded beyond any hope of producing grounding action. To add insult to

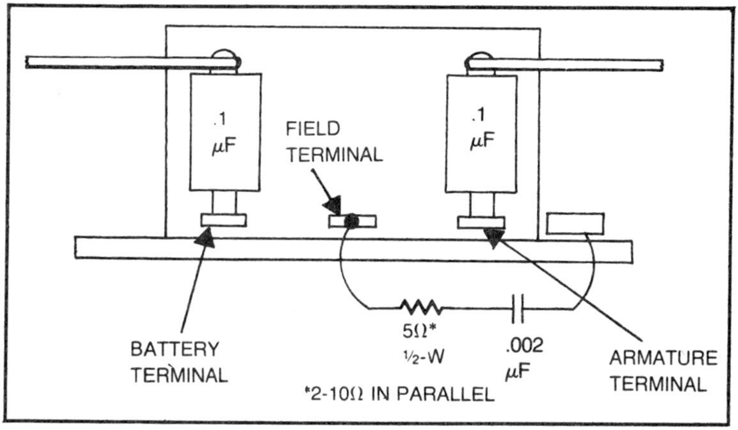

Fig. 17-6. Noise suppression on a mechanical regulator.

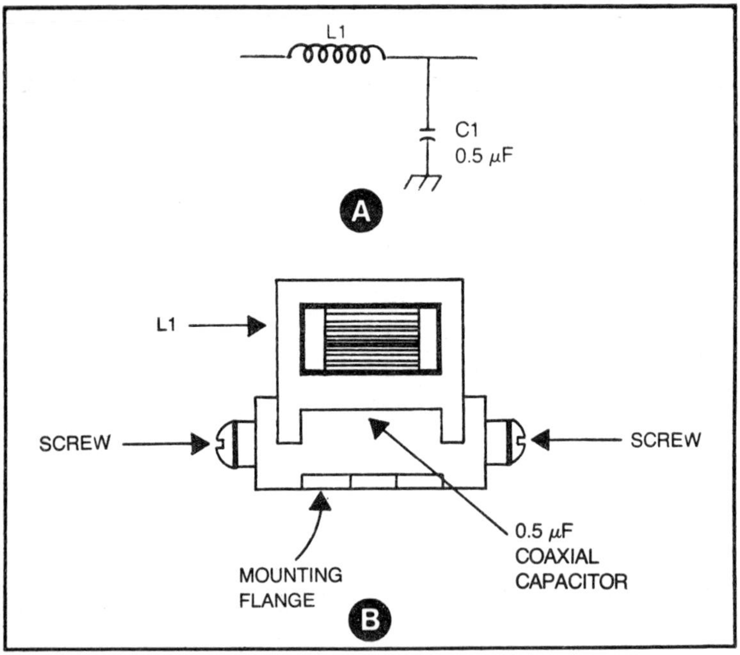

Fig. 17-7. L-section noise filter circuit (A); and mounting(B).

injury, the strip was almost a quarter wavelength on 11-meters, so the noise problems were aggravated even further. The chrome strip was picking up the noise and reradiating it because of its resonant length. Cleaning the metal surface on the underside of the chrome strip, and on the roof where it was mounted, and renewing the fasteners solved the problem.

Inductive and capacitive pick-up, coupled with reradiation problems similar to the extreme case cited above, cause some of the most difficult to solve motor noise problems in mobile installations. Noise suppression is *usually* a simple matter of rerouting a wire, shielding something, or adding a ground wire somewhere. The big problem is finding the source.

Figure 17-8 shows a very common source of reradiation—the engine exhaust pipe. It is a conductor, and hangs from insulated supports. Furthermore, it often has a length that is one-eighth, one quarter, or half wavelength at one or more of the high-frequency Amateur bands and so may not act like a grounded object even though it is attached to the engine block.

Even in the best of circumstances, however, the exhuast pipe is not well grounded because of rust and the types of fittings

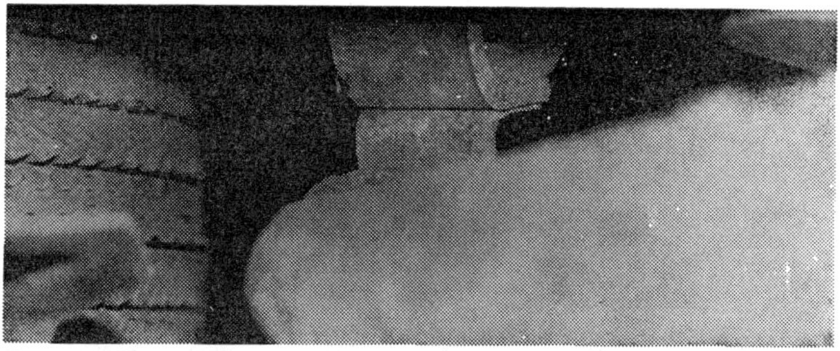

Fig. 17-8. Exhaust-pipe hangers are a frequent cause of reradiated noise.

normally used. To cure this type of problem, it is often necessary to ground the tailpipe to the frame of the automobile at intervals of every few feet, taking care not to select one or more intervals that are electrical lengths of the frequency being used.

"SNIFFERS" AND "SLEUTHS"

Trying to locate the source of inductive or capacitive pick-up, or reradiation, is difficult at best. Almost any conductor that passes through the firewall of the vehicle is a prime suspect. The emergency-brake cable and the accelerator linkage are but two examples. Also causing problems are the wires passing through the firewall. They will pick up noise impulses from nearby ignition system components inside of the engine compartment, and then couple these pulses to the radio A-line, antenna, or speaker leads via the methods discussed above. A very frequent contributor to the noise problem is the antenna, speaker, or (in the case of remotely controlled trunk-mounted models) control leads running under the carpet to the trunk. The car manufacturer generally provides a channel, or certain openings, to permit the cable run to

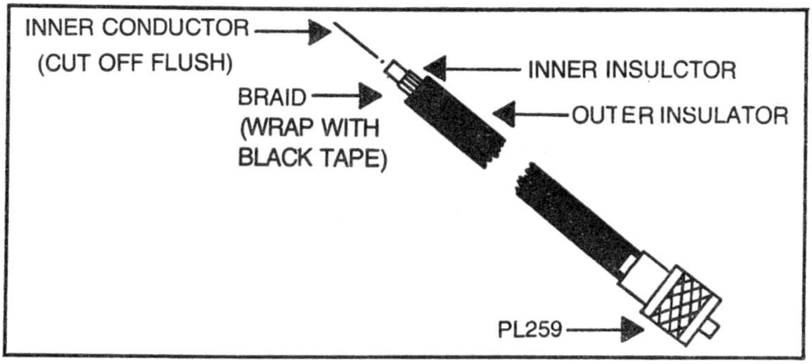

Fig. 17-9. Coaxial cable "sniffer."

Fig. 17-10. Channel Master "Sleuth" finds noise spots.

the rear lights, gasoline gauge, etc. It is all too easy to also run the mobile radio cables along this same path! This procedure can cause noise pick-up. A far better approach is to run such wiring down the opposite side of the car from the power wiring; i.e., if the power wires run (as usual) down the driver's side of the car, then run the mobile-radio antenna, loudspeaker, or control wires down the passenger's side!

An rf "sniffer" is a tool used by auto electronics technicians to locate the source of noise pick-up. An example of this simple tool is shown in Fig. 17-9. It consists of a piece of coaxial cable (preferably RG-58 or 59) stripped so that an inch or so of the inner insulator is exposed. The other end is fitted with a PL-259 UHF connector, or whatever connector is compatible with your receiver. The device is used as an antenna to probe in the vicinity of suspected noise sources, using the receiver output as an indicator.

A commercial version is the Channel Master *Sleuth* (Model 5270 rf noise detector) shown in Fig. 17-10. This device is actually superior to the sniffer, because it is more directional. The cost is less than $20, so it makes a handy tool for almost any commercial radio shop. Although the Channel Master Sleuth was designed for the CB service technician, I have found that it also works well in high-frequency Amateur installations, in AM broadcast radios, and moderately well in FM-broadcast applications.

Chapter 18
Operational Amplifiers

Although operational amplifiers, or "op-amps," have been in existence for several decades, they only came into their own during the 1960s. Some of the first linear integrated circuits were operational amplifiers, and the popularity of the op-amp has grown by leaps and bounds.

Until recently, much of the Amateur monthly literature and books ignored the op-amp, but now more and more new Amateur equipment and homebrew projects use them. Amateur use of the op-amp has been stifled, perhaps, by the attitude of one of the leading Amateur magazines, which required all operational amplifiers used in projects they published to use a single-polarity power supply. This short-sighted rule was based on the face that the project could not be used in mobile applications if a bipolar power supply were used. But, it failed to recognize the fact that mobile operation was one of the smallest Amateur categories, and most of it was with factory built 2-meter FM rigs, and a few factory-built HF rigs. This policy tended to deny Amateurs both a proper understanding of the working of the op-amp and the full use of its potentials . . . which could be realized only when used in a dual-supply circuit.

In this chapter, I will briefly cover op-amp theory, and discuss some of the ways you can troubleshoot op-amp circuits.

INTRODUCTION TO OP-AMPS

The operational amplifier has been available in IC form for the past dozen or more years, and in that time has become one of the principle linear IC devices sold. The orignal vacuum-tube op-amps were designed for use in analog computers, and were used to peform mathematical operations, hence the name operational amplifier. The range of circuit applications today, however, far exceeds that rather limited view of the original inventors.

Although it is quite possible to describe the operational amplifier in terms that are mathematically quite complex, we may also provide an adequate working description using just Ohm's law, Kirchoff's current law, and the basic properties of the op-amp itself.

One of the profound beauties of the modern IC op-amp is its utter simplicity when viewed as a "black box," or circuit-building block. We may build our knowledge of the behavior of the op-amp on the universal transfer function for all voltage amplifiers, namely gain $(A_v) = E_{out}/E_{in}$.

There are six basic properties ideal to all op-amps. The first five of these are:

1. Infinite open-loop voltage gain
2. Infinite input impedance
3. Zero output impedance
4. Infinite bandwidth
5. Zero noise generation.

Of course, it is not possible to obtain a real op-amp with these ideal properties, but the approximations that are possible are quite exciting. In the list above, replace the word "zero" with "very, very, low," and "infinite" with "very, very high," when talking about real, kind-you-can-buy operational amplifiers. Most real op-amps, for example, have open-loop gains in the 20,000 to 2,000,000 range, input impedances over 1 megohm, and output impedances under 100 ohms.

Differential inputs are found on most IC op-amps. Note that the schematic diagram symbol for the op-amp, shown in Fig. 18-1, has two inputs. One is labeled (−) and is the *inverting* input while the other is labeled (+) and is the *noninverting* input. These inputs produce output signals that are 180 degrees out of phase with each other; the noninverting input produces signals in-phase with the input signal, while the inverting input produces output signals that are 180 degrees out of phase with the input signal.

The two differential inputs of the op-amp have equal, but opposite-polarity, effects on the output signal for any given input signal.

At this point, now that we have introduced the concept of differential inputs, let us add the sixth basic property of the op-amp to the list above:

6. Differential inputs follow each other and must be treated as if they were at the same potential.

This property implies that the two inputs will behave as if they were at the same potential, especially under static conditions. In Fig. 18-2 we see a simple, inverting, follower circuit in which the noninverting (+) input is grounded. Any point that is grounded in this circuit, incidentally, has a potential of zero volts. The sixth op-amp property tells us that we must now treat the other input (inverting input), *as if it were also grounded*. Many magazine articles and textbooks, including some by this author, have labeled this concept a "virtual" ground, but, that term tends to confuse people. It is better to merely accept as a basic rule, for purposes of calculations and voltage measurements, that the inverting input will act as if grounded when the (+) input is really grounded, and that it will take on the same voltage as the (+) input if the (+) input has a certain voltage applied to it. The reasons why this is so are due to the internal operation of the device.

Before proceeding to the analysis of the simple op-amp circuit, let me once again direct you to Fig. 18-1. Note the power supply terminals. For normal operation, the op-amp requires a dual-polarity power supply (in reality, two power supplies). The Vcc power supply is positive with respect to ground, while the Vee supply is negative with respect to ground.

CIRCUIT ANALYSIS

We know from Kirchoff's current law that the algebraic sum of all currents entering and leaving a point in a circuit must be zero.

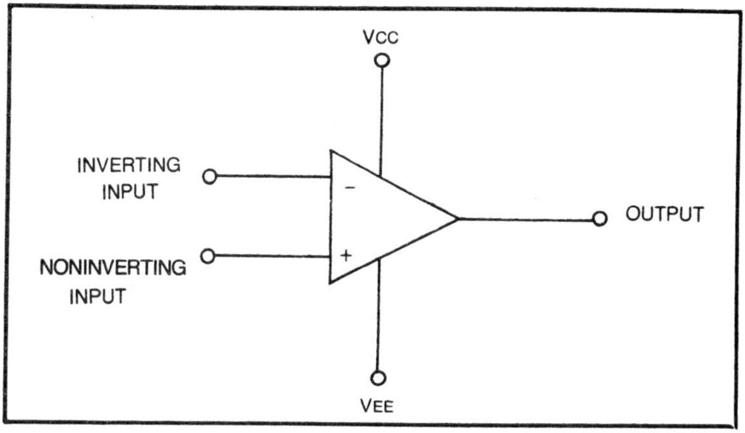

Fig. 18-1. Basic operational amplifier.

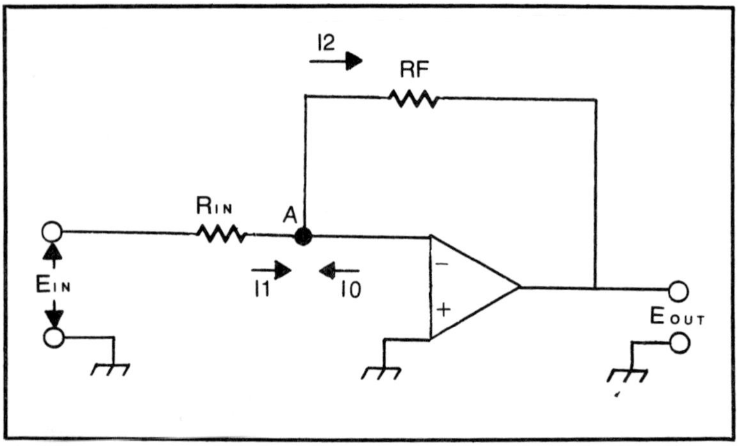

Fig. 18-2. Inverting follower.

The total current flow into and out of point "A" in Fig. 12-2, then, must be zero. Three possible currents exist at this point: input current I_1, feedback current I_2, and any bias currents flowing into or out of the op-amp, I_0. But according to the second ideal property, the input impedance of this type of device is infinite, and, by Ohm's law:

$$I_0 = E/Z_{in} \tag{18-1}$$

Since Z_{in} is infinite, then current I_0 must be zero. So if current I_0 is equal to zero, then we must conclude that only I_1 and I_2 are present in the external circuit. This means, by Kirchoff's law, that $I_1 + I_2 = 0$, so

$$I_2 = -I_1 \tag{18-2}$$

We also know that

$$I_1 = E_{in} / R_{in} \tag{18-3}$$

and,

$$I_2 = E_{out}/R_f \tag{18-4}$$

By a little algebraic manipulation, we may substitute Equations 18-3 and 18-4 into Equations 18-2, and obtain as our result:

$$\frac{E_{out}}{R_f} = \frac{E_{in}}{R_{in}} \tag{18-5}$$

Solving Equations 18-5 for E_{out} gives us the transfer equation normally found in textbooks and magazine articles for the gain of an inverting follower:

$$E_{out} = \frac{-E_{in} R_f}{R_{in}} \tag{18-6}$$

The term "R_f/R_{in}" is the voltage-gain factor, and is usually designated by the symbol A_v, which is written as

$$A_v = R_f/R_{in} \tag{18-7}$$

we sometimes see Equation 18-6 written using the symbol for voltage gain,

$$E_{out} = -A_vE_{in} \tag{18-8}$$

When designing homebrew projects using the operational amplifier, use Equations 18-7, or 18-8. Let's look at a practical example. Suppose that we wanted an amplifier that had a voltage gain of 50. We further want to drive this amplifier from a voltage source that has an output impedance of 100 ohms. A standard rule of thumb, used by most designers, is to make the input impedance of the circuit at least ten times the source impedance, so we would need a minimum input resistor (R_{in}) of 10 × 100 ohms, or 1000 ohms. To make the example simple, let's use this figure as a selected value for R_{in}. The question now becomes: what value resistor for R_f would yield a voltage gain of 50?

$$A_v = R_f/R_{in} \tag{18-9}$$

$$50 = R_f/1000 \tag{18-10}$$

$$R_f = 50{,}000 \text{ ohms}$$

Our gain-of-50 amplifier would look like Fig. 18-3.

NONINVERTING FOLLOWERS

All inverting follower circuits suffer badly from low input impedance. If the (+) input is grounded, as is almost always the case, then the maximum impedance seen by the driver of the amplifier is R_{in} (remember that "virtual" ground). The problem becomes espe-

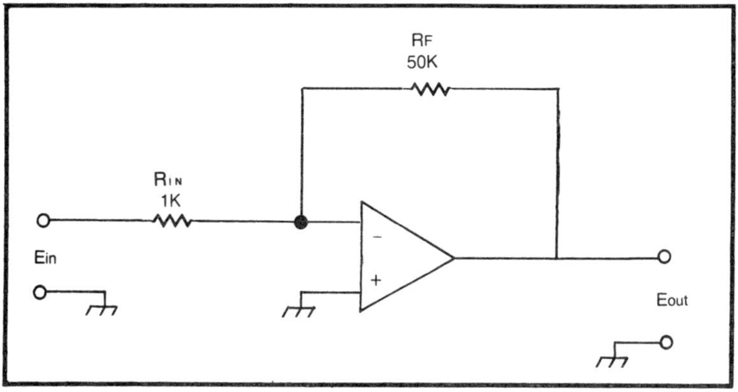

Fig. 18-3. Gain-of-fifty inverting follower.

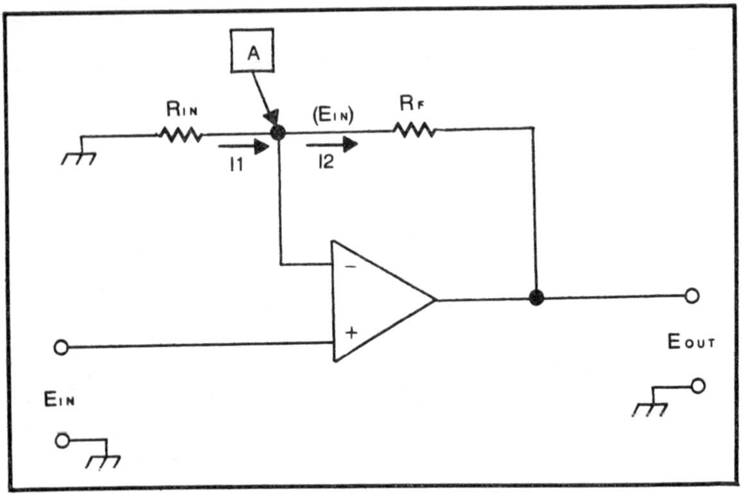

Fig. 18-4. Noninverting follower.

cially acute at higher gains because we must use lower and lower values of input resistor in order to obtain values that we can actually go out and buy for the feedback resistor.

The noninverting follower circuit of Fig. 18-4 solves this problem nicely. The input impedance is typically the input impedance of the op-amp device, which is very high (as high as 1.5×10^{12} ohms, that's 1.5 *terraohms*), especially in those op-amps with a MOSFET input stage.

We may use the same general method for deriving the transfer equation of the noninverting amplifier as we used in the case of the inverting amplifier. Remember that the op-amp inputs track together (property no. 6). If a potential is applied to the noninverting input, then we must treat the inverting input, point "A" in Fig. 18-4, as if it were also at that potential. In this case, the noninverting input is at potential E_{in}, so we must assume that the inverting input is also at E_{in} (as, in fact, a voltmeter would verify!). By the same process as before, slightly modified:

$$I_1 = I_2 \quad (18\text{-}11)$$

$$I_1 = E_{in}/R_{in} \quad (18\text{-}12)$$

$$I_2 = (E_{out} - E_{in})/R_f \quad (18\text{-}13)$$

Again, using the same concept as before:

$$I_1 = I_2 \quad (18\text{-}14)$$

$$\frac{E_{in}}{R_{in}} = \frac{E_{out} - E_{in}}{R_f} \quad (18\text{-}15)$$

Using a little algebra on Equation 18-15 leads us to the transfer equation normally given for the noninverting follower circuit:

$$E_{in} = \left[\frac{R_f}{R_{in}} + 1\right] \times E_{in} \qquad (18\text{-}16)$$

There is a special case of the noninverting amplifier in which the voltage gain is unity (1). If the output terminal is shorted to the inverting input, then the "R_f/R_{in}" term of Equation 18-16 is zero, and the equation reduces to $E_{out} = (1) \times E_{in}$.

OP-AMP POWER SUPPLIES

Earlier in this chapter, I told you that the op-amp prefers to use bipolar power supplies (Fig. 18-5). The Vcc supply is positive with respect to ground and the Vee power supply is negative with respect to ground. Keep in mind that, although batteries are shown in Fig. 18-5, the actual power supply in AC-powered equipment will be + and − DC power supplies.

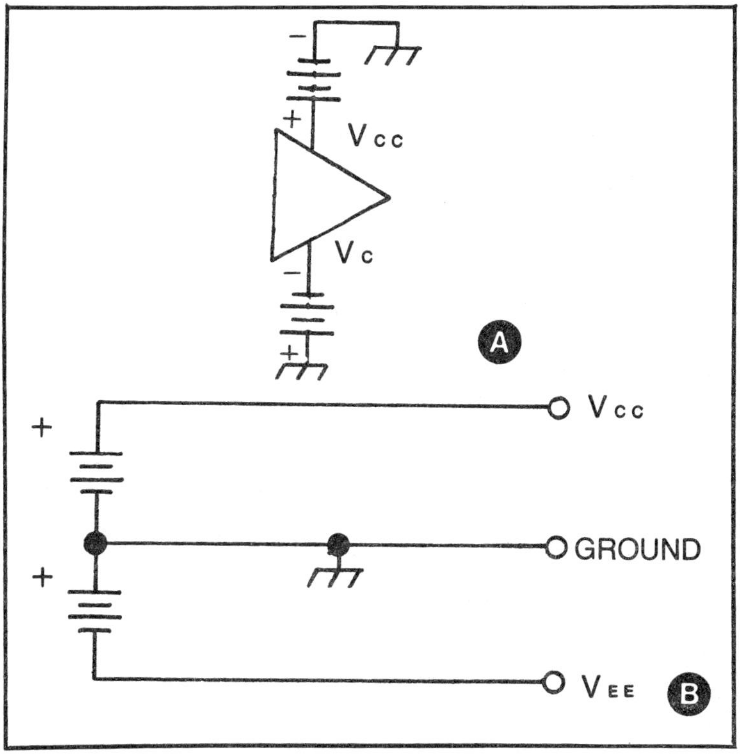

Fig. 18-5. (A) Op-amp power-supply connections; (B) typical dual-polarity power supply.

Most op-amps will accept power supplies in the ± 3 to ± 22 volt, with the vast majority being in the "less than ± 18 volt" range.

There are two constraints placed on op-amp users with respect to power-supply voltage selection. One is the minimum allowed voltage for a given maximum output voltage. Some op-amps require a supply voltage that is 3.7 volts higher than the maximum expected output voltage. If you expect the output voltage to swing to ± 10 volts, then the supplies must be at least (10 + 3.7), or 13.7 volts dc. Other devices, such as the RCA CA3140 BiMOS op-amps, will allow operation with outputs that are just 0.5 volts less than the supply voltages.

The second constraint is that the maximum value of Vcc to Vee must not be exceeded, and this potential is not always merely twice the maximum value of Vcc or Vee alone. An op-amp such as the 741, for example, might be able to operate with Vcc *or* Vee values up to ± 18 volts, but the Vcc to Vee value is only 30 volts. Consider an example. If we require a Vcc supply of ±18 volts, and the Vcc to Vee rating is only 30 volts, the maximum allowable value for Vee is (30) – (18) = 12 volts. This is true even though the Vee rating may be –18 volts! If the maximum voltage for both Vcc and Vee were used, then the difference voltage would be (18) – (–18) = 36 volts, which is 6 volts greater than is allowed by the Vcc to Vee rating.

PRACTICAL OP-AMPS: SOME PROBLEMS

Before we can successfully build projects with op-amps in them, or troubelshoot existing op-amp circuits, we must learn that the ideal op-amps of textbook analysis simply do not exist. We, who write the books, use the ideal case because it simplifies the job of writing the book . . . but, there is still a duty to the reader to let you know some of the problems in the real-world devices.

In general, the lower the cost of the device, the less like the ideal it will be. The 50-cent, 741-family devices, for example, look a lot less like op-amps than do certain 50-dollar models. But, the fact that the problems accompanying being "less than ideal" are manageable is shown by the fact that the 741 is the most popular op-amp in industry and hobby markets alike.

Three main problems are found in op-amps: offset current, offset voltage, and frequency response. Of less importance to our discussion, but critical in scientific instrumentation applications, is the generation of noise inside the op-amp.

In real op-amps, the input impedance is something less than infinite, and this implies that a small bias-current flows into, or out

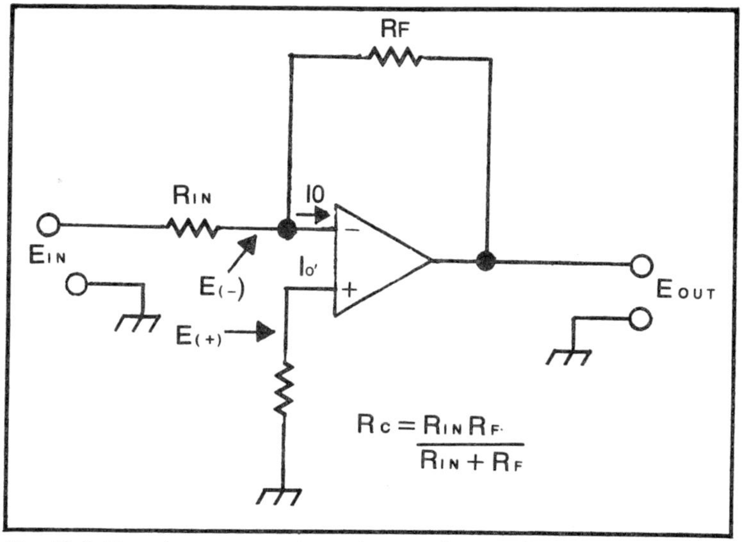

Fig. 18-6. Use of a compensation resistor to change offset.

of, the inputs of the device. In other words, current I_0 of Fig. 18-2 is not actually zero, but will have some value. The output voltage produced by the existence of I_0 will be equal to $-I_0 \times R_f$, and this represents an output offset-voltage due to an input offset-current.

One cure for this problem that works pretty well, for the expense of but one resistor, is the compensation resistor shown in Fig. 18-6 between ground and the noninverting input of the op-amp. The value of the compensation resistor R_c is the parallel combination of R_{in} and R_f.

The compensation resistor works well because the input bias currents flowing in the two inputs are approximately equal. The bias current from the (+) input, then, flows through R_c to create a small voltage $E_{(+)}$ at the noninverting input. This voltage counteracts the voltage applied to the inverting input due to the current I_0 flowing in R_f and R_{in}.

The input bias-current for the transistors of the op-amp are only one source of an output offset-voltage. There are actually several such voltages. An output offset-voltage is the voltage at the output terminal in cases where the input voltage $E_{in} = 0$. In the ideal op-amp of the textbook world, the outut voltage would be zero anytime E_{in} was also zero. But real op-amps don't always work that way.

Almost any output offset voltage will be cancelled by one of the two circuits shown in Fig. 18-7. The circuit in Fig. 18-7A uses a pair

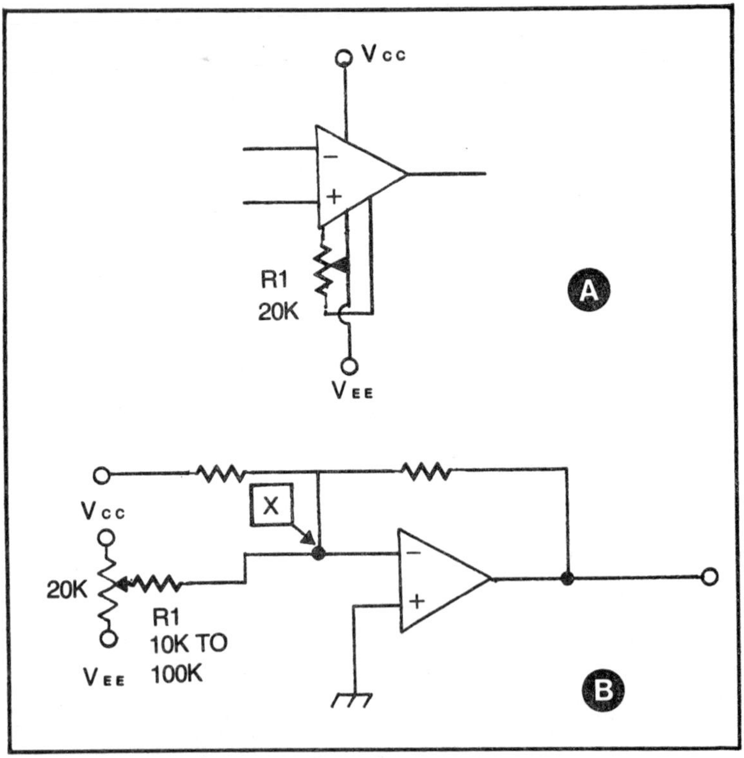

Fig. 18-7. Onfset-null circuits:(A) when null terminals are used; (B) when no null terminals are available.

of special *offset null* terminals found on many IC op-amps. If the op-amp that you select lacks these terminals, or if the range of control that it provides is insufficient for the null observed (often the case with bargain op-amps bought for a song at mail-order or retail hobby stores), then use the circuit of Fig. 18-7B.

The offset-null circuit of Fig. 18-7B creates a small current flowing in resistor R1 to the summing junction of the op-amp (point "X"). Since the offset current may flow *into* or *out of* the input terminal, the null circuit must be able to supply current of both polarities. This means that the ends of the potentiometer (R2) must be connected to Vcc and Vee, instead of just one supply.

In many cases, it is found that the offset null is small compared with normally encountered signal levels. This is especially true in low-gain applications. In those cases, the offset may be negligible, so you might not want to take any action at all. In still other cases, where a cascaded chain of stages is used, then the cumulative offset might be big, even though the individual offsets are small. In that case, apply the null circuit only to one stage late in the chain.

In some cases, especially when dealing with low signal levels and/or high gains, the simple circuits of Figs. 18-7A and 18-7B will not offer enough resolution to completely null out the offset to the fine degree needed. Figs. 18-8A and 18-8B offer simple solutions to these types of problems. In Fig. 18-8A a pair of resistors are connected between the ends of the potentiometer and the Vcc and Vee power supplies. The values of these resistors are usually equal to each other, and are 5 X to 20X the total value of the potentiometer. If the potentiometer is a ten-turn model, then a very small change of voltage at the wiper is possible for any given turn of the adjustment screw.

In Fig. 18-8B, we have essentially the same thing, except that a pair of Zener diodes are used to regulate the voltage at the ends of the potentiometer to values substantially less than either Vcc or Vee. This, again provides a smaller increment of voltage at the wiper per turn of the adjustment screw.

DC DIFFERENTIAL AMPLIFIERS

The existence of two complementary inputs on the op-amp make it very easy to make differential amplifiers. These circuits produce an output voltage that is proportional to the difference between the voltages applied to the two inputs.

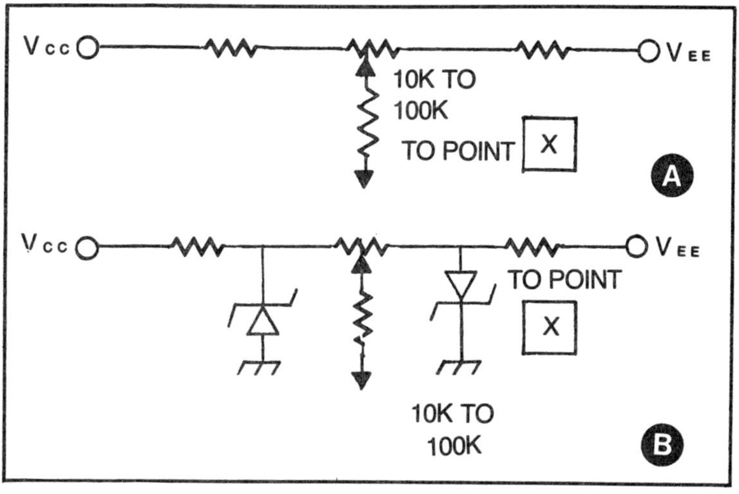

Fig. 18-8. Medium-resolution offset control (A), high-resolution offset control (B).

Recall, from our previous discussion, that the two inputs have equal, but opposite-polarity, effect on the output voltage. This means that the output voltage will be zero if the two input voltages are equal. If the same voltage, or two equal voltages, are applied to the inputs (the *common mode* voltage, E3, in Fig. 18-9), then the output voltage will be zero. The transfer equation for the differential amplifier is

$$E_{out} = A_v (E_1 - E_2) \qquad (18\text{-}17)$$

so, if $E_1 = E_2$, then $E_{out} = 0$.

The circuit in Fig. 18-9 shows a simple DC differential amplifier using just one operational amplifier. The voltage gain of the circuit is given by

$$A_v = R3/R1$$

provided that R1 = R2 and R3 = R4.

The main appeal of this circuit is that it is very simple, requiring only one operational amplifier. It will reject common-mode voltages reasonably well (including 60-Hz pick-up from nearby power wiring), and is reasonably well-behaved. But, it suffers from low input impedance.

In this brief introduction to operational-amplifier theory, we cannot even scratch the surface of the whole story. For those who want to know more let me direct you to my book, *OP-AMP Circuit Design & Applications* (TAB book No. 787).

Troubleshooting Op-AMP Circuits

For the most part, troubleshooting op-amp circuits is the same as troubleshooting any other electronic circuit. Be careful,

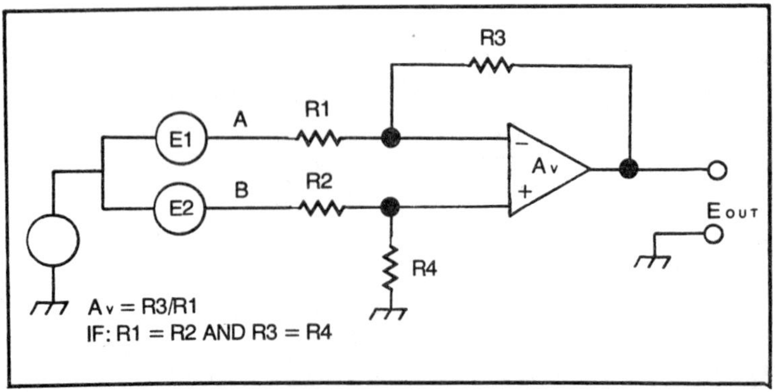

Fig. 18-9. Signal voltages affecting an op-amp.

though—there are some pitfalls. For example, when using a signal tracer or signal generator, do not look for, or inject, a signal directly into the inverting input of an inverting follower amplifier. This is tempting because, after all, it is the input. But it is also at a ground, or near ground,potential.

Use the DC voltmeter to measure the potentials on all the terminals, and then reason out the problem using the six basic properties and the transfer equations.

In most cases, the fault will cause the output voltage to rise to the level of one of the supply voltages. If this happens you need to find out whether the problem is inside of the device (a bad op-amp) or whether it is externally caused. Unless there is circuitry, that could be damaged, following the op-amp, the best way to see if the op-amp is good or bad, in the absence of a tester, is to simply short together the (–) and (+) input terminals. This should cause the output voltage to drop to zero, or near zero. If it does not, then suspect a bad op-amp. In that case, troubleshooting by replacement is then justified.

Chapter 19

Testing Components

There may be several reasons why an Amateur might want to test electronic components. One might be to verify that the hamfest or auction "grab-bag" specials purchased for only a song and a prayer are actually good. It is surprising how many people sell the proceeds of cleaning out their junk box at hamfests, only to include large numbers of bad parts. Have you ever tried to get a construction project working properly using untested junk box specials? Or have you purchased assortments of components, allegedly "tested," from a national mail-order or retail-chain merchandiser, only to find out that they were what is euphemistically called "industrial surplus" (often as not, translatable as "rejects."). Another reason for wanting to test individual components is to troubleshoot electronic circuits. Whether newly constructed, or a well-used piece of equipment, it is possible that the troubleshooting procedures will only serve to get you "in the ballpark," and you will have to troubleshoot the rest of the way by testing individual components.

There are two levels of testing any components. One is a quick test, qualitative in nature and often subjective. The other is a quantitative test that tells how many μF, ohms, or whatever. Both are valid, although the qualitative tests are used more in troubleshooting because, a) they tell the story of go/no-go, and, b) they are simple to perform.

QUALITATIVE SEMICONDUCTOR TESTS

A simple tester for transistors and diodes can be made from only a switch and a battery/resistor combination. The circuit so resembles an ohmmeter as to suggest to the clever that an ohmeter can be forced to make simple transistor tests. While it is true that we cannot easily measure the beta or the frequency response of a transistor, we can tell whether it is good for DC . . . or at least predict with high certainty that it is a good device. This is all that we need in troubleshooting, most of the time.

A solid-state diode consists of a single PN junction of semiconductor material. Such a junction will pass current in only one direction, and there lies the basis for an ohhmeter test of diodes.

Most ohmeters uses a DC voltage source, in VOM's and most electronics VMs, a dry-cell battery, and some calibrating resistors. The meter will pass a current through the unknown resistance, and the current level allowed to flow by the resistance of the unknown is inversely proportional to the resistance. Because the voltage source is DC, the ohmmeter probes are polarized; one is positive and the other is negative.

When the ohmmeter forward biases the PN junction, as in Fig. 19-1A, then the ohmmeter will register a low resistance (on the order of 50 to 1000 ohms, depending upon the diode and the current flowing). When the ohmmeter probes are reversed, exactly the opposite occurs (see Fig 19-1B); a high resistance is measured.

Note well that the terms "high" and "low" in this context are used in a relative sense. When I say high, this means something in the over-1000-ohms range, and possibly up to the resistance of an open circuit. Similarly, low means from 0 to about 1000 ohms.

When testing PN junctions made of germanium, the ratio between the high and low readings obtained when using the ohmme-

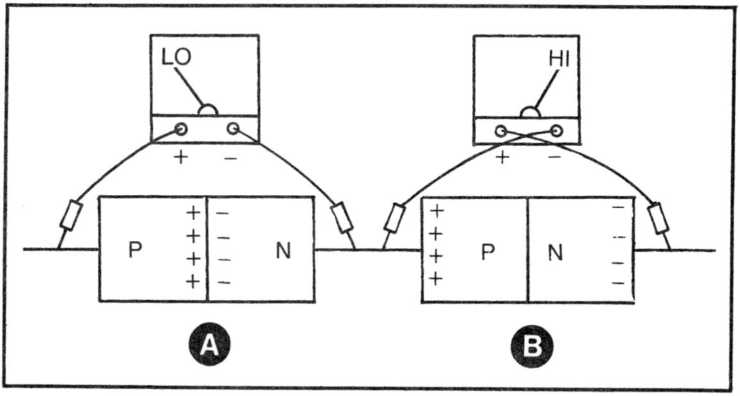

Fig. 19-1. In testing a pn junction with an ohmmeter (A), measure the reverse resistance, and then (B), measure the forward resistance.

ter as above will be approximately 5:1 or 10:1 in older devices, and maybe 20:1 in newer devices. Silicon devices typically have forward/reverse resistance ratios of 50:1, or more.

It is very important to remember which ohmmeter scale to use when measuring these resistances. Failure to observe this will either negate the results (a fancy way of saying that you don't really know what is going on!), or may destroy the device due to excessive current flow.

As a general rule, when using an ohmmeter that has a 1.5-volt battery for the voltage source, use the X1 and X10 resistance scales for rectifier diodes (power supply types) and power transistors, but use the X100 or X1000 scales for small-signal transistors and small-signal diodes (1N60, 1N34, 1N914, 1N4148, etc.)

Also, be careful of using the ohmmeter sections of some modern electronic voltmeters, especially the digital voltmeter instruments. The voltage source used in these ohmmeters is an electronic source from the main supply. Because of the sensitivity of these instruments, it is possible to use very low voltages in the ohmmeter section. This is often hailed as a benefit (which is true in some instances) because the user can make measurements in-circuit without fear of forward biasing diode and transistor PN junctions. They are also safer to use in-circuit because they will not damage any PN junctions (remember the admonition concerning the right scale given above).

However, the very feature that makes the low-voltage ohmmeter useful in those instances also tends to make it *useless* for making simple semicoductor tests: the low voltage will not forward bias the PN junction!

Some models, however, have both a low and high voltage ohmmeter capability. The markers of these instruments seem to have a good eye on their market, because they include a switch that permits the customer (i.e., you) to select either a low-voltage ohmmeter position, or a high-voltage (0.8 or more) position. They are obviously thinking of semiconductor testing in one such company, because the hi/lo ohmmeter switch is marked with the diode symbol for the "hi" position!

A bipolar (NPN or PNP) transistor can be viewed naively as a pair of diodes back-to-back. We may therefore use the same test procedure as on diodes to obtain a good prediction of whether the transistor is good or bad. The test procedure for transistors is given in Fig. 19-2.

There are three factors that must be checked on the transistor: *base-emitter junction, base-collector junction*, and the *collector-to-emitter leakage resistance*. These tests are shown in the various parts of Fig. 19-2. The test of the b-e junction is shown in Figs. 19-2A and 19-2B, the test of the b-c junction is shown in Fig. 19-2C and 19-2D, while the test of leakage is the subject of Fig. 19-2E.

Both PNP and NPN transistors may be tested in this manner, but expect the hi-lo readings to be reversed (after all, the junction polarities are reversed!). Also, it is a good idea, when testing Germanium transistors, to make the leakage test twice, reversing the leads of the ohmmeter as in the junction tests, and then accept the higher reading as valid. This seemed to be a problem mostly in PNP Germanium devices made in the mid 1960s.

We obey the same rules for transistors as for diodes: Use the X1 and X10 scales *only* on power transistors, and the X100 or X1000 scales for small-signal transistors. Use of the lower-range scales on the small transistors could cause excessive current to flow, and damage the junction.

When using an ohmmeter to test semiconductors, it is very important—let me repeat that, *very important*—to make sure of the battery being used in the ohmmeter! Use only those ohmmeters that use a 1.5 volt battery (C, D, or AA cell). Do not ever use an ohmmeter with a higher-voltage battery. There are many instruments, mostly older stuff made in the 1950s, that use batteries in the 4.5- to 22.5-volt range, and these will blow the transistor!

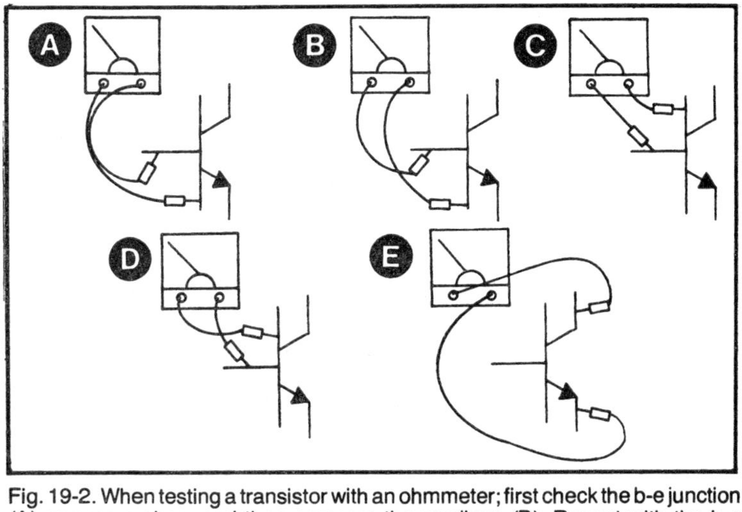

Fig. 19-2. When testing a transistor with an ohmmeter; first check the b-e junction (A) reverse probes, and then compare the readings (B). Repeat with the b-c junction (C, D), then measure the c-e leakage (E).

The forward and reverse resistances of the PN junctions in a transistor tell us their relative condition, and are a reasonable predictor of the ability of the transistor to operate properly, at least at DC. A better test method would tell us whether or not the base is capable of controlling collector current, and this is not shown by the simple ohmmeter tests above.

Figure 19-3 shows us whether the base is in control of the device, and uses only a simple ohmmeter. The ohmmeter probes are connected across the c-e terminals of the transistor in the polarity that would be normal for that transitor. This means that the positive terminal goes to the collector in an NPN transistor, but in a PNP transistor the negative terminal of the ohmmeter goes to the collector.

With the base lead of the transistor open, as in Fig. 19-3A, the ohmmeter will read a high resistance. But, in Fig 19-3B, the c-e resistance drops low because the base lead is connected to the collector potential.

A slightly better approach, more complicated than the first due to the need for an extra resistor, is shown in Fig. 19-3C. Here the base is connected to the collector through a resistor of such a value that will cause forward biasing of the transistor. This resistor will be 50 - 100 ohms in most power transistors, and 10 kohms to 500 kohms in most small-signal transistors.

The ohmmeter tests discussed thus far are generally useful in audio transistors, rf small-signal transistors well into the UHF region, and almost all of the transistor. They do not work well on high-collector-voltage transistors and VHF/UHF power transistors.

TRANSISTOR TESTERS

Once you have mastered the art of testing transistors using only a simple ohmmeter, you may want to become more elegant and use a genuine transistor tester. There are actually several varieties of transistor tester on the market: *conduction* (or *leakage/gain*), *oscillator, beta*, and the *curve tracer*.

The different types of transistor tester vary in their validity, but all find some use under the right circumstances. With the exception of the simple conduction tester (which is an ohmmeter in disguise), all of them are considered a notch above the ohmmeter tests given earlier. The conduction type of transistor tester merely formalizes the ohmmeter tests, and makes them seem more valid by using a switch to make the connections.

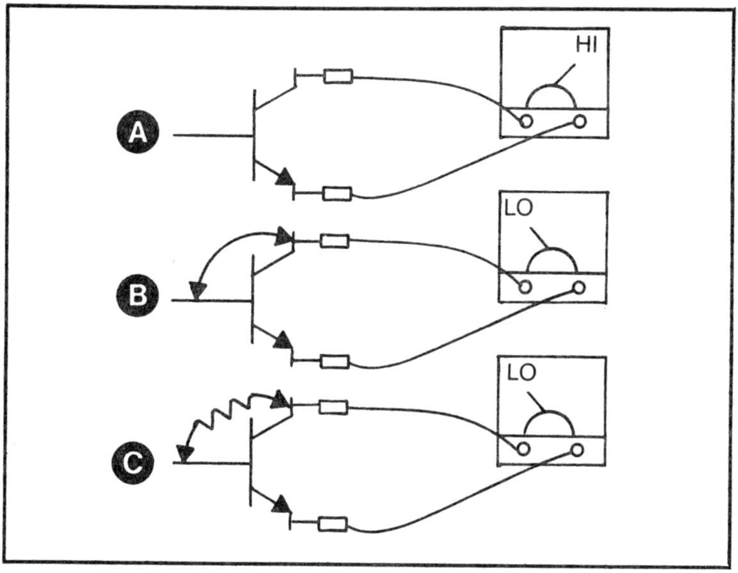

Fig. 19-3. More active method of measuring the transistor with an ohmmeter.

Conduction Tester

An example of the simple conduction tester is shown in Fig. 19-4, and , as you can easily see, it is very similar to an ohmmeter circuit . . . with modifications. When the *test* switch (S1) is in the open position, the meter will read the leakage resistance between the collector and the emitter terminals of the transistor. But, when S1 is closed, you have a rough, but purely qualitative, test of something like the beta of the transistor (one would find it difficult to actually place numbers on this test, but it is a form of beta test).

The scale of meter M1 is usually marked with two sections. One will read leakage, although the numbers tend to be either dimensionless or fake, and the other will read gain. Some use color codes instead of numbers; i.e., green for good, red for bad and yellow for "your guess is as good as mine."

Oscillator

The *oscillator* type of transistor tester is sort of a dynamic type of device, because it places the transistor in a real oscillator circuit to see if it will oscillate. An AC-coupled detector on the output of the oscillator drives a meter that indicates whether or not oscillations are present. If they are present, then the transistor is deem "good." The transistor is deemed bad if the device will not oscillate. Such

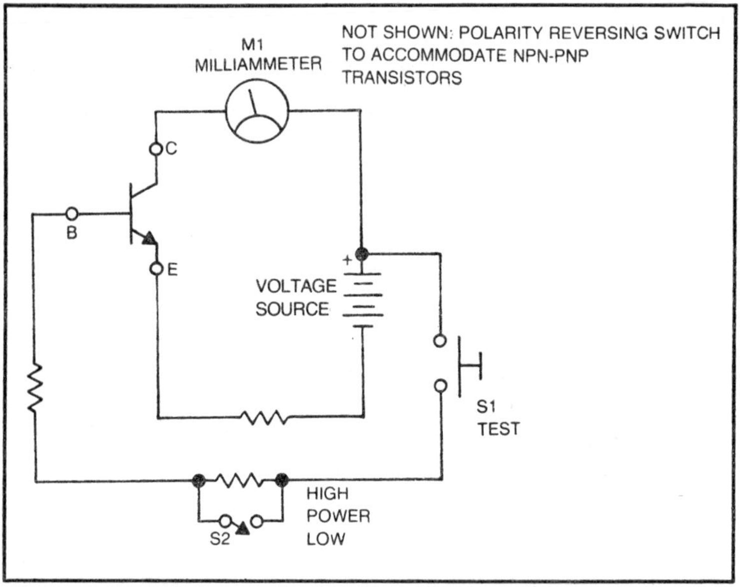

Fig. 19-4. Simple transistor tester.

transistor testers are also capable of testing crystals, so are sometimes combined as a single unit in a low cost test set. Many models of these have been made for the CB service market (and the CBer) over the past 15 or so years. Although some Amateurs tend to look with disdain on the CBer, there are some good, low-cost pieces of test gear floating around in that market!

Note that the oscillator tester is often easy to recofigure (using a single rotary switch) so it can test MOSFET and JFET devices, as well as bipolar (NPN/PNP) devices.

Beta Tester

A real, live, honest-to-gosh beta transistor tester will yield both qualitative and quantitive information on the transistors being tested. Some models are automatic and cost an arm and a leg. Others are simple, manually operated, and can be bought for very low cost.

Beta is defined as the ratio of collector current to base current (I_c/I_b). We can use this fact to create a beta tester. An example is shown in Fig. 19-5. Note that it is similar to the circuit used to make the conduction tester. The main difference between this type of tester and simpler types is that the base current is fixed and is constant (and is *known*). If we measure the collector current, then, we will find that it is proportional to the transistor beta. We are therefore justified in calibrating the scale of the collector current meter (M1) in units of beta rather than units of current. This is allowable because the beta is the variable being measured by an indication of the collector current, and the only other variable is not variable at all; the base current is held constant.

The relative accuracy of any given beta tester is dependent upon how accurately the collector current is measured and, most importantly, how constant the base current is held. At least one transistor tester is known to use a differential-current-meter scheme to make the measurement of the base current, and then crank the result into the collector-current measurement as an accuracy control.

In some so-called "dynamic testers," i.e., those that purport to measure h_{fe} instead of H_{FE}, a small AC signal is applied to the base of the transistor, in series with the bias resistor, and the collector meter is AC-coupled to the collector circuit. If the base current is held constant, and the AC signal amplitude at the base is constant, then the amplitude of the AC signal produced across the collector load resistor is reflective of the device beta.

There are many claims for all transistor testers on the market. In fact, one well-known company wrote glowing descriptions of their cheapie conduction device that made it sound like their higher-priced true-beta tester (a darn good instrument), so let the

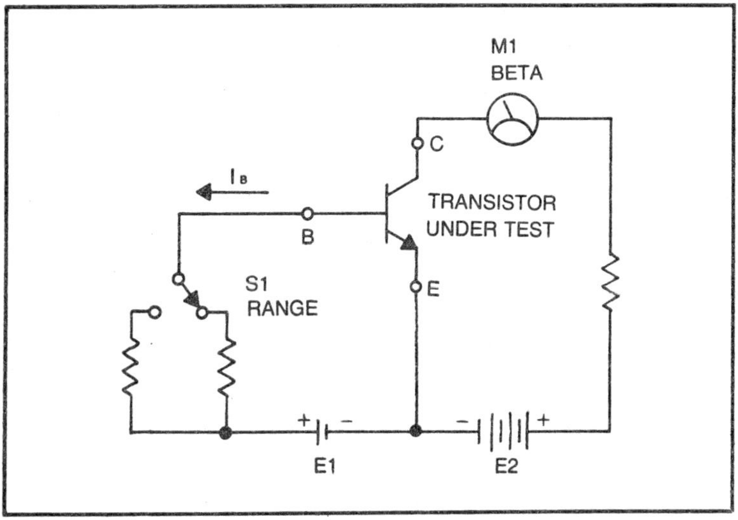

Fig. 19-5. Beta tester.

buyer beware. There are, however, some hallmarks of a good tester as generally agreed upon by most professional servicers:

1. Measure I_{cbo}
2. Measure I_{ceo}
3. Measure at least two ranges of *beta* values
4. Measure super-beta (up to 20,000) for Darlington transistors
5. Test PN junctions in both signal and rectifier diodes
6. Test MOSFETS (enhancement and depletion)
7. Test JFETs
8. Test Zener diodes
9. Test unijunction transistors (UJTs)
10. Wash the dishes for the XYL
11. Repair the rig for you.

Actually, all but 10 and 11 can be found in some testers designed for the professional servicer, at costs less than $200. Lower-cost instruments, however, will usually offer features 1-5 as a matter of course.

Some years ago, the first in-circuit transistor testers produced so many false negatives that they fell into disrepute in a real hurry. Too often, professional servicers would waste time and bucks replacing a good transistor because they were led astray by the transistor tester. But, today, some of the in-circuit testers are capable of approximating the performance of out-of-circuit testers. There will still be some false negative results, so it is wise to use the n-circuit feature to screen a large number of devices, and then the out-of-circuit tester to confirm the result. While often more expensive, the in-circuit tester has an appeal for the pro because of the time saved.

Curve Tracers

The *curve tracer* is an instrument that has come into its own in the service industry in the past few years. They were used by engineers for a much longer time, but costs and unfamiliarity kept the instruments out of the hands of service technicians until the 1970s.

Most curve tracers will plot, on an oscilloscope, the characteristic curve of a transistor or diode. For transistors, they will plot the I_c *–vs* $–V_{ce}$ curve at several different levels of base current. This results in a family of curves that describe the operation of the device under differing circumstances. Most of these instruments use from one to ten different levels of base current, step-switched to automatically select the I_b levels. A variable-voltage supply will cause the collector potential to change V_{ce}.

Some of the more expensive curve tracers, designed mostly for engineering laboratories, have the oscilloscope built-in. Interestingly enough, some of the older "large mainframe" transistor curve tracers, such as those by Tektronix, have been seen on the hamfest market and in electronic surplus stores. Most of the lower-quality instruments, those designed for the workday service technician, require an external oscilloscope. This approach to design keeps costs low.

CHECKING CAPACITORS

Capacitors can be perverse little animals to test! Yet, they seem too simple when described in textbooks! For most capacitors in the 0.01 to 1.0 μF range, the testing is relatively simple and straightforward . . . especially the type of testing a book on troubleshooting would detail. But, at very low capacitance values, or at very high capacitance values, or when electrolytics are being tested, some of the simple and old-fashioned quick checks simply fall short.

Ohmmeter Tests

When you connect a discharged capacitor across an ohmmeter, the capacitor will charge up because of the current flowing from the ohmmeter. The initial current flow is large, so the ohmmeter will (momentarily) show a low-resistance value. But, as the capacitor charges, the current flow becomes progressively less, so the ohmmeter reading tends to rise. When the capacitor is fully charged, the capacitor voltage and the open-terminal-voltage of the ohmmeter are equal, so no current will flow. This makes the ohmmeter appear to read an open circuit.

In most cases, it is wise to use the highest ohmmeter scale available, or the higher current levels of lower ohms scales will charge the capacitor too fast for convenient reading. This becomes really important at low capacitance values, because the *RC* time constant of the capacitor and the internal resistance of the ohmmeter is too short for the heavily damped meter pointer to follow.

It takes a relatively small amount of practice to become proficient in the use of this "test," but some practice is in order so that you will know what good and bad capacitors should look like. Take a handful of different value capacitors, and an ohmmeter, and test them just for the sake of learning. This is called "getting experience."

The ohmmeter, again on the highest range, will also tell you whether or not the capacitor is shorted, or has a high leakage (i.e., a high-resistance short rather than a dead short). If the ohmmeter pointer does not rise, or fails to rise all of the way to open-circuit (so-called infinity), then the capacitor is shorted or leaky. Take care, though, most aluminum electrolytic capacitors will appear leaky, especially if the wrong ohmmeter polarity is used.

Voltmeter Tests

We may also test both the quality and the leakage of a capacitor using a voltmeter.

In the leakage test, we require a circuit with at least a few volts of potential across the capacitor. The "cold end" of the capacitor is disconnected from the circuit, and the voltmeter is used to measure the potential from the cold end to ground. If the capacitor is good, then there will be no potential (or at least very little . . . but that discussion takes us deeply into capacitor theory).

If the capacitor is shorted, the above test will reveal a voltage . . . and that voltage will be the highest when the short is the worst.

We may use the voltmeter, and a DC supply, to test capacitors in a qualitative manner . . . and two methods are available. In one, we use a series resistor between the capacitor and the power supply, and a voltmeter across the capacitor (the higher the voltmeter input impedance, the better). The capacitor is discharged, and then the power supply is turned on. The reading on the voltmeter should rise exponentially as the capacitor charges.

If the reading on the voltmeter fails to rise at all, then the capacitor is shorted. If the voltmeter rises, but stops below the terminal voltage supplied by the power source, then the capacitor is leaky. In the last case, the capacitor's series resistance and the resistance you used between it and the supply form a simple voltage divider.

We may also qualitatively measure the "goodness" of a capacitor by charging the capacitor to some voltage, and then connecting a very high input-impedance voltmeter across the capacitor. When the supply is disconnected,a perfect capacitor would hold the charge indefinitely, but a real capacitor will allow the charge to *slowly* bleed off.

The last technique is another one in which some practice is needed before you can do it well . . . and avoid false results. A little practice goes a long way.

Measuring Capacitance

Two methods have been traditionally popular for measuring the capacitance of capacitors. One of these is to place the capacitor in one leg of an AC bridge circuit, and then operate the bridge's calibrated controls to find the capacitance. This usually involves a null technique. The other method has been the so-called "Q-meter" in which an unknown reactance (either L or C) is placed in a series-tuned tank circuit. If we measure the AC voltage drop across the L or C element being tested, then we will be able to make a very accurate measurement of the value of the capacitance (C) or inductance (L).

Both the bridge and the Q-meter methods, however, are difficult to make cheaply and accurate at the same time. Further, unless they are really costly automatic models, they tend to be too time-consuming to operate. Because of the cost, and the time involved in using them, most servicers shun these instruments. A simpler approach, that overcomes these problems, is shown in Fig. 19-6. This is the block diagram of a simple digital capacitance meter. The circuit consists of a period counter in which the on-time of the main gate is set by a monostable multivibrator. The capacitor being measured is used to set the timing of the multivibrator. The RC time constant sets the length of the output pulse, and the output pulse determines how many clock pulses are admitted to the counter. A higher capacitance makes the pulse longer, so more clock pulses are admitted. In practice, it is very easy to achieve better than 5-percent accuracy in low-cost instruments.

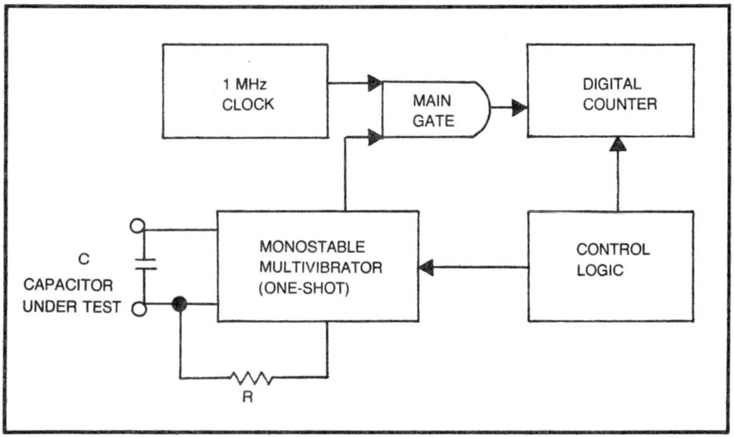

Fig. 19-6. Digital capacitor-tester.

TUBE TESTERS

Everything is coming up solid-state . . . the vacuum tube is dead! Really? My ham rig, as of this wiring, is a Heath HW-101, it still works well and is all tubes! While new equipment in the under-300-watt class tends to be all solid-state, many not-too-old rigs are tube types, and most high-power amplifiers use vacuum-tube devices.

An awfully large number of vacuum tubes change hands at hamfests and in surplus stores. Many more are traded, sold, or given as gifts between individual Amateurs. This is especially true of power tubes. I wish I had a dime for every 4X150 that I have seen offered at hamfests! The problem with actually trying to *use* tubes of uncertain origin is that they are often bad . . . that's why they are on the market, sad to say. When you acquire a collection of these "bargains," it is wise to test them (at least on a screening basis) to determine whether or not they stand a good chance of working in your rig.

In troubleshooting, also, it is a good idea to be able to test tubes. This is not too important with small-signal tubes, or even moderate power transmitter tubes, because it is economically feasible to "test by substitution." But, one may be a little hesitant to buy a $150 power-amplifier tube on the off-chance that the one in the linear amplifier is bad!

This leads us to a discussion of tube testers. For most small tubes, you can easily go to the local drug store and use the "Harry-&-Harriet-Homeowner-You-Fix-Your-TV-Yourself" tube tester. These are not the best instruments in the world, but are very useful for spotting gross defects.

There are three basic forms of tube tester: *short-circuit tester, emission tester*, and *transconductance tester*. These different instruments provide different levels of testing ability, but are all useful in at least some cases.

Short-Circuit Testers

Figure 19-7 shows the circuit for a simple short-circuit tester. It will allow you to check for interelectrode shorts, a very common form of failure. In many instances, a short circuit test will be built into emission or transconductance testers. Some have only one lamp, and a switch to select between the various elements.

The primary use for the short tester is when the suspected failure is an interelectrode short circuit, or when you wish to quickly screen a large number of grab-bag tubes of uncertain origin. A short-circuit test should be a part of both other types of test because

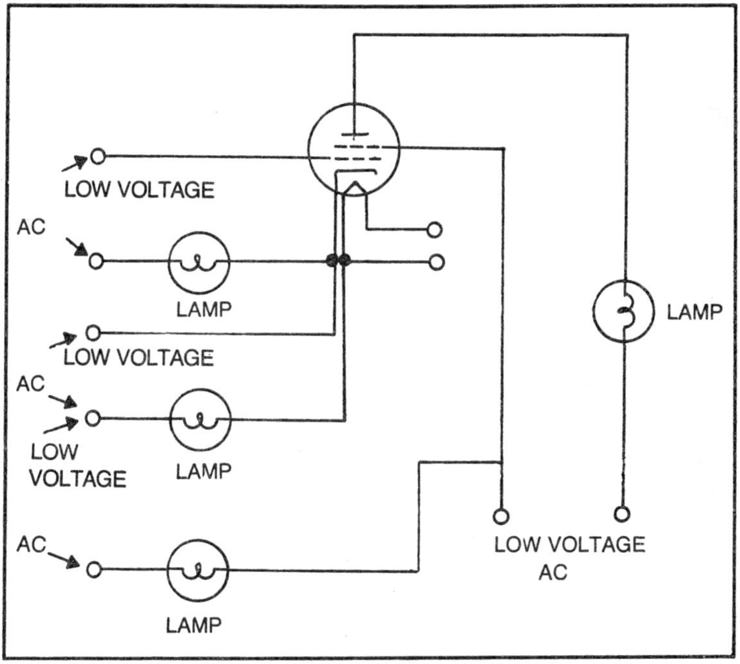

Fig. 19-7. Vacuum-tube-shorts tester.

some tubes will seem to function in a tube tester, even with a short that renders them inoperative in the real circuit.

It is prudent to wait several minutes when testing for shorts because many interelement shorts do not occur until the tube is hot. Unlike transistors, tubes may require 2 or 3 minutes of warm-up time before they can operate properly. It may take this same length of time for the short to occur. This is why a simple ohmmeter is not used to detect shorts . . . it will say a tube is good when, in fact, a heat-related short exists.

Emission Testers

The simplest type of tube tester that yields any real qualitative results is the emission tube-tester circuit shown in Fig. 19-8. This circuit tests the tube for the emission of electrons from the cathode. The tube is connected in a diode configuration, that is, all grids and the plate are tied together. Declining cathode emission is one very common form of failure in tubes, especially those that have been in service for a long time.

Note that the meter calibration on the emission tube tester is meaningless as far as measurement of tube parameters is con-

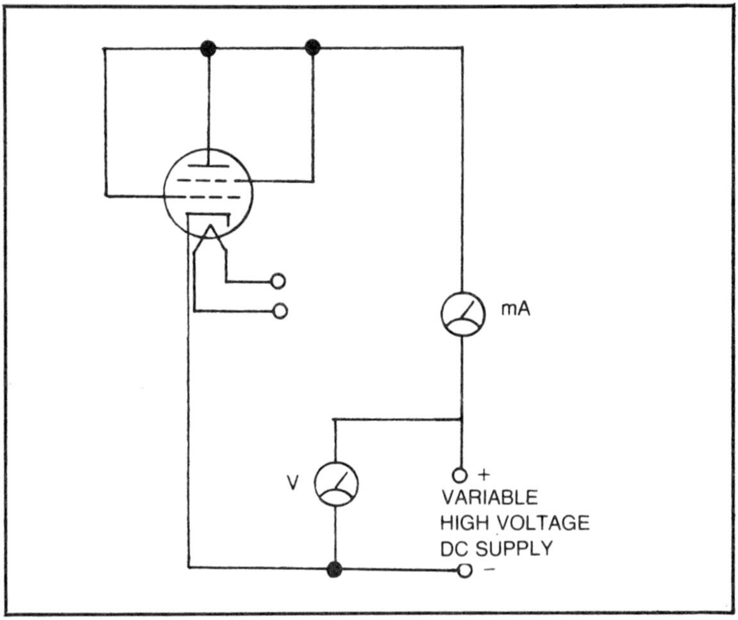

Fig. 19-8. Emission tester.

cerned. In fact, most use the red-green-yellow subjective markings.

The emission current of the tube is defined as the saturation current when the tube is connected as a diode. If the plate potential is increased from near-zero volts, we will see the plate current is also increase in a very nearly linear manner, except at very low potentials. The current will continue to increase with increased plate-cathode voltage until a certain critical saturation point is reached. Past that point we get very little, if any, increase in plate current for increases in plate voltage. The tube is then said to be saturated, and we call this current the emission current.

The emission tube checker tests for the correct emission current for the type of tube checked. If the emission current is substantially lower than the rated current, then the meter will read in the red, or "reject," region.

In tubes where the emission current is inconveniently high, or if the current is higher than the allowed plate current of the tube, we may test the tube at a lower current that is well within the allowed range. If the current produced by the tube under test falls to less than 70 or 80 percent of the test current for a known-good tube, then the tube is rejected.

Note that most drug store and other "self-service" tube testers are emission types because of the simplicity of operation (and possibly the few extra false negative results they generate!).

Transconductance Testers

There are some defects in tubes that will prevent normal functioning in the circuit, yet will not show up on an emission-type tube tester. This is why most professional servicers prefer to buy a *transconductance* (also called *mutual conductance*) tube tester. There are two different types of transconductance tube tester, static and dynamic. The simple circuit of Fig. 19-9 is a static transconductance tester. It uses what is called the grid-shift method of measuring the transconductance.

Recall the definition of transconductance in a vacuum tube. It is the ratio of a small change in *plate current* (I_p) for a small change in *grid voltage* (E_c):

$$gm = \frac{\Delta I_p}{\Delta E_c}$$

Switch S1 is in position no. 1 at the beginning of the test. This position will make the grid voltage E_c equal to the value of voltage E_2 alone. Both E_1 and E_2 are adjusted to produce a convenient and safe level of plate current in the tube being tested. When these voltages are adjusted to produce a reasonable current in the tube plate circuit, note (and write down!) the values of E_b (plate-cathode voltage), I_p, and E_c.

Next, place switch S1 in the position no. 2. This adds a small voltage to E_2, making E_c equal to $E_2 + E_3$. If the plate voltage changed (it might), readjust E_1 to the previously recorded value.

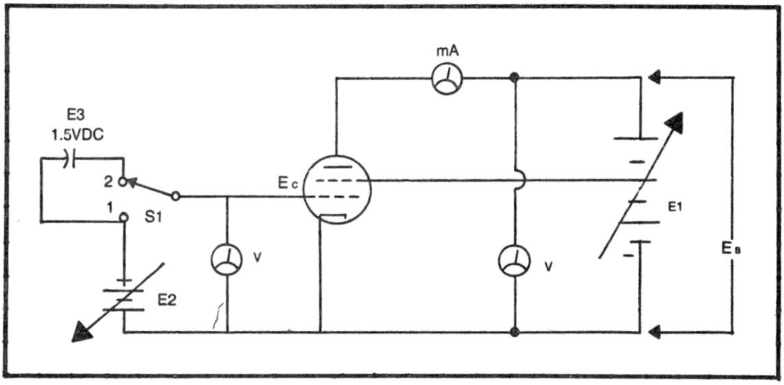

Fig. 19-9. Grid-shift transconductance tester.

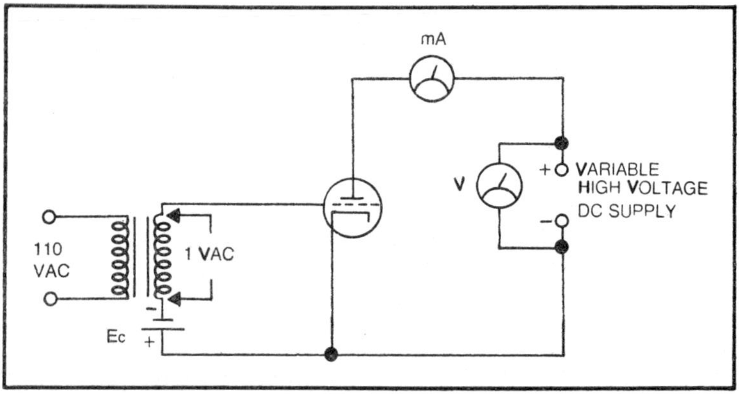

Fig. 19-10. Dynamic transconductance tester.

The plate voltage must be equal in both parts of this test. Now, read the new value of plate current, and calculate the transconductance using the above formula. Voltage E_3 is used as ΔE_c.

The answer yielded by the formula will be in mhos (or the new term, Siemens) if volts and amperes were the units. But most vacuum-tube data books list the transconductance in micromhos. Make the correction by multiplying the answer by 1,000,000 (there are 10^6 μmhos in one mho).

A dynamic-transconductance tube tester is shown in Fig. 19-10. This is the basic circuit used in most professional tube testers. A low voltage AC signal is applied to the grid of the tube, and this voltage becomes the ΔE_c term. The actual value of the transconductance in μmhos (μS) is measured on the plate meter.

Chapter 20

The Phase-Locked-Loop

Phase locked loops have been with us for many years. Originally intended as a new type of amplitude-modulation detector (not FM, as is currently the case), the phase-locked loop has come into its own in the past decade or so due to advances in IC technology. What had been a relatively difficult circuit to design and build is now a simple matter of correctly connecting an IC device.

PLL ICs have been used in many different applications. One, of course, is the FM demodulator mentioned above (their use as an AM demodulator is not largely gone by the wayside). But, they also find use as stereo decoders in FM hi-fi receivers, SCA (background music on subcarriers of FM broadcast signals) demodulators, frequency control in receivers, transmitters, etc., and in decoding frequency-shift keying in teletypewriter communications and computer data channels. In this chapter we will discuss some of the basic theory of the phase-locked-loop, some of the more common applications (as above mentioned), and provide some service information.

WHAT IS A PLL?

Good question. Simply put, a phase-locked-loop is a feedback system in which a local oscillator locks onto, and then tracks, any changes of input frequency. Figure 20-1 shows the basic block diagram for a phase-locked-loop circuit. This circuit consists of four basic sections: two-part phase detector, low-pass filter, DC amplifier, and a voltage-controlled oscillator.

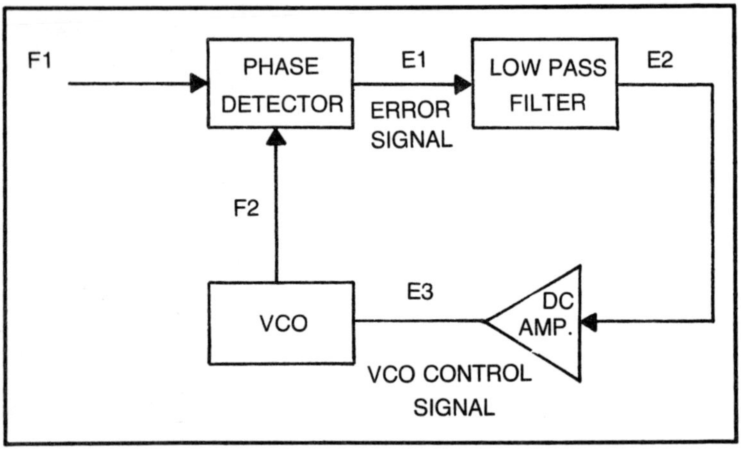

Fig. 20-1. Elementary block diagram of a PLL.

Two frequencies are involved in the PLL circuit. An input frequency (F1) comes from the outside world, or another local circuit, while the other (F2) is generated by the voltage-controlled oscillator (VCO). The VCO is in the feedback loop.

If no input signal exists, then the VCO will oscillate at its natural resonant frequency, or *free-running frequency*. This frequency is determined, as in most oscillator circuits, by the *RC* or *LC* time constants in the oscillator. When F1 is applied to the input, however, the phase detector compares F1 and F2, an issues and *error signal* that is proportional to the difference between F1 and F2. The error signal is fed to a low pass filter, so that it becomes a DC level. The actual difference signal is a DC level represented by voltage E1 in Fig. 20-1.

Voltage E2 may be fed directly to the VCO in some cases, but it is usually necessary to scale it to the range proper for the VCO by passing it through a DC amplifier. Regardless of which technique is used in any given circuit, though, the error voltage is applied to the DC control input of the VCO, where it used to pull the VCO frequency to correct the discrepancy between F1 and F2. This particular circuit seeks the condition where F1 = F2, and E2 = 0.

PLLS IN FM DEMODULATION

One application for a circuit such as Fig. 20-1 is FM demodulation. Recall the basic FM theory which states that the frequency of the carrier varies at a rate that is dependent upon the frequency of the modulating audio signal. Also, please note from the above discussion that the DC error voltage varies directly as the input-frequency varies. If an FM I-F signal is applied to the input (as F1), then the error voltage, E2, will be the audio signal that was the modulation.

Several integrated-circuit manufacturers offer IC phase-locked-loops that are suitable for FM-demodulation applications. Three types seen often in the Amateur Radio literature are the 560, 561, and 565 devices from Signetics.

Figure 20-2 shows an FM-demodulator circuit using the 565 IC device. This circuit will not work above about 500 kHz, but is suitable for use in demodulating the narrow-band FM signal from the low-frequency I-F (455 kHz) used in many mobile receivers. It will also operate in other FM applications, such as SCA demodulation and certain telemetry tasks.

DIGITAL TONE DECODING

The PLL is a natural tone decoder. We can, for example, use the PLL IC to produce a digital-logic level depending upon the existence or nonexistence of a certain frequency tone. This allows for applications such as tone calling in communications systems, and for the transmission of digital data via radio channels. In some areas, the local VHF or UHF repeater club will put a microcomputer on the repeater for member's use. The digital information is transmitted as tones, one each for logical states 1 and 0.

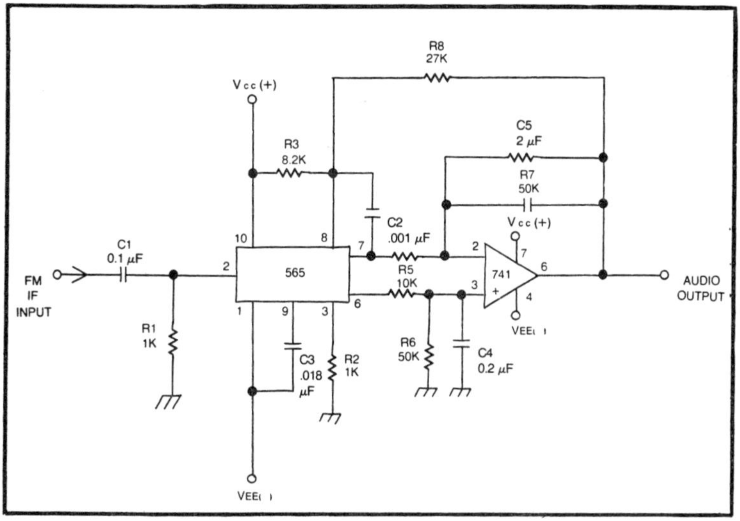

Fig. 20-2. PLL FM detector.

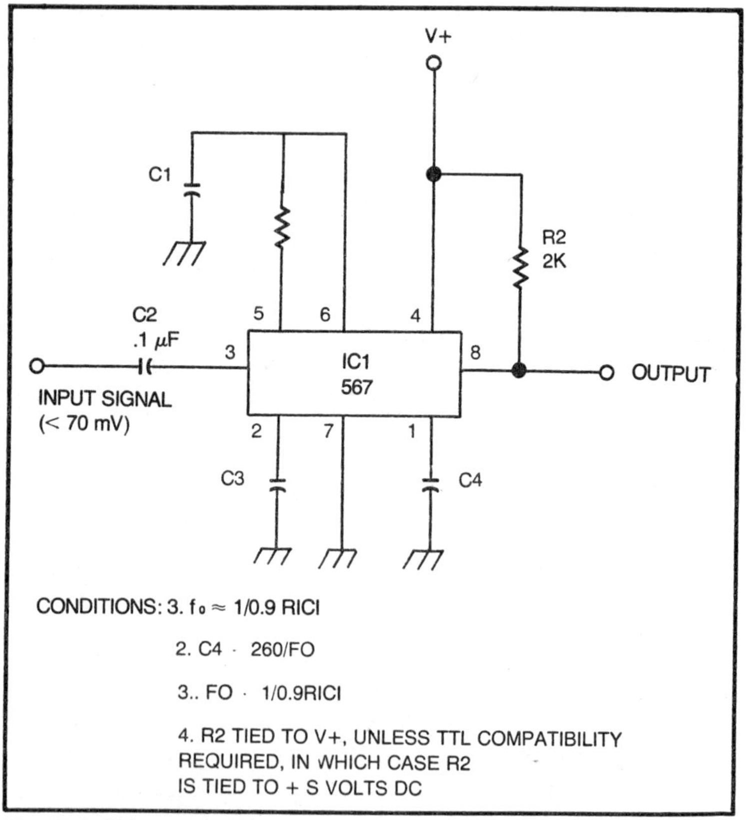

Fig. 20-3. PLL tone decoder.

A typical tone decoder would be a circuit such as Fig. 20-2, but with the operational amplifier replaced by a level comparator.

The 567 is a PLL integrated circuit that is designed especially for applications such as tone decoding, and contains an internal comparator (Fig. 20-3). The component values shown are typical for tones in the 400- to 2500-Hz range, but other values may be selected by using the formulas in the figure.

Tone decoders find applications in communications where they may be used to actuate tone-operated squelch systems in private-call radio communications.

A tone decoder may also be found in electronic instrumentation applications where it is necessary to tape record low-frequency analog signals that are too low in frequency for the normal passband of the tape recorder. Such signals can be recorded on audio tape if they are used to frequency modulate an audio-frequency carrier. In the playback mode, therefore, a tone decoder will be needed to recover the recorded audio signal.

FREQUENCY CONTROL BY PLL

Two-way radios must be crystal controlled for proper, on-channel, mobile operation. This is true of Amateur, CB, marine, land-mobile, and aviation transmitters. In past years, a 24-channel, 2-meter mobile rig would cost approximately $250, and then the owner would have to go out and spend another 150 - $200 for crystals to fill all of the sockets in the receiver and transmitter portions of the rig! In some cases, the cost of crystals would be greater or equal to the cost of the rig itself!

The main reason for the popularity of the PLL in transceiver frequency control is that it allows full coverage for a lot less money. Several 2-meter Amateur transmitters are on the market, which cover as many as 600 - 800 channels. This would cost several thousand dollars, not to mention make the radio as big as a house, if done by using an individual crystal for both receiver and transmitter on each frequency.

Figure 20-4 shows the basic block diagram of a simple PLL receiver- or transmitter-frequency generator. This circuit is essentially the same as the circuit we presented earlier, except that the input frequency is not from the outside world, but from a locally generated crystal reference oscillator.

You might think that a PLL is capable of producing only a single VCO frequency when there is only one reference frequency, but this situation is modified somewhat by using a programmable

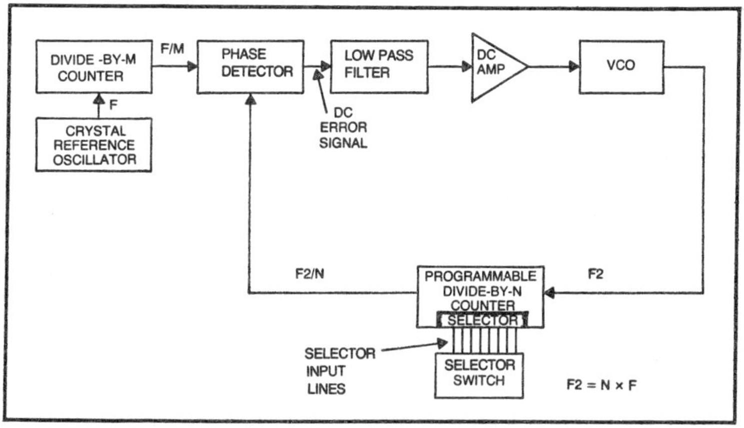

Fig. 20-4. PLL frequency synthesizer.

divide-by-N counter between the output of the VCO and the input port of the phase detector.

The VCO output is also sent to the rest of the receiver and/or transmitter where it is used as the master oscillator. The VCO will produce a stability that approaches that of the reference oscillator, which, being crystal controlled, can be quite good.

The VCO signal may be used directly, may be used to heterodyne to another frequency (as in SSB transmitters/ receivers), or may be used as the master oscillator in a multiplier or divider chain.

The key to multi-channel operation lies in the programmable divide-by-N counter. The frequency-division ratio is set externally by the operator. When you select a channel on a PLL controlled transmitter, then you are manipulating the digital word applied to the control-line inputs of the divide-by-N counter. The selector switch will produce the needed binary codes. This type of frequency control also makes it easier to use LED digital readouts to indicate either frequency or channel number, without using a counter circuit especially for the dial.

Chapter 21
Servicing Untuned Amplifiers

An untuned amplifier is a stage that will pass a wide range of input frequencies. There are no tuning circuits in the amplifier to limit frequency response. By tuning circuits we specifically mean *LC* tuned circuits. Certain audio amplifiers, for example, might use *RC* frequency shaping to achieve a certain passband, but are still considered "untuned" amplifiers. Untuned amplifiers, in general, are audio amplifiers and video amplifiers, as well as any amplifier that could conceivably be used—except those that are equipped with an *LC* circuit—to pass, or reject, certain frequencies.

BIASING

One of the most important aspects of troubleshooting any circuit is to learn how it works. In most untuned amplifiers, this means learning the operation of the transistor (or vacuum tube) and then learning the different ways in which proper bias is established. We will consider only the transistor case here, because they are current, and also, tubes have relatively few different bias schemes.

Most of us are probably familiar with the more-or-less standard approaches to biasing a transistor. Some of the most popular are shown and explained in Fig. 21-1. These circuits, and their variations, are used in most Amateur equipment.

The bias method shown in Fig. 21-1A uses a resistor returned to the collector supply in order to create the base-bias current. This method offers simplicity, but at the cost of input impedance and stability. In Fig. 21-1B we see a variation on the same scheme, but

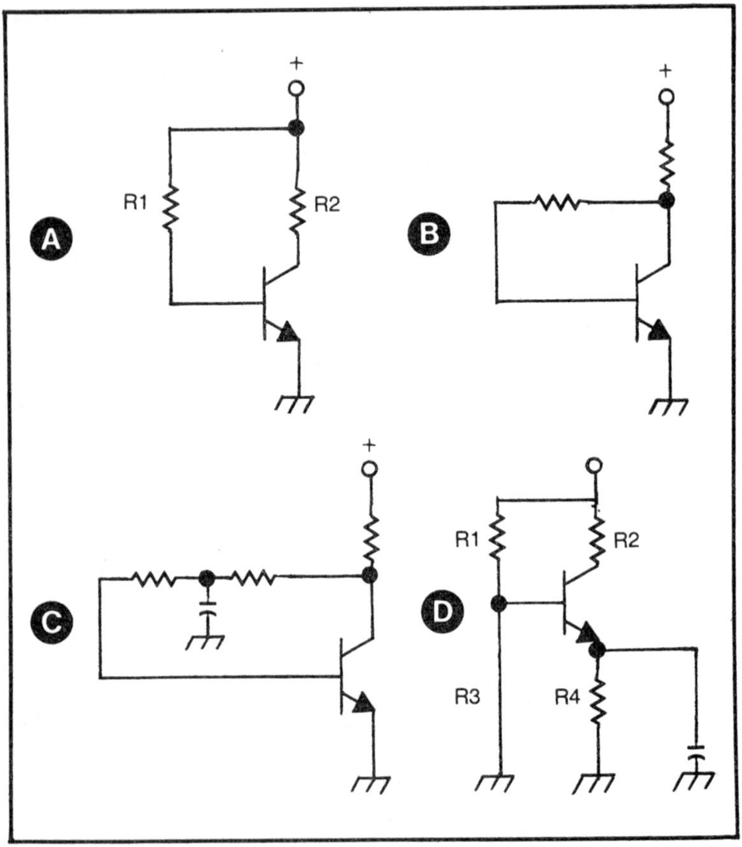

Fig. 21-1. Transistor biasing networks.

in which the base-bias resistor is returned to the collector to take advantage of a little negative feedback. This method is actually a form of self-bias, because any change in collector current will cause a proportional change in the transistor bias. The method shown in Fig. 21-1C is still another variation, but of the collector-feedback technique. Here we see the bias resistor split into two components, and an AC decoupling capacitor placed to ground in order to smooth out the variations normally found in the collector current (due to signal). The scheme in Fig. 21-1D is the most stable, and is considered to be the best of these schemes. Resistors R1 and R3 forms a voltage divider to set the bias current on the transistor, while R2 serves as the collector load. Emitter resistor R4 is used to improve the thermal stability of the circuit, and also increases the input impedance of the circuit. An emitter-bypass capacitor, with a capacitive reactance at the lowest frequency of operation of one-tenth the value of R4, is used to place the emitter at AC ground, while keeping the DC level above ground.

DUAL SUPPLY OPERATION

Much fixed-station equipment is now using dual power supplies, long familiar in operational-amplifier, and other IC, circuits. Now that many linear ICs are being used in communications equipment, it has become necessary to incorporate such supplies. Of course, mobile equipment still operates from a 12- to 16-volt negative-ground power system, but fixed-station equipment (which derives power from the AC mains) is not so constrained. Transistor circuits sometimes operate better from dual supplies.

Figure 21-2A shows a transistor amplifier stage that is operated from a dual-polarity power supply. In the NPN case, the emitter resistor is returned to the V – supply, while the collector is returned to the V+ supply. In PNP stages, exactly the opposite scheme is used. The base is returned to ground, not to either of the supplies.

An example of a dual-polarity supply is shown in Fig. 21-1B. Note that we mean dual-voltage, dual-polarity supply. The V– is

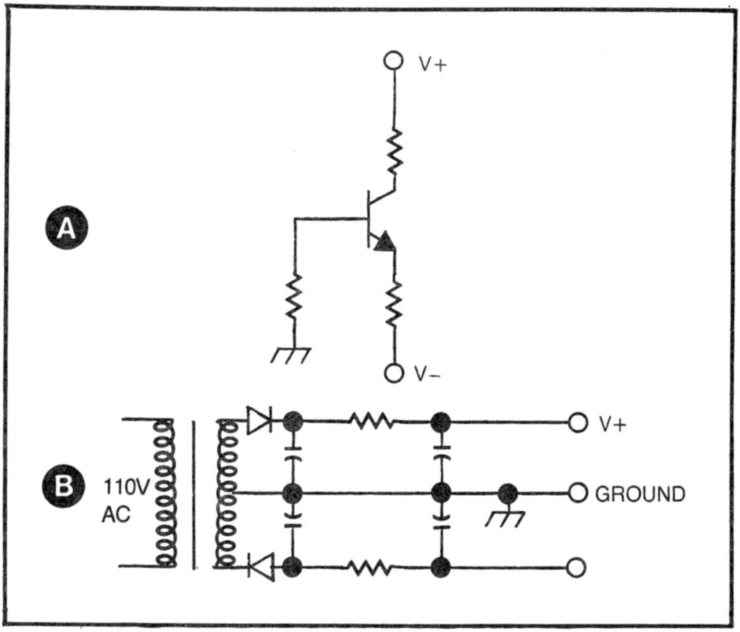

Fig. 21-2. Dual polarity power supply operation of transistor, (A), and a dual supply (B).

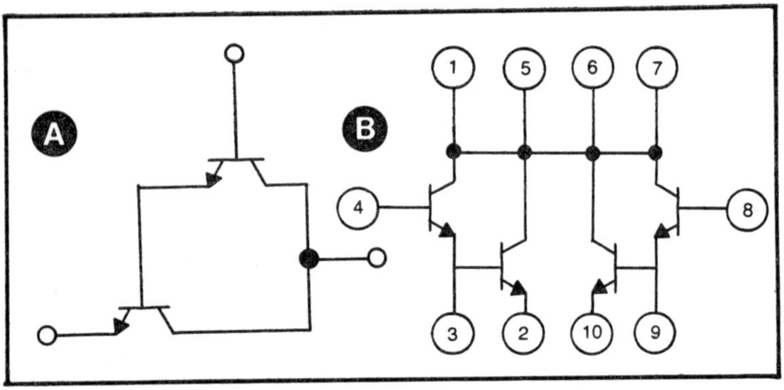

Fig. 21-3A. Darlington pair. B. IC dual-Darlington pair.

not the other side of a V+ supply, but a completely separate supply of opposite polarity. The V+ supply produces a potential that is positive with respect to ground, while the V− supply produces a potential that is negative with respect to ground.

The advantages of the dual-polarity circuit is that a larger output voltage swing is possible. Additionally, some immunity from power-supply hum is improved. This less of a problem is later equipment, because of the availability of low-cost voltage regulator ICs and discrete circuits, but in equipment only a few years older, they may use a dual supply for this reason.

DARLINGTON PAIRS

Another type of circuit that has seen increased use is the Darlington amplifier, or Darlington pair as it is sometimes called. An example of this configuration is shown in Fig. 21-3A. Notice that both of the transistor collectors are tied together. Also note that the input transistor is used as a direct-coupled emitter follower to drive the base of the second transistor. This produces a much higher current gain and a much higher input impedance than is possible with only one transistor. The beta of the Darlington pair is approximately equal to the product of the betas of the two transistors, i.e., h_{fe1} X h_{fe2}. In the usual case, where identical transistors used for both in the pair, then the total beta is the square of the beta of one. This multiplication of beta is the reason why these are called super-beta circuits.

A number of manufacturers use the Darlington pair inside of a single package. These transistors are often called *superbeta* transistors. It is not uncommon to find superbeta transistors that boast beta figures in the 1000 to 20,000 range.

Figure 21-3B shows an RCA integrated circuit often used in communications equipment. This IC has a pair of Darlington pairs sharing a common power-supply connection at the collectors. In one two-way-radio circuit, the CA3036 was used as a squelch-controlled amplifier. The Darlington pair on the left side was used as the audio amplifier, while that on the right was a squelch switch. When the squelch circuit sensed no signal, the voltage at pin 8 was high, causing it to saturate, and drop the collector voltage to zero. Bringing this point low restored operation, allowing audio to pass.

POWER AMPLIFIER DESIGNS

There seems to be an almost endless variety of power amplifier circuits, especially in the lower end of the power range where we find most Amateur-Radio audio circuits. In this section we will deal with several of the most common circuits. Keep in mind that many variations will be found, and that these circuits are intended to be guidelines only.

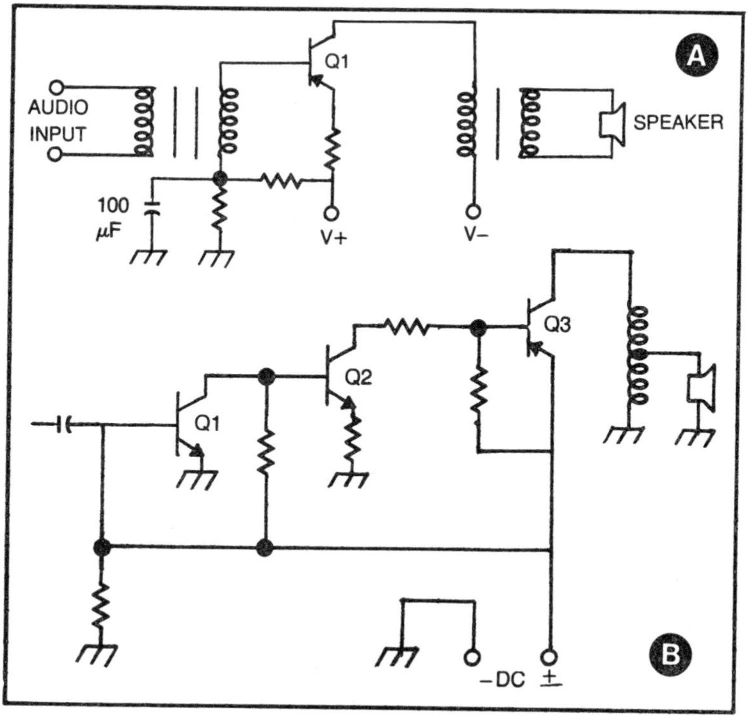

Fig. 21-4A. Transformer coupled audio output stage (class-A). B. Direct-coupled audio output stage (class-A).

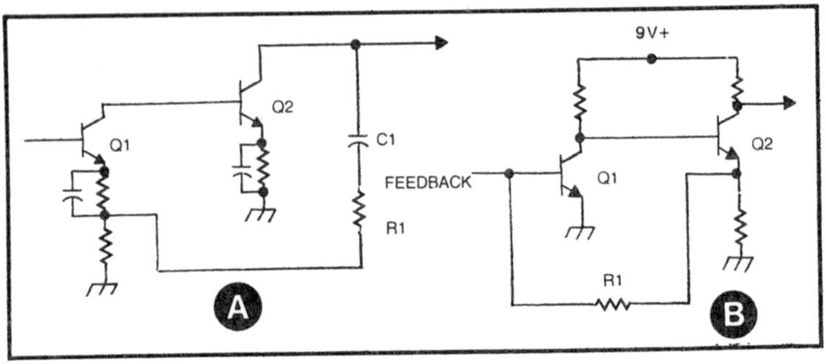

Fig. 21-5. Typical feedback circuits.

Single-ended, transformer coupled. One of the oldest solid-state power-amplifiers is the single-ended type of circuit shown in Fig. 21-4A. In this circuit, a single power transistor usually a PNP Germanium device, is transformer coupled to both the driver circuits and the output load (i.e., loudspeaker). Resistors from the DC power supply set the bias point to class-A, and signal was coupled in from previous stages by an interstage transformer. This design was in the first sets, probably because the designers, new to transistors, were mostly vacuum-tube devotees. In fact, most had little experience in the solid-state world.

But, transformers are bulky and expensive. Whenever you can get rid of a transformer, you are money and aggravation ahead (or so it seemed). The power amplifier of Fig. 21-4B is a more recent version, and may still be found in currently offered designs. This particular circuit is common to mobile equipment and automobile radios, but is also found in some fixed station equipment. Note that the output transformer is now an autotransformer. In some cases, only an output indicator, or choke is used. This stage is a direct-coupled device, and problems in any of the stages can cause output-stage symptoms. We discussed at length the problems that can crop up in this type of amplifier, so will not repeat them here.

Feedback

There are two basic kinds of feedback systems normally found in untuned-amplifier circuits. One, shown in Fig. 21-5A is called the "second-collector-to-first-emitter" system. With correct values of components, this circuit can make a relatively mediocre amplifier behave like a much more expensive circuit. An open or shorted capacitor, or a change in the value of a resistor in this network will create problems that range from mildly irritating distortion to a runaway thermal or oscillatory condition that can render the circuit either ineffective, or dead. Transistors and integrated circuits have been destroyed by problems in feedback loops. A step-by-step inspection of the feedback loop is generally in order whenever you encounter strange, unusual symptoms that seem difficult to trace to a single stage or component.

Figure 21-5B shows the second widely used feedback system. This one has been dubbed the "second-emitter-to-first-base" system. This circuit often employs only one resistor to supply feedback voltage.

Push-pull Circuits

The push-pull amplifier is widely preferred over other types, for both its ability to handle power and its overall fidelity. The push-pull amplifier obtains its fidelity from the fact that it will cancel the even order harmonics (2, 4, 6, etc), producing a much more flat output-frequency response.

Figure 21-6 shows the standard push-pull power-amplifier circuit which has been used in almost every audio-amplifier application since the introduction of commercially feasible transistors. Everything from a five-dollar transistor portable radio to multi-kilobuck transceivers have used this design, or one closely related to it. This circuit, however, is far from being efficient when compared with later circuits.

The circuit in Fig. 21-7 is a more recent addition to the family of solid-state push-pull amplifiers. It has often been called the "split-

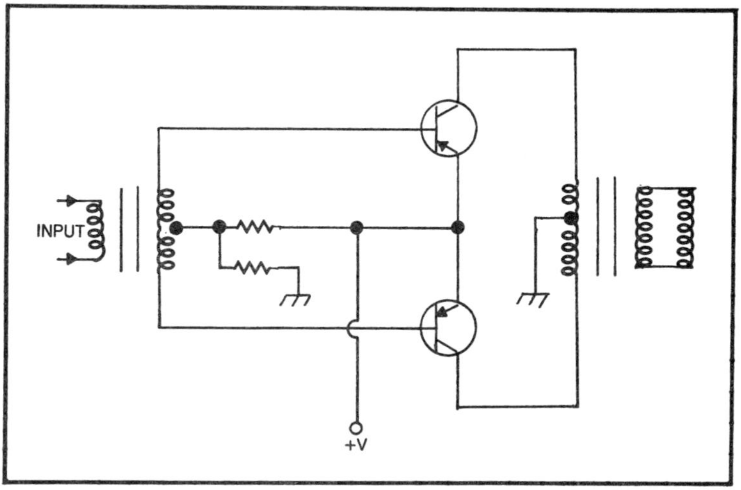

Fig. 21-6. Conventional push-pull amplifier.

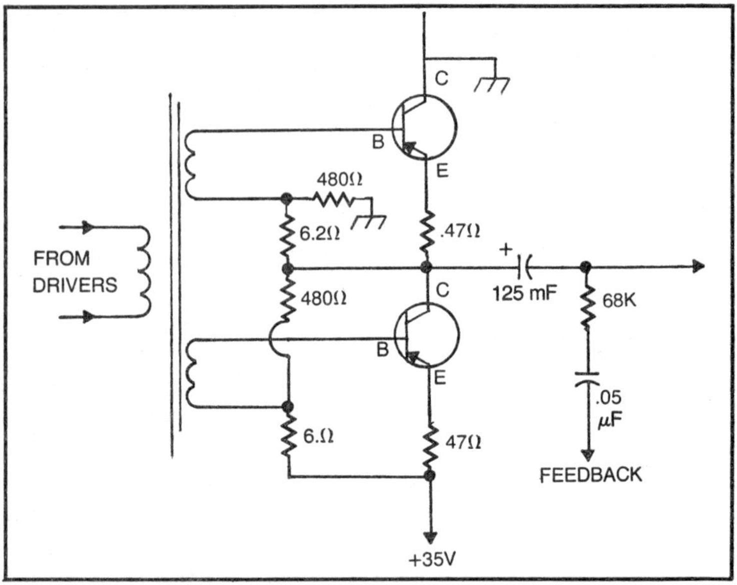

Fig. 21-7. Improved push-pull amplifier uses a series-connected transistors. Phase inversion is by twin-secondary interstage transformer. This circuit is sometimes called the split-secondary totem-pole push-pull amplifier.

secondary-totem pole" amplifier circuit, and has been used extensively in Amateur equipment, imported car radios, tape players, and the like. In fact, it is so common that you probably own several pieces of equipment that use this circuit. The series connection of the power-amplifier transistors and the transformer with two secondary windings are the chief identifying features of the circuit.

One feature which all push-pull amplifiers have in common is the necessity for phase-splitting the input signal to provide two signals 180 degrees out of phase with each other. One signal each is used to drive the two halves of the push-pull circuit. In older designs, this was accomplished by either a center-tapped interstage transformer (Fig. 21-6) or a split-secondary interstage transformer (Fig. 21-7). In many modern designs, however, the interstage transformer has been left out entirely. Of course, while this reduces cost, it does nothing for fidelity unless some other means is provided for splitting the input signal into the two required out-of-phase components.

The transistor phase inverter circuit is one possible replacement for the interstage transformer. These circuits are very similar to their vacuum-tube counterparts. They have one driving signal taken from the collector circuit, and another, of opposite polarity,

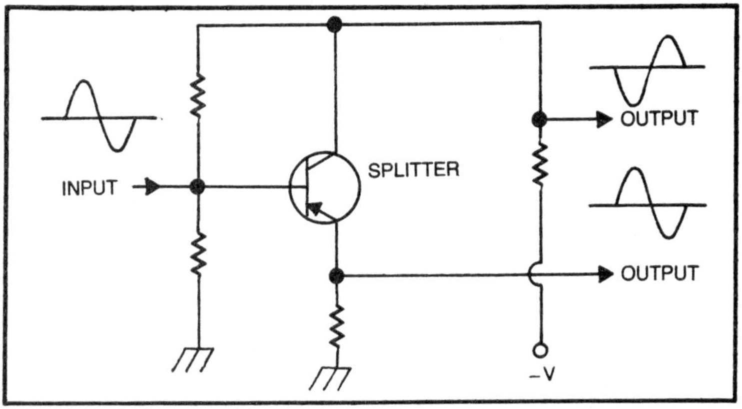

Fig. 21-8. Transistor phase-inverter produces equal amplitude, but opposite phase outputs.

taken from the emitter. An example of solid-state phase inverter is shown in Fig. 21-8.

Another approach to the design of this type of circuit is to use a special integrated circuit that provides out-of-phase output signals. A few ICs on the market have both inverted and noninverted output terminals, so would be well suited to this application. Most of these devices are referred to as "wide-band" or "video" amplifiers with push-pull or "balanced" outputs. An example of such a circuit is shown in Fig. 21-9.

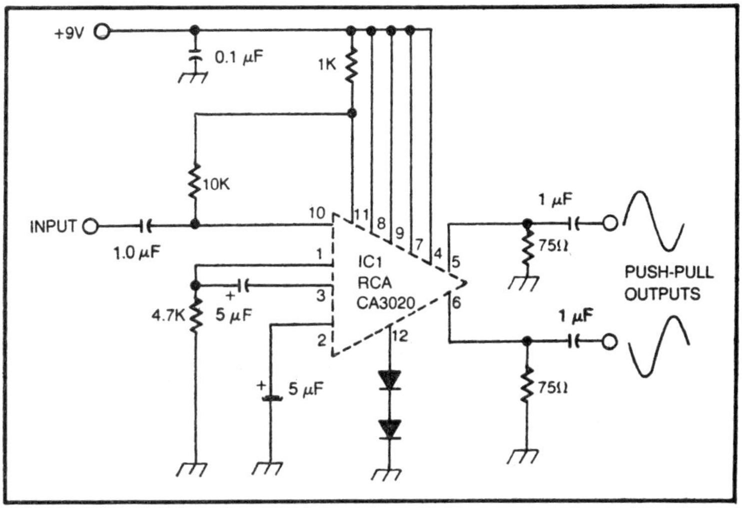

Fig. 21-9. IC phase-inverter stage.

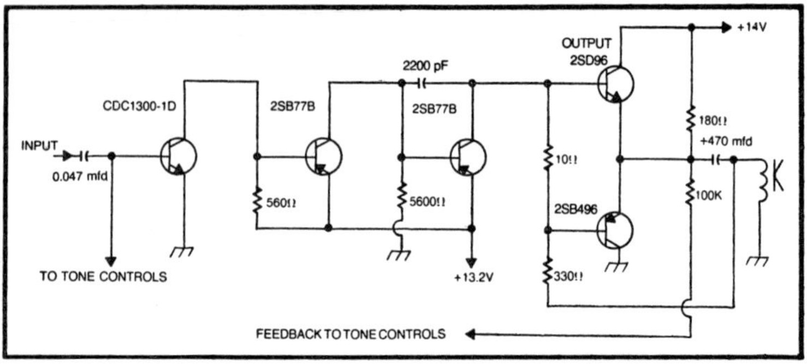

Fig. 21-10. Direct-coupled complementary push-pull amplifier.

Designers have another method for accomplishing phase inversion that often is more economical than either of the methods mentioned above. It is called the ***complementary symmetry*** power amplifier circuit. This design, shown in simplified form in Fig. 21-10, takes advantage of the fact that PNP and NPN transistors require signals of opposite polarity to perform the same function. Notice that the bases of the two transistors are fed in parallel by the input signal. Also note that the loudspeaker, minus the output transformer, is connected to the midpoint of the two series transistors. Versions of this circuit which use a single assymmetrical power supply usually employ a capacitor to block DC from the loudspeaker. Dual-supply versions, however, often do not use the output capacitor.

Complementary symmetry circuits once had a significant disadvantage: it was often difficult to find matched pairs of NPN and PNP transistors. Most of the pertinent electrical specifications of the two devices had to be nearly identical, so (unless the transistor manufacturer intended it that way) matching devices was difficult. Manufacturer's spec sheets of only a few years ago reveal that only a few selections in complementary power transistors were available. As the required power level went up, as in high fidelity amplifiers, then the number of selections available to the designer declined drastically. The problem to the troubleshooter was even more acute because even fewer of these filtered down to the service industry ranks. Only expensive factory replacements would work, and the home constructor (including many Amateur-Radio people) were just stuck.

It was, however, relatively easy to obtain matched pairs of the same transistor, whether the power level could be described as low, medium, or high. It seemed that NPN power transistors in silicon and PNP devices in germanium were particularly prevalent. Even the universal-replacement lines (which often offered less than "universal" choices) could offer reasonably well matched NPN power transistors of the 2N3055 class.

The lack of easily available matched complementary transistors led to an interesting modification of the complementary symmetry power-amplifier design. It seems that it was a lot easier to obtain complementary pairs at low and medium power than at higher powers. It was quite possible to order transistors with collector dissipations rated in milliwatts, where pairs rated in watts or dozens of watt were unavailable. The circuit shown in Fig. 21-11 is a quasi-complementary amplifier in which low-power, complementary drivers are used ahead of the same-polarity, push-pull, power-amplifier transistors. In this circuit, the output transistors are connected in series after the fashion of the totem-pole amplifier discussed earlier. The drivers are complementary. For the designer, this meant a larger selection of possible power transistors, and, for the servicer, a wider selection of possible replacement transistors. You are generally safe using one of the replacement-line transistors offered to professional service technicians (such as HEP, SK, ECG, GE, etc).

Both complementary and quasicomplementary power amplifiers can provide a host of their own unique troubleshooting problems. For example, these amplifiers have some characteristics

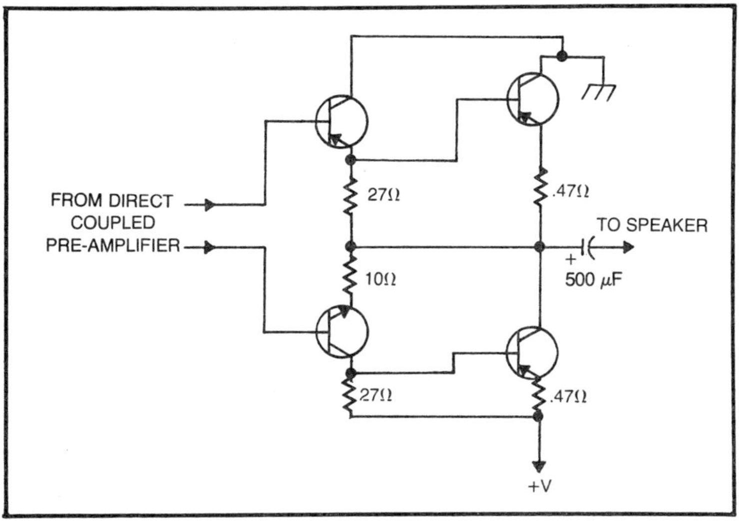

Fig. 21-11. Quasi-complementary push-pull amplifier.

similar to rf or wideband-video amplifiers. One characteristic is the wide bandwidth required of the transistors used in the circuit. Because these types of circuit tend to be fed from low-level signal sources (such as a receiver detector, microphone, etc), they must have very high gains. This, coupled with the high-frequency transistors used, can create instability problems.

High-frequency transistors in high-gain circuits are quite capable of oscillating at a frequency that extends from supersonic audio to well into the VHF region. The result can be a high level of distortion, "lisping," etc. Lisping is an added "sss" sound whenever sounds like "p", "s", "z", are pronounced. A squarewave test of the amplifier will usually show if the lisping is due to ringing of the circuit, i.e., oscillations that occur on fast-attack signals and then die out shortly thereafter. The visual indication of ringing will be extra cycles on the trailing edge, or a blurring of the trace if a low-frequency oscilloscope is used. Such oscillations can be caused by open capacitors, incorrect replacement transistors (too much beta or, more often, too high an operating frequency, F_t), or improper lead dress.

Word of Wisdom. Don't assume, just because your rig is a factory made kilobuck wonder machine that they didn't make lead dress errors. Both erroneous placement of leads by nonalert or sloppy production people, and improper design has been noted. I can't count the number of manufacturer's "engineering change notices" and "service tips" that I have seen in which some problem is cured by changing the lead dress someplace inside of the equipment.

GENERAL TROUBLESHOOTING TECHNIQUES

Most technicians agree that the newer circuits are more difficult to service than some older designs. It seems that design engineers have gotten cleverer than before. . . and that new devices are now available. Just take a look at the complexity of modern transceivers and receivers compared with what we called the "ultimate design" only a few years ago. Part of the reason why troubleshooting is more difficult is that many of the newer circuits are direct-coupled, and that makes it harder to isolate the stage.

Most troubles, however, can be eliminated by one of two methods: "shotgunning," and a logical analysis procedure.

The term "shotgunning" is used throughout this book, and is used to mean replacement of several, or a large number, of parts to correct a fault in one of them. Shotgunning is not terribly elegant, but it is easy to justify unless you have some super hang-ups about finding the particular bad component using circuit-analysis techniques.

In general, shotgunning is justifiable whenever it saves you time and money. Even commercial servicers, who are quite ethical by any reasonable standard, will shotgun some circuits. Why charge the customer for an hour of troubleshooter's time (at $20 - $40 per hour) to find which of five or a dozen $1 transistors is bad? In a direct-coupled circuit, it may take the entire hour (or more) to "analyze" which is bad. True, you have the "pride factor" (i.e., "me Superman") when you find a single bad component, but, while you are basking in self-admiration, I will be back on the air using my equipment!

But, let's not knock logical procedures too much, after all they do have their uses. The first step is a preliminary inspection using sight, sound, smell, and touch. Notice, for example, whether any fuses are blown, or if any fuse resistors used are overheating (fuse resistors, i.e. fusistors, are often used as the emitter resistor in power amplifiers. They then serve the dual function of protection—as a fuse—and to provide a small amount of negative feedback to the power amplifier, thereby stabilizing it).

Although not used too much in Amateur Radio equipment, some lower-cost Japanese transistor amplifiers, especially those used in car-radio power amplifiers, use the lead wires from the printed-circuit board to the emitters of the power transistors as the fuse resistors. This wire is special resistance wire, and replacements must be either identical wire, or a special "fusistor."

Notice whether fusistor insulation appears charred or cracked. Also note whether any printed labels on the transistors, or fusistors, have been erased by overheating. This could easily point you to which of several transistors are shorted, saving time. A transistor that heats up before shorting can often be detected by your sense of touch. I occasionally touch several transistors in a single section to determine if any of them are heating quicker than others.

Shorted and leaky transistors cause a large percentage of failures in solid-state power amplifiers. In small-signal amplifiers, and those power amplifiers that use plastic power transistors, it is the open emitter that seems to predominate.

One good way to locate defective transistors is to measure the bias and supply voltages with a VTVM. Generally speaking, though, such tests only confirm what an observant person would have already guessed.

Locating a suitable replacement transistor might take more time than an actual diagnosis. A good rule of thumb is to replace a transistor with an original (identical) part. In years gone by, the replacement lines were regarded as second best, and somewhat optimistic. Today, we find a lot more numbers in each line, so the crossover guide can find fits closer to the original. In fact, since most such lines are offered by transistor makers, they may be originals in disguise.

It is, however, good policy to be forever suspicious of crossover guides. The problem is not that the companies are trying to rip your off by selling a ringer, but that information flow is not always the best. The crossover-guide compiler will often select a replacement that seems well-suited to the original, but, the application for which you need a replacement transistor might not be the same application that the original transistor maker had in mind.

It is also possible that the guide is just plain wrong on their notion of what a good replacement is supposed to be. In one case, for example, I was supplied a replacement (from a well known replacement line trusted and used by professional servicers) transistor in a TO-3 diamond-shaped power-transistor case that was supposed to "sub" for an original in a tiny TO-5 case! Another time, I was sent a drift-field PNP "oscillator/rf amplifier" transistor to replace an NPN audio-power amplifier! In still another case, I received a transistor with a collector-emitter specification of 80 volts to replace a transistor operating at 350 volts. All of these examples come from, not a fly-by-night outfit that tries to rip off their customers (or victims), but a well-known, established semiconductor manufacturer all readers would recognize immediately.

Your best protection is to look up the replacement number, and then check its specs. These are given in tabular form in the front or back of the crossover guide. Compare the proposed replacement with what you know about the circuit it must operate in. If the replacement seems flaky to you, try another brand. The replacement may well be of high quality, but is unsuited to the job.

Remember, it is no longer possible for a handful of different transistor types to replace all of the transistors that a servicer will encounter. Whether you are building a new project, or repairing existing equipment, there might be dozens of different types of transistors needed. Chapter 32 covers transistor substitution in more detail. That material should be well worth your time reading.

Important to remember, when replacing power transistors, even in low to medium power applications, is proper heat sinking of the new device. If the new transistor is not heat sinked well enough, I can guarantee another failure in the near future (hmmm, there must be *something else* wrong in the circuit. . . this is the second time it failed!). Careful tightening of the mounting screws and the use of a silicone heat-transfer grease will do wonders for your "callback rate". If you do not have a tube of heat-sink grease, then buy one when you buy the replacement transistor. Do not rely on "salvaged" grease from the old transistor. It has long since lost its ability to do the job.

The admonitions given in the last paragraph are especially needed when dealing with plastic power transistors, especially if they have anodized finish on the metal part. Use a generous amount of heat-transfer grease, and then be sure to tighten the screw properly.

Defective output transistors often result in a complaint of low volume and lots of distortion. One way to spot this defect in a hurry is to measure the DC voltage at the junction where the two output transistors and the output capacitor are connected together. Compare this voltage with the overall supply voltage. Note that the output transistors in several classes of amplifier are connected in series, so the voltage at this junction should be approximately one-half of the overall supply voltage. If the measured voltage is significantly higher or lower than one-half of the supply voltage, then suspect a fault.

Unfortunately, many types of distortion problems are not so simple to locate. In some cases, for example, a small transistor with a defect in an early stage of a cascade chain will cause massive failure of the output transistors. It is always good practice to check all of the other transistors in the direct-coupled chain to find out whether any gross defects are in evidence. This can be done using the in-circuit transistor checks discussed in Chapter 19.

IC DESIGNS

It has become increasingly popular among Amateur-equipment designers to use IC power amplifiers in the audio stages. These devices can produce up to several watts of audio power, yet take up only small amounts of space. In most cases, if a voltmeter check of pin potentials shows discrepancies, you are safe in just replacing the IC. Fortunately, most of these are made by only a few companies. They are "standard" types, so replacements are easily obtained either from the equipment maker, the device maker, or the universal-replacement IC lines offered by the same companies who offer the replacement transistors.

Warning. Many IC audio power amplifiers use a balanced output to the loudspeaker, instead of the chassis-ground referenced system of other designs. If you follow "standard practice" and ground one side of your test or bench speaker to the chassis, then expect to blow out the audio IC! It has happened to me too many times, so take a word of wisdom, and avoid an expensive mistake.

MEASUREMENTS ON UNTUNED AMPLIFIERS

There are only a few measurements used to evaluate untuned-amplifier circuits. You will want to use these techniques for two reasons. One is to test out new designs or new projects, to see if they do what you want them to do. Secondly, it will also be true that you will occasionally use these techniques to troubleshoot a problem in existing equipment, or "proof of performance," after a troubleshooting job is finished. For example, when you used that replacement line transistor in the audio preamplifier section of your modulator, does it really have pretty much the same gain and frequency response as the original? These tests could help you find out.

The meaurements which we will consider in this chapter include voltage gain, power gain, output power, input sensitivity, frequency response, squarewave testing, total harmonic distortion (THD), intermodulation distortion (IMD), slew rate, full-power bandwidth, input offset voltages, and common-mode rejection ratio.

But first, let us cover the concept of decibel notation. Many tests will express results in either dB, or in a decibel derived scale (such as volume units, etc.).

DECIBELS

Many measurements that result in specified levels of voltage, or current, are best expressed in *decibels*, usually denoted "dB." The decibel causes many people to anguish over its meaning. They seem to think that it is a "something," or a unit of some kind. They become confused because they do not seem to understand how it is derived, or just how it related to voltages, currents, and powers. After all, we see amperes, voltages, and watts expressed as dB, and we *know* that these are mutually exclusive categories. So how come dB describes them all?

The answer is so simple that it bothers people: decibels express *ratios* . . . and nothing more. This makes it suitable for expressing gains and losses in amplifiers and systems, frequency-response plots, and almost any other place where we are required to make *comparisons* between two levels.

For electrical *power* we express decibels as

$$dB = 10\ Log_{10}\ \frac{P1}{P2}$$

Where: db is the gain or loss in decibels P1 and P2 are the power levels involved Log_{10} refers to the base-10 logarithms

Example:

Calculate the power gain of an audio amplifier used to amplitude modulate a radio transmitter if 100 milliwatts at the input produces an output power of 25 watts.

Solution:

$$dB = 10\ Log_{10}(25/0.1)$$
$$dB = 10\ Log_{10}(250)$$
$$dB = 10\ (2.4) = 24 \text{ decibels}$$

Similarly, we can express both *voltage* and *current* gains and losses in decibel notation. The multiplication constant, however, is 20 instead of 10:

$$dB = 20\ Log_{10}\ (E1/E2)$$
$$dB = 20\ Log_{10}\ (I1/I2)$$

Example:

Calculate the voltage gain of a preamplifier if the input voltage is 56 mV, and the output voltage is 9 volts.

Solution:

$$dB = 20\ Log_{10}(E1/E2)$$
$$dB = 20\ Log_{10}(9/0.056)$$
$$dB = (20) Log_{10}(161)$$
$$dB = (20)\ (2.21) = 44 \text{ decibels}$$

The examples given above show how to express *gains* in dB notation. In both cases, the numerator in the ratio was larger than the denominator. Losses can be expressed by dB notation also. But to distinquish losses from gains, and to allow them to be mixed in the same system (as in antenna design), we have adopted a convention regarding the decibel equations. The ratio is always expressed in the form "output/input". If this convention is followed, then we find that gains result in positive numbers, and losses result in negative numbers! The numerical values of dB are the same, regardless of whether the larger number is in the denominator or the numerator.

If you doubt this, take out your calculator or rusty old slide rule and try the following:

a. $dB = 10 \, Log_{10}(3/7) = -3.68$ dB
b. $dB = 10 \, Log_{10}(7/3) = +3.68$ dB

In both cases, the numerical result was the same, but the loss (version a), was negative and the gain (b) was positive. The same holds true for both current and voltage ratios.

Notation: *Gains* are positive and *losses* are negative when you use the output/input rule.

Example:

A piece of coaxial cable will demonstrate a certain amount of loss, usually expressed in terms of dB/100 ft. Find the loss of 100 feet of a certain coaxial cable operated at 2-meters. The transmitter applies 10 watts at the input, while the dummy load at the other end of the 100-foot run receives only 4.2 watts.

Solution:

$dB=10 \, LOG_{10}$ (output/input)
$dB=10 \, LOG_{10} (4.2/10)$
$dB=10 \, LOG_{10} (0.42)$
$dB=(10)(-0.378) = -3.78$ decibels

Special dB Scales

Decibel scales represent ratios between two sets of numbers, i.e., voltage, current, or power ratios. It is sometimes convenient to reference all measurements in a system, or an entire industry, to some common standard level that we *define* (that is, agree to use) as 0 dB. If this is followed, then we may simply add up all of the gains and losses, and then note how the result compares with some standard. We can use this system to predict signal levels, determine needed gains, and to compare two pieces of equipment or systems. In short, use of a standard reference scale based on the decibel notation is a way to solve many problems.

There are several commonly employed decibel-based scales used in various areas of electronic communications:

1. dBm (used in rf measurements). 0 dBm defined as 1 mW dissipated in a 50-ohm resistive load.
2. dB. Obsolete scale used in telephone and public-address work. 0 dB defined as 6 mW dissipated in a 500-ohm resistive load.
3. Volume Units (VU). Newer version of the dB scale. Defines 0 VU as 1 mW dissipated in 600 ohms, and is calculated using the power dB formula.

4. dBmv. Used in TV antenna systems. 0 dBmv is 1 mV in 75 ohms.

The user of a special dB scale may be required to convert from dB notation back to power or voltage notation, i.e., milliwatts or millivolts, etc. If we know the reference standard for 0 dB, and the gain or loss is dB, then we can find the information needed. All that we need do is a little algebra on the exponential 10^x. For example, how do we find the actual power level in milliwatts, if the dBm system is being used? Simple:

$$dBm = 10 \text{ Log}_{10}(p/1 \text{ mW})$$
$$dBm/10 = \text{Log}_{10}(p/1 \text{ mW})$$

Raise 10, the base of the common logarithms, so dBm/10. Doing the same procedure on the right side of the equal sign simply erases the Log_{10}, leaving us with the term (P/ 1 mW):

$$10^{(dBm/10)} = P/1 \text{ mW}$$
$$(1 \text{ mW})\ (10^{(dBmB)}) = P$$

Example:

How many watts are applied to the input of an rf amplifier if the signal generator used as the source is set to produce an output of 35 dBm?

Solution:

$$P = (0.001)\ (10^{(dB/10)})$$
$$P = (0.001)\ (10^{(35/10)})$$
$$P = (0.001)\ (10^{3.5})$$
$$P = (0.001)\ (3162) = 3.16 \text{ watts}$$

VOLTAGE GAIN MEASUREMENTS

The gain of any electronic circuit is defined as the output over the input. This is known as the "transfer function" of the circuit. For voltage amplifiers, then, the transfer function is

$$A_v = \frac{E_{out}}{E_{in}}$$

Where:

A_v is the gain of the circuit, and is dimensionless
E_{out} is the output potential in volts
E_{in} is the input potential in volts

We can measure the gain of the circuit by measuring the input signal voltage, the output signal voltage, and then taking the ratio expressed by the transfer function. Similarly, we can make the measurements and then plug the data into the voltage/dB expression given earlier.

Figure 21-12 shows the basic equipment set-up needed to make a large number of measurements on amplifiers. Of course, for the voltage-gain measurement, we only need part of it, i.e., a means for determining input and output voltages. This equipment set-up is essentially the same for all amplifiers from DC to daylight. Whether they are in high-fidelity equipment, communications equipment, or television receivers, the basic principles are the same; only the nature of the instrumentation needed to make the voltage measurements will change.

Procedure:

1. Adjust the attenuator, oscillator-output level, and amplifier-gain controls for a convenient output level (*note*: Most authorities insist that the amplifier gain control be set to maximum—or to some specified point determined by the manufacturer).
2. Measure the input and output signal-voltage levels.
3. Make the gain calculation given earlier, or convert to dB; $dB = 20 \text{ Log}_{10}(A_v)$.

The voltage-gain measurement is usually made with the amplifier gain set to maximum. In most cases, a gain-control setting of less than maximum results in a meaningless measurement. If the amplifier is driven into clipping, then reduce the input-signal level or the gain control until clipping disappears.

OUTPUT POWER MEASUREMENTS

Power amplifier measurements have been subject to a lot of abuse in the consumer electronics field. Fortunately, amateur radio equipment is usually specified in a relatively straight forward manner. But, for background, let us describe some of the ways in which Hi-Fi manufacturers and sales people have abused the power measurement. This situation became so bad before the Federal Trade Commission (FTC) stepped in that professional servicers and non-audio electronic engineers and technicians often felt that "taffy was being distributed" when talking to Hi-Fi salesmen! There seemed to have been several different ways to measure "power."

It would seem, on casual inspection of the problem, that a measurement as basic a power would be cut and dried, with little room for improvisation. But this was not the case. It seems that $P = E \times I$ can take on a lot of different meanings when you begin to define your terms. I have seen the following types of audio-power rating:

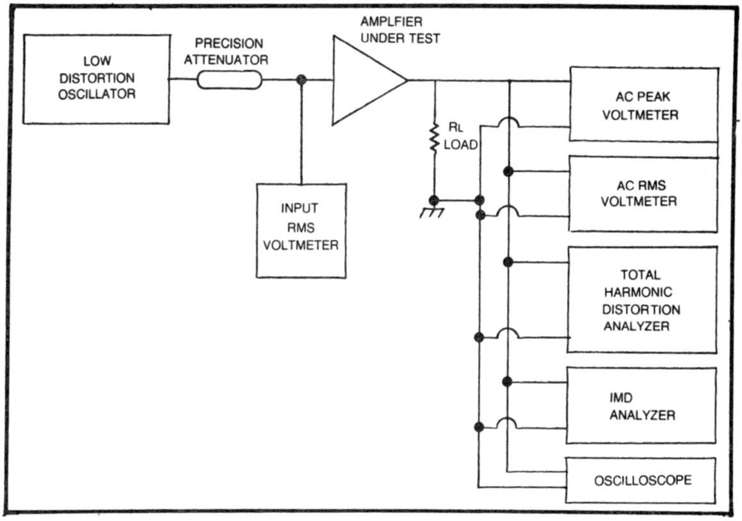

Fig. 21-12. General set-up for most tests of untuned amplifiers.

a. Rms watts at 1% THD
b. Rms watts at 5% THD
c. "Peak watts"
d. Peak-to-peak watts
e. Something called "music power"

Of these, only rms watts has any business being quoted. Rms watts tells us something of the amplifiers ability to do work, i.e., pump air through a loudspeaker system, modulate an rf carrier, etc.

Notice that there are two different rms wattage ratings! It seems that a lot of manufacturers were fond of measuring the THD at 1 watt output (where any reasonable amplifier was decent), and then measuring the output power at a THD of 5 percent. So what, you say? Well, it seems that most amplifiers will produce a lot more output power at 5 percent THD than they will at 1 percent. This is a standard phenomenon that can be noticed on almost any amplifier capable of low THD percentages.

Note that, in an AM radio transmitter, the THD content of the modulating signal will affect the bandwidth of the signal. Splatter is greater when the harmonic distortion is increased. To be sure of less splatter, one would want to create the needed power at a lower THD figure.

Let us consider some of the types of "power" measurement, with respect to the proper measurement of audio power at 1 percent THD. An amplifier is nominally rated at 25 watts rms at 1%

THD, when connected to an 8-ohm resistive load. At 1-percent THD we find a potential of 14 volts rms across the load. Using the standard methods for obtaining the various power measurements presented earlier, we arrive at output power figures of:

1. Rms watts at 1% THD.
   $P = (E_{rms})^2/R$
   $P = (14 \times 14)/8$
   $= 24.5$ watts rms at 1% THD
2. Peak watts.
   $P = (E_p)^2/R$
   $P = (1.414 \times 14 \text{ volts})^2/8$
   $P = (392)/8 = 49$ watts
3. Peak-to-peak watts
   $P = (2.83E_{rms})^2/R$
   $P = (2.83 \times 14)^2/8$
   $P = (39.6 \times 39.6)/8$
   $P = (1570)/8 = 196$ watts!
4. Music Power
   $P = 2 \times$ rms power
   $P = (2)\ (14 \times 14)/8$
   $P = (2)\ (196)/8 = 49$ watts

Note well the wide range of numerical results . . . and then consider that all of these figures have been advertised by various manufacturers. You will soon see why consumer complaints forced the FTC to take some action. Manufacturers might make use of these different measurements, but are required to prominently display the rms wattage at 1 percent THD. The use of these techniques is one explanation why a $10 portable phonograph could have the same power output statistics as a piece of quality high-fidelity equipment. In those days, professional servicers, when asked the output of a customer's amplifier, would quote (tongue in cheek) so many "watts IYL". . . If You're Lucky!

Measurement procedure:

1. Adjust the amplifier for maximum gain, increase the oscillator for either a) clipping begins (and then back off until the clipping disappears), or b) until the THD increases to 1%, or to the level specified by the equipment maker or the FCC
2. Measure the rms output voltage across the load resistor.
3. Calculate the rms power from $P = E^2/R_L$.

In most cases, the measurement is specified at some particular frequency; 100 Hz is used in hi-fi work, and 400 Hz in many communications applications. One authority feels that it is best, and most reasonable, to measure the power at 1000 or 400 Hz, and then again at the points at the high- and low-frequency ends of the band, where the power output drops to −3 dB of the 1000 (400) Hz power. This will always yield a lower power if the three are averaged, but the figure is more nearly equal to that which the amplifier produces under all circumstances.

Input Sensitivity

The sensitivity rating of an amplifier tells us how much signal is required to drive the amplifier to a specified output level. Two methods seem generally applied: a) drive level required for full output power, b) drive-level required to produce 1 watt of output. Procedure:

1. Adjust the amplifier gain to maximum.
2. Increase the input signal level to produce the specified output power.
3. Measure the rms input voltage.
4. The sensitivity will be expressed as "mV at 1 watt," or "mV for full output power."

FREQUENCY RESPONSE

The frequency response of an amplifier refers to its ability to amplify signals of different frequencies. Ideally, a perfect amplifier

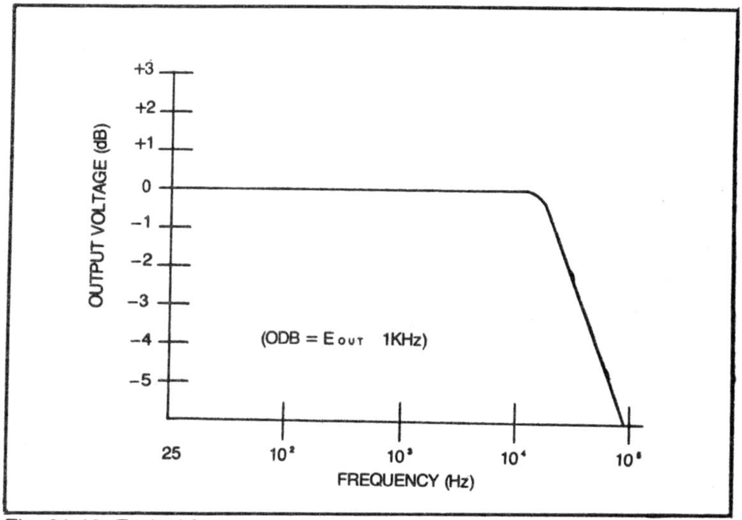

Fig. 21-13. Typical frequency-response graph.

will amplify *all* frequencies equally, but in practical amplifiers this ideal is never met.

The response plot of an amplifier (Fig. 21-13) graphs *gain-vs-frequency* on semilogarithmic paper. The relative gain is not actually important, but is inferred from observing the output voltage (assuming that the input voltage is kept constant).

The gain, then, is expressed in terms of output volts (usually converted to decibel notation). The 0-dB reference level will be the gain of the amplifier at some low or medium frequency, usually 400 or 1000 Hz. The reference point must be far removed from the frequency at which the gain drops −3 dB.

The most practical way for Amateurs to measure the frequency-response data is to make the measurements manually by readjusting the frequency of the oscillator, or other signal source, and then reestablishing the input voltage (remember it must be constant), using the precision attentuator. The output voltage and frequency are then recorded.

In some cases, though, it might be possible to buy or build a voltage-controlled signal generator that will sweep the spectrum. If a sawtooth sweep-input voltage is applied, and is also used as the oscilloscope time base, then the amplitude of the amplifier output signal (applied to the oscilloscope vertical input) represents the frequency-response data. Note that care must be taken to insure the constancy of the input signal voltage.

Square Wave Tests

We can examine the frequency response of an amplifier in a qualitative way by using a squarewave input signal. It is the nature of square waves to possess a large number of harmonics of the fundamental signal. These harmonics cause the shaping of the waveform, and if removed by an amplifier or other circuit, will alter the waveform shape. It seems that the ideal square wave will contain an infinite number of odd-order harmonics.

The square wave test lacks something in quantitative data but is a quick method for obtaining a qualitative view of the circuit under test.

Figure 21-14 shows several different waveforms obtainable in squarewave testing. The first example, Fig. 21-14A, shows the input squarewave. Note that the output waveform in a very-wideband amplifier would not look significantly different from this illustration. In most practical wideband amplifiers, you will notice only slight rounding of the upper corners of the waveform. It is the higher harmonics that give the waveform its "sharpness."

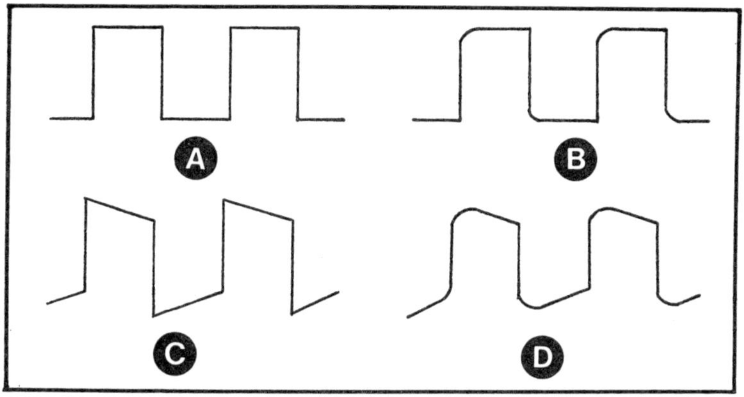

Fig. 21-14. Waveforms: (A) Normal squarewave; (B) loss of high frequencies; (C) loss of low frequencies; (D) loss of both high and low frequencies.

Loss of significant numbers of upper harmonics produces the trace of Fig. 21-14B. Note that loss of even more harmonics has cost the trace much of its sharpness.

In Fig. 21-14C we see the result of loss of the lower frequencies. This trace is characteristic of a squarewave passed through an amplifier or circuit with poor low-frequency response. In this case, the leading edges of both positive and negative excursions are overpeaked, and the flat portion has become tilted.

TOTAL HARMONIC DISTORTION (THD)

All waveforms (and in fact most continuous mathematical functions) can be described in terms of a Fourier series of sines and cosines of the fundamental frequency. The specific harmonics that are present, and their relative phases and amplitudes, determine the shape of the waveform. Only the sinewave is pure, i.e., contains no harmonics, all other waveforms have harmonic content.

When an amplifier, or any other circuit, distorts a waveform, it is creating harmonic components in the output waveform that did not exist in the input signal. The measure of this type of distortion is called *total harmonic distortion*, or THD.

Total harmonic distortion is measured by using a harmonic distortion analyzer, block diagrammed in Fig. 21-15A. The oscillator supplying the input signal to the amplifier being tested must have a very pure sine-wave output. The THD analyzer does not care where the distortion originates—if the distortion is present it is measured.

An rms AC voltmeter is used to measure the distortion. We first measure the rms voltage of the amplifier output potential E, and then measure the rms voltage due to harmonics E_n. This voltage is measured by passing the signal through a high-Q *notch filter* that is adjusted to pass all frequencies except the oscillator's output frequency (the fundamental frequency). The response of a typical notch filter is shown in Fig. 21-15B. The percentage THD is expressed by the equation below:

$$THD = \frac{E_h}{E} \times 100$$

Where:

THD is the total harmonic distortion in percent
E is the rms output voltage before using the notch filter
E_h is the rm output voltage with the notch filter in place

Example:

Calculate the total harmonic distortion if the amplifier output voltage is 13 volts rms without the notch filter, and 256 mV with the notch filter in place.

Solution:

$$THD = (E_h / E) \times 100$$
$$THD = (0.256/13) \times 100$$
$$THD = 1.97\ \%$$

Note well that THD for any given amplifier varies markedly with output power. An amplifier in which the THD is measured at 1 watt may produce an entirely different (usually worse) figure when producing full rated output power. This matter is usually of more concern to hi-fi enthusiasts than radio amateurs, however, because our communications audio amplifiers typically produce THD figures in the 2 - 10 percent range. An example of a commercial THD analyzer is shown in Fig. 21-15C. This is the Sound Technology, Inc. model 1700B (a hi-fi industry standard).

INTERMODULATION DISTORTION (IMD)

When two signals are combined together in a linear network, they will both pass through the system without affecting each other. Figure 21-16 shows the result of mixing two signals in such a network. In this case a high frequency (2 - 4 kHz) is mixed with a low frequency (30 - 100 Hz). If the output signal of the network is examined on a spectrum analyzer we would find that only the two original signals are present in the output.

However, if these same signals were passed through a nonlinear network, such as an amplifier that produces distortion, then mixing, or "heterodyning," will result. The lower frequency will

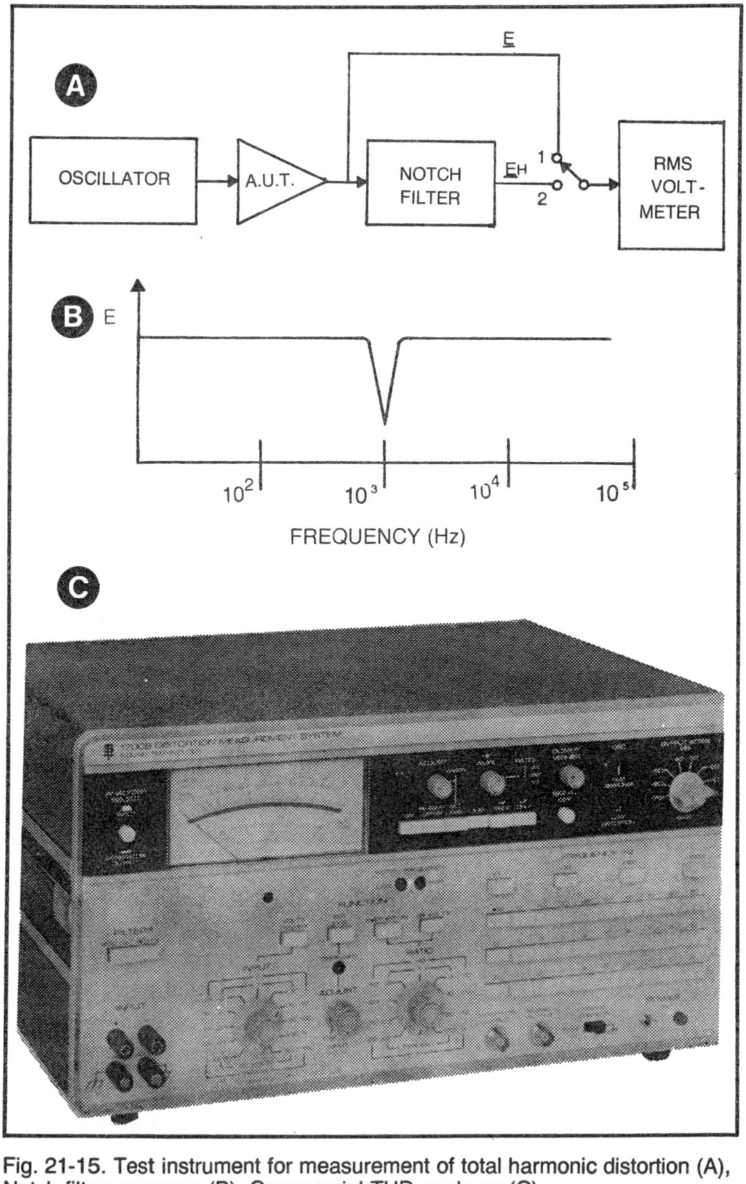

Fig. 21-15. Test instrument for measurement of total harmonic distortion (A), Notch filter response (B), Commercial THD analyzer (C).

tend to amplitude modulate the higher frequency. If F1 and F2 are the two original frequencies, then the output spectrum will contain at least the following frequencies: F1, F2, F1 + F2, and 1 – F2. There might also be present several different sum and difference

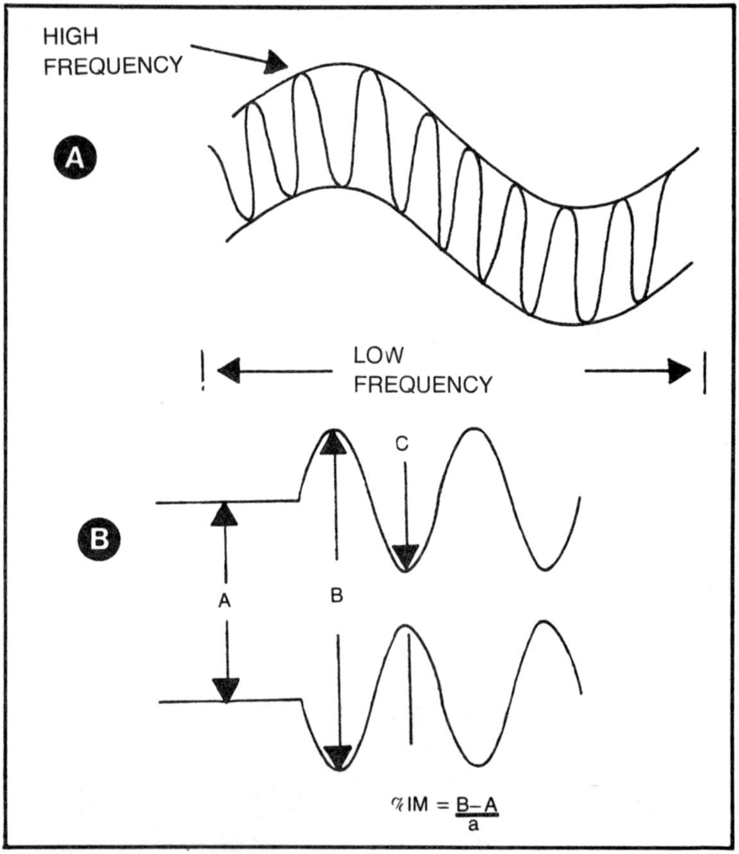

Fig. 21-16. Mixture of high and low-frequency components in a linear network is the test signal for IMD tests (A). The resultant output signal due to nonlinearity in the amplifier resembles amplitude modulation (B).

products of the harmonics of the two fundamental frequencies (i.e. $2F1 \pm 3F2$, etc).

Note that distortion that tends to mix these frequencies will produce an output signal that is more like the familiar amplitude modulation waveform (Fig. 21-16B) than the linearly mixed signal previously shown. The percentage of intermodulation distortion is given by:

$$IMD = \frac{(b - a)\,(100)}{a}$$

Where:

IMD is the intermodulation percentage
terms "a" and "b" refer to Fig. 21-16B.

A block diagram to a simple IMD analyzer is shown in Fig. 21-17. This instrument can measure IMD by simply rectifying the output signal. A standard test signal is applied to the input of the amplifier. The higher frequency must be not less than 50 times greater than the lower frequency. For a 60-Hz lower frequency, then, the upper frequency must be at least 3000 Hz. The two signals are combined in a linear network (such as a Wheatstone bridge).

The test circuit uses a high-pass filter to reject the low frequency component of the output signal, and pass only the high frequency component plus any modulation products (i.e. F1 ± F2) that lay near the higher frequency. A low-pass filter is used to average the rectifier output (a low-pass filter acts like an integrator) to remove any residual high-frequency information present. The remaining signal is a rectified and filtered dc potential that is proportional to the IMD, that is, the low-frequency component mixed with the high frequency. Since the original high-pass filter removed the low-frequency signal, it will not affect this voltage level. The output voltage is proportional to intermodulation distortion, so can be calibrated in percent IMD.

In untuned amplifiers the IMD is also sometimes called *cross-modulation*. When dealing with radio communications equipment this term is misleading. In receivers, for example, the two terms intermodulation and crossmodulation have similar meanings, but there is a slight difference. In some hi-fi circles, however, you will hear these terms used interchangeably.

OTHER AMPLIFIER MEASUREMENTS

Many of the tests discussed thus far are particularly useful in hi-fi equipment, and when dealing with the modulators in radio

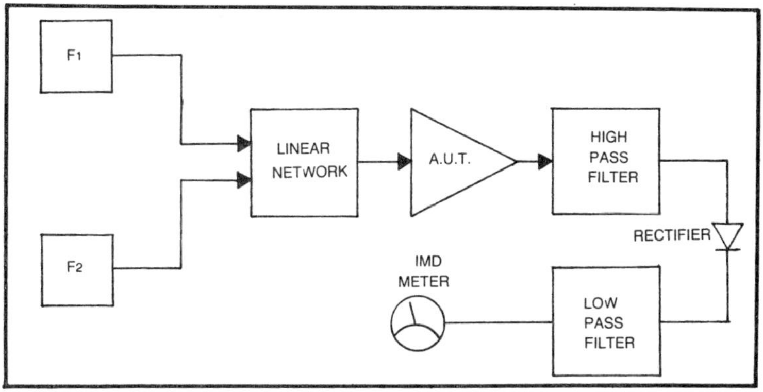

Fig. 21-17. IMD test set-up.

transmitters. The tests presented in this section are used primarily in other types of amplifier. Some of these tests are specified for *differential* amplifiers, while others are applicable to all amplifiers in general. We will use the operational amplifier for our examples, but please be well aware that they are applicable to all forms of amplifier.

Slew Rate

The slew rate (usually denoted S_r) of any amplifier is the measure of its maximum rate of change of output voltage dE/dT; while the amplifier is producing its rated output current. Figure 21-18 shows the test configuration for this test. In operational amplifiers, the unity-gain, noninverting follower configuration is specified because it produces the worst case results. Gain followers can also be used, but the results obtained may not be as valid in all cases.

The input signal E_{in} is adjusted to severely overdrive the amplifier to both positive and negative limits (Fig. 21-8B). The slope of the output voltage waveform is viewed on an oscilloscope, and represents the slew rate. Mathematically,

$$S_r = \Delta E_o/\Delta T$$

Slew rate is usually expressed in terms of a voltage or current per unit of time. In voltage amplifiers, you will see the specification written as ***volts per microsecond*** or ***microvolts per second***, depending upon whether they want the number to look large or small. Current-output amplifiers will use amperes or milliamperes per unit of time. In those circuits the millisecond seems popular.

Full Power Bandwidth

The *full power bandwidth* of any amplifier is the frequency at which the open-loop voltage gain drops to unity (1), while the amplifier is delivering its maximum rated output power. This frequency will usually be less than the gain-bandwidth product (F_t) of the same amplifier.

It is usually difficult, or impossible, to measure the full power bandwidth on operational amplifiers because the open loop gain is tremendous. In that case, an input signal in the microvolts range may tend to saturate the output of the amplifier. In those devices, we approximate the full power bandwidth measurement by using a closed-loop gain in the X100 to X1000 range.

The input-signal amplitude must be adjusted to a level that drives the output signal to its maximum power level. Load resistor

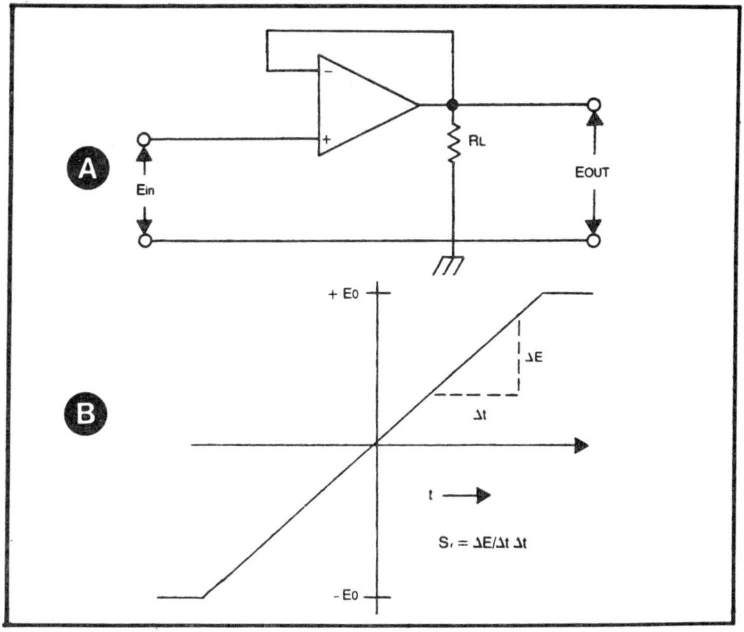

Fig. 21-18. Test circuit for slew-rate determination uses the unity gain configuration because this is often a worst-case situation (A). Mathematical and graphical definitions of slew rate (B).

R_L can be set to absorb the power. The frequency of the input signal is then increased (the input signal voltage is held constant) until the output voltage is equal to the input voltage, i.e. $A_v = E_o/E_{in} = 1$. The frequency at which this occurs is the full-power bandwidth for the amplifier.

Input Offset Voltage

The input-offset voltage of a DC amplifier is defined as the potential across the inverting and noninverting inputs that is required to force the output potential to zero at a time when the normal input-signal voltage is also zero. We would normally expect the output voltage to be zero any time the input signal is also zero, but certain internal amplifier defects sometimes cause an offset to exist.

The measurement circuit is shown in Fig. 21-19. You should recognize the circuit as the inverting follower circuit with a gain of X100 to X1000. The "input" end of resistor R1 is grounded to insure that the input voltage is, indeed, zero. Then a precision, high-resolution DC power supply is connected across the input terminals

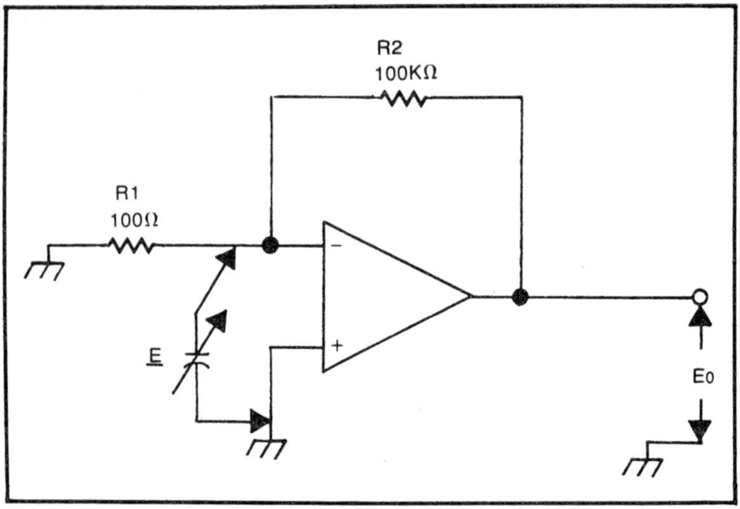

Fig. 21-19. Test circuit for determination of offset voltage.

of the amplifier. This supply is adjusted until the output of the amplifier drops to zero volts. The input voltage from the supply that causes the output voltage to drop to zero is the input-offset voltage of that amplifier.

Common Mode Rejection Ratio

Common-mode rejection is a property of differential amplifiers by which such amplifiers will reject, that is, not recognize as valid input signals, those signal voltages that are applied equally to both the inverting and noninverting inputs. The common mode rejection ratio (CMRR) is defined by:

$$CMRR = \frac{A_d}{A_{cm}}$$

Where:

A_d is the differential gain
A_{cm} is the common mode gain

CMRR can also be defined as $((E_o/E_{in})/(E_o/E_{cm})$ and (E_{cm}/E_{in}), in which case the output voltage is taken to be a constant.

In actual measurement, the input-signal voltage, and the output voltage (E_o) that it produces, are measured first. Then the two amplifier inputs are tied together, and the common-mode voltage is applied to the input. This potential is raised until the output voltage previously obtained exists. The common mode voltage is then measured, and used in the above equations.

Chapter 22

Testing Tuned Circuits

Tuned circuits are used almost everywhere in Amateur Radio equipment. In fact, most of the circuits in your receiver and transmitter are resonant, or tuned, circuits. These circuits consist of various combinations of inductance (*L*) and capacitance (*C*), and seem relatively simple. But, wait until you try testing such circuits, and try to obtain valid results without changing the circuit too much. If, for example, you try to use a signal generator and an rf voltmeter in the simplest way, then the capacitance of the cable and voltmeter-input will detune the circuit such that the result will be totally useless. Consider the ordinary shielded coaxial cable often used on signal generators and rf voltmeters. This cable has a capacitance of 30 pF/ft (or thereabouts). If you connect a signal generator to a circuit through three feet of cable, then you will add 3 × 30, or 90 pF to the circuit. Most amateur circuits use less than this to resonate the inductor, so you cannot measure the circuit using this method.

TUNED CIRCUITS - A REVIEW

In this chapter, we are going to consider many of the test instruments and techniques used in the measurement and troubleshooting of tuned circuits. But first, let us review some of the basics of resonant *LC* circuits. Note that antennas are also tuned circuits, and some of the techniques used in antenna measurements are also used in testing circuits where the *L* and *C* constants are lumped in the form of inductors and capacitors, instead of being

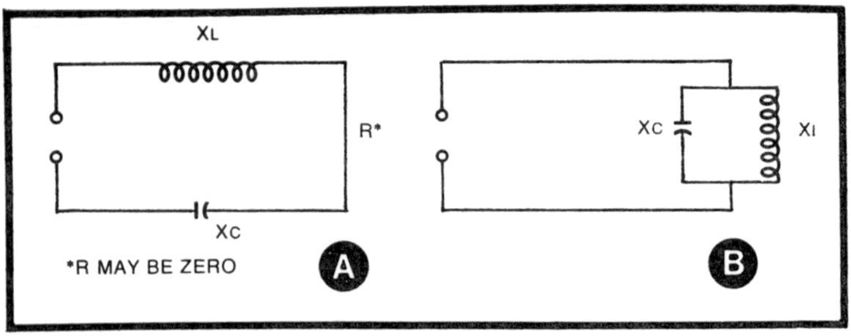

Fig. 22-1. (A), Series-resonant tank; (B), parallel-resonant tank.

distributed along the length of the antenna. Chapter 26 covers antenna systems, so you may wish to skip ahead to that material after you finish reading this chapter.

All tuned circuits consist of various series and parallel combinations of inductors and capacitors. Resistance in a tuned circuit alters the performance of the circuit somewhat, but resistance is not part of the definition of the resonant condition needed for the tuned circuit. Figure 22-1 shows the two basic forms of tuned circuit, series and parallel. The series-tuned circuit is shown in Fig. 22-1A, while the parallel-tuned version is shown in Fig. 22-1B.

Tuned circuits differ in behavior from untuned circuits because of the phenomenon known as resonance, a property of *LC* circuits. Resonance occurs in an *LC* circuit whenever the reactance of the capacitor (X_c) and the reactance of the inductor (X_L) are equal. This condition is described mathematically below.

$$X_L = X_c$$
$$2\pi FL = 1/2\pi FC$$

The capacitive and inductive reactances produce equal, but opposite phase, effect on the AC signal, so when the magnitudes of these reactances are equal (occurs at only one frequency), the next two effects cancel each other totally, leaving only any resistance in the circuit to oppose the flow of current. All tuned circuits, then, become resistive at their resonant frequency, and will be inductive or capacitive at frequencies other than resonance.

We find the formula for the resonant frequency of a simple *LC* circuit by solving the equation for F:

$$F = 1/2\pi\sqrt{LC}$$

Where:

F is the resonant frequency in hertz (Hz)
L is the inductance in henrys
C is the capacitance in farads.

Example:

Find the resonant frequency of a tuned circuit in which a 100 pF capacitor and a 0.1 uH inductor are connected in parallel.

Solution:

$F = 1/((2)\ (3.14)\ LC^{1/2}$

$F = 1/(6.28)\ ((0.1 \times 10^{-6})\ (100 \times 10^{-12})^{1/2})$

$F = 1/(6.28)\ ((10 \times 10^{-18})^{1/2})$

$F = 1/(6.28)\ (3.3 \times 10^{-9})$

$F = 1/(21 \times 10^{-9}) = 0.047 \times 0.047 \times 10^{9} \times 4.7 \times 10^{7}$

$F = 47$ MHz.

There is a fundamental difference between the behaviors of the two different types of resonant circuit. The graph showing impedance-vs-frequency for the two types of circuit is shown in Fig. 22-2. The curve for the series-resonant circuit shows us that it presents a high impedance at all frequencies removed from resonance, and a low impedance at the resonant frequency.

The behavior of the parallel resonant circuit is exactly the opposite. It presents a low impedance at frequencies that are removed from resonance and high impedance at the resonant frequency.

Note that, in both cases, the response falls off gradually as the frequency departs from resonance. The curve does not contain any really sharp edges. The relative sharpness, however, is a function of the circuit quality factor, or "*Q*." The *Q* of an *LC* tank circut is given by:

$$Q = F_r/BW = X_L/R$$

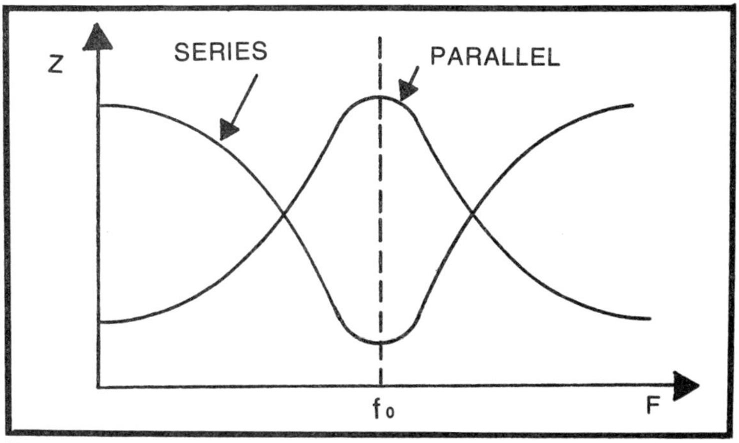

Fig. 22-2. Impedance-vs-frequency curve for series and parallel resonant circuits.

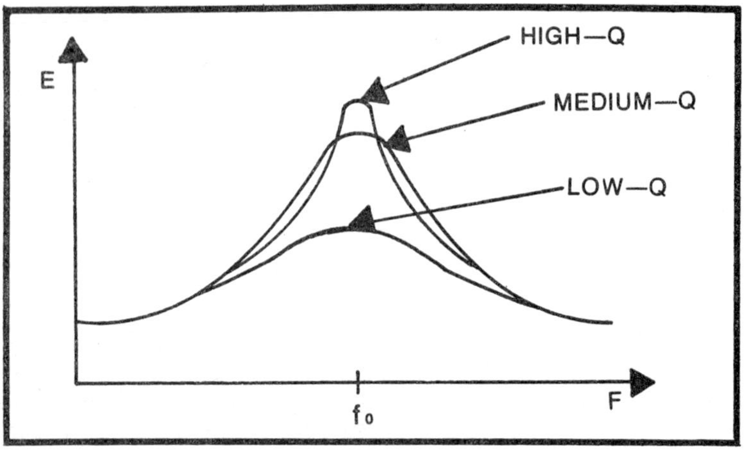

Fig. 22-3. High, medium, and low-Q responses.

Where:

Q is the quality factor

F_r is the resonant frequency

BW is the bandwidth in the same frequency units as F_r

X_L is the reactance of the inductor at F_r

R is any resistance in the circuit (the winding resistance of the inductor, or a lumped resistance).

Note that it is not critical which units of frequency you select, so long as both F_r and BW are in the *same* units.

Figure 22-3 shows the frequency response curves of a resonant *LC* circuit at various values of Q. Note that we are plotting the voltage across the circuit (parallel-tuned), *vs.* the frequency.

Note how the bandwidth of the circuit decreases as the Q increases. The sharper the tuning, then, the higher the Q factor. It is known that shunting resistance across the inductor will cause the Q to drop (remember, X_L/R), so one sometimes find a potentiometer in a tuned circuit, used to vary the bandwidth by varying the Q.

The moderate Q curve is marked with the point where the voltage has dropped to 70.7 percent of the value at the resonant frequency. This is known as the ***half-power point*** or the *– 6 dB point*. It is usual to specify the bandwidth in the Q formula as the difference in frequency between the points on the high and low sides of resonance at which the response drops off 6 dB.

Tuned transformers are used extensively in all forms of Amateur receivers and transmitters. In some cases, the primary and the secondary windings are resonated by a capacitor, while in other cases only one of the windings (usually the primary) will be resonated. Figure 22-4 shows several examples of tuned transformers.

The apparent Q of a tuned transformer is affected by the degree of coupling between the primary and secondary tank circuits. Figure 22-5 shows how we may obtain different values of Q by changing the coupling.

The curve in Fig. 22-5A shows the result of subcritical coupling between the primary and secondary tank circuits. In this condition, only a few of the flux lines from the primary cut across the secondary winding. Only those signals close to the resonant frequency have the strength to induce a secondary current.

The result of critical coupling is shown in Fig. 22-5B. In this case, we have more primary flux lines cutting the secondary winding, and the curve looks like a normal bell-shaped distribution.

When we overcritically couple the primary and secondary tank circuits, we obtain a response such as Fig. 22-5C. In this case, there is a dip in the middle of the passband. Most of the flux from the primary will cut across the secondary in an overcritically coupled transformer.

TESTING TUNED CIRCUITS

All resonant tank circuits are a little more sensitive to the methods used to test them than are untuned circuits. The problems become especially acute as the frequency increases.

The most critical problem is the loading of the tank circuit by the test equipment used to measure it. We discussed the problem of cable-capacitance loading at the beginning of this chapter, and found

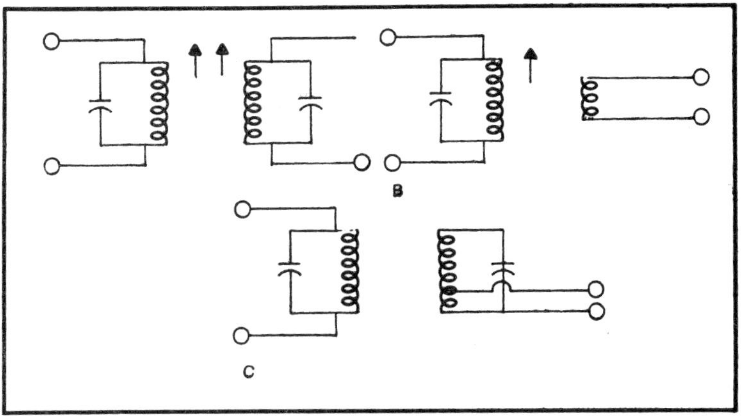

Fig. 22-4. Various types of tuned transformer.

that significant measurement errors resulted when the cable capacitance was a significant fraction of the total tank capacitance, and they became unbearable when the capacitance approached or exceeded the circuit capacitance.

A signal generator, and other test equipment, may be used only when the tank circuit is isolated properly from the instrument. In circuits where the tank is part of a cascade amplifier, for example, then we may sometimes use the ordinary input and output terminals of the amplifier to make the measurements. By injecting the signal at a point preceding the tank, and then measuring the response at a point further down the chain from the tank, then we may infer the properties of the tank from overall performance.

Sometimes, however, we may have to use a "gimmick" to excite the tank. Alternatively, we may use a very-low-capacitance probe, such as are often used for oscilloscopes and rf voltmeters. A "gimmick" is a short piece of wire connected to the "hot" lead from the rf signal generator. The "gimmick" is placed near the tank circuit inductor, and signal is thereby coupled by induction.

The response curve can be measured by plotting the voltage across the tank *vs.* frequency settings. We measure the voltage at certain frequencies above and below resonance, and then hand plot the curve on a piece of graph paper.

SWEEP GENERATOR

But there is a better technique . . . use a sweep generator.

A sweep-signal generator uses a voltage-controlled oscillator (VCO) driven by either a sawtooth or sinewave voltage. This will

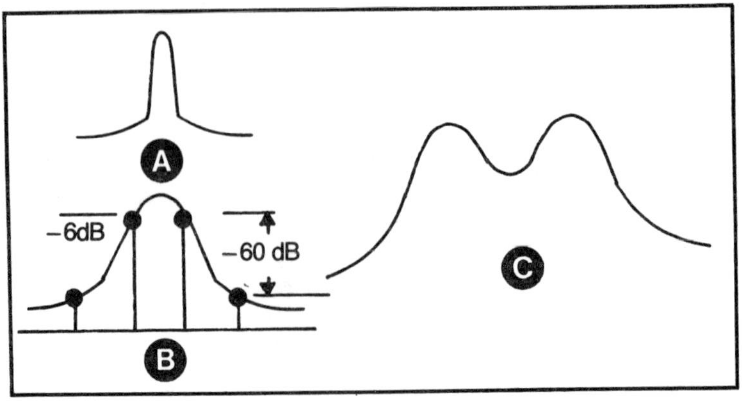

Fig. 22-5. (A), subcritical coupling; (B), critical coupling; and (C), overcritical coupling.

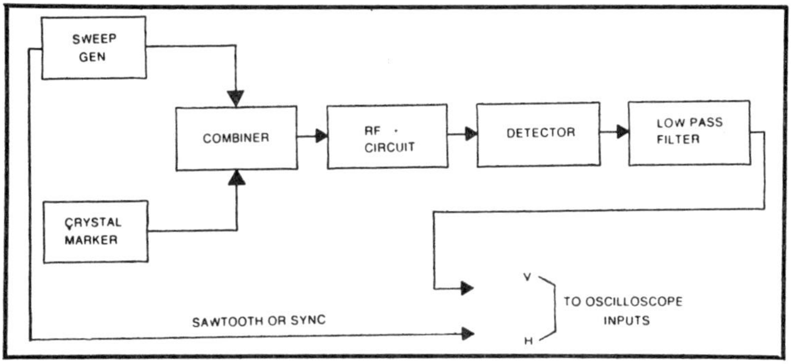

Fig. 22-6. Use of sweep generator and oscilloscope to view response of tuned circuit.

create a varying frequency that sweeps back and forth across the dial at a rate determined by the sawtooth or sinewave.

The test set-up for a sweeper is shown in Fig. 22-6. The signal from the sweep generator, and a signal from a crystal-controlled calibrator, are mixed in a network before being applied to the circuit under test. The output of the rf circuit is passed through a detector that rectifies is to DC, and a low-pass filter that provides smoothing for the pulsating DC produced by the rectification process. The sweep generator also controls the horizontal sweep of the oscilloscope by sending either the sawtooth itself (which becomes the horizontal deflection signal of the scope) or a synchronization pulse that starts the sweep of a triggered-sweep oscilloscope.

One problem, when using the sweep technique to test an rf circuit, is the matter of *ringing*. If the sweep rate is too high, then the circuit will behave as if a pulse were applied. It will, therefore, ring, or oscillate, at its resonant frequency. This is avoided by keeping the sweep rate low. In most cases, the sweep will be 60 Hz (or 50 Hz, in countries where the AC line frequency is 50 Hz.).

Other Methods

We may also use the dip oscillator to measure certain aspects of the tank circuit. Unfortunately, the dipper is often very inaccurate, so often produces erroneous readings. The best way to use a dipper, if you must, is to find the point on the dial at which the meter dips, and then measure the oscillator's frequency on a communications receiver.

At this time, you may also want to read Channel 26, which details antenna tests, and Channel 14, which discussed troubleshooting to oscillator circuits.

Chapter 23

Measuring Frequency and Period

Frequency is defined as the number of *events per unit of time*. In radio communications, we consider the proper units for frequency to be cycles (events) per second (unit of time). We define 1 *hertz* as 1 cycle per second.

We also recognize certain units based on multiples of 1 Hz, such as the *kilohertz* (kHz), which is 1000 Hz, and the megahertz (MHz), which is 1,000,000 Hz.

The basic period of time in all of our measurements is the second (s). But, seconds usually prove to be too large for most practical purposes, so we also use the *millisecond* (1 mS = 0.001 s), microsecond (1 μS = 0.000001 s), and the nanosecond (1 nS = 0.00000001 s).

The concept of *period* in electrical circuits assumes that the waveform is periodic, that is, it repeats itself at regular intervals. A sinewave, for example, can repeat itself quite regularly. Examine the AC waveform from a sinewave signal generator, or the AC mains power in your house or office. The period, then may be defined as the time interval between identical features on adjacent waves. Figure 23-1 shows a "squareish" waveform. The period is the time interval (T_2 - T_0), which are the points where the successive waveforms begin to go positive. We could measure the period from any other two points that could be readily identified as being the same on both waves (i.e., T_3 - T_1).

If the waveform were a sinewave, then the only time measurement, that we would make would be period. But if the

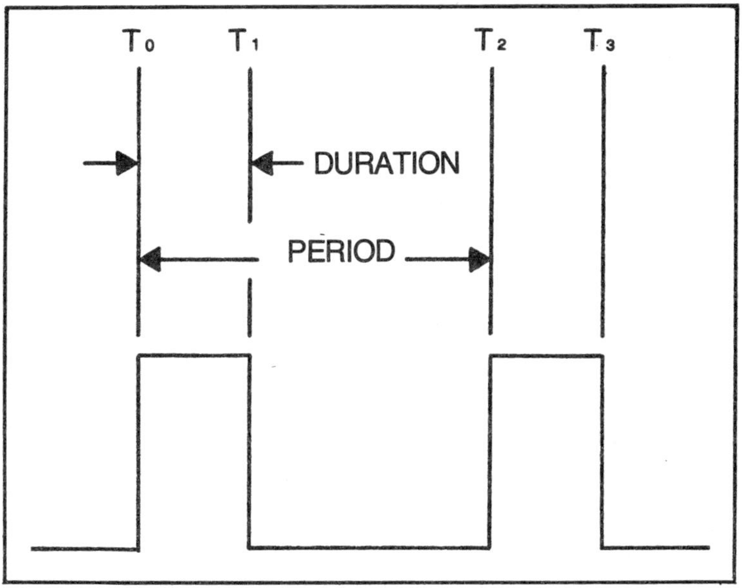

Fig. 23-1. Period and duration of a pulse.

waveform is a pulse, or a squarewave, then we may also be required to consider the *duration* of the wave. Duration is defined as the time during which the pulse is high, or in the case of Fig. 23-1, the time periods designated as (T_3 - T_2) and (T_1 - T_0).

Frequency and period are related to each other. Recall that frequency is events per unit of time, or cycles/second. Period (the "seconds" part) is the reciprocal of frequency, or seconds/cycle. We may calculate the period of a waveform by the following

$$P = 1/F$$

Where:

P is the period in seconds
F is the frequency in cycles per second (Hz)

PERIOD MEASUREMENT

The period is measured by comparing the event being tested-with a known time-base, or clock. We could, for example, use a digital frequency/period counter. In the period mode, the decimal counting circuit counts pulses from the time base, while the gate is turned on and off by the input waveform. The trick is to get the trigger circuits to respond to the same points on successive waveforms.

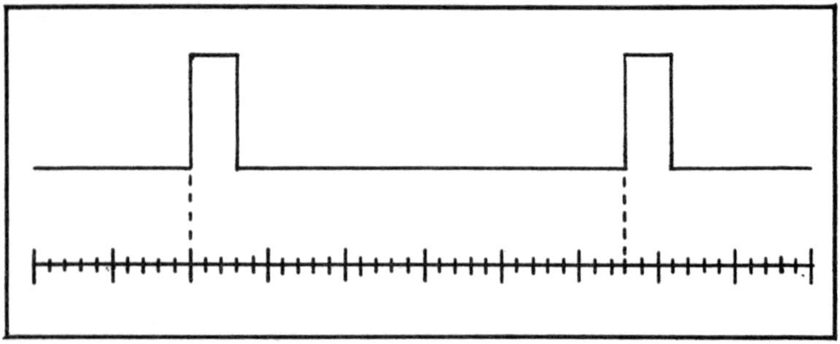

Fig. 23-2. Checking the period on a scope.

Alternatively, you could also use an oscilloscope time base to graphically display the information required. There are two different methods, although they are closely related. In the most common, we use the time base of the oscilloscope to make the meaurement. If the horizontal sweep control is calibrated in tie/division (such as mS/cm), then we may use this technique. In Fig. 23-2, we see a waveform that occupies 5.6 horizontal divisions. If we noted that the time base selector was set to 5 μS/div, then we could calculate the period as

$$\frac{5\ \mu S}{div} \times 5.6\ div = 28\ \mu S.$$

This technique is quick and simple, but is limited by the fact that most low-cost, or older high-cost, oscilloscopes do not possess a well-calibrated time base. Of course, the high-cost oscilloscope can be calibrated, but the calibration will be valid for only a few months.

A second technique is to use a time-base marker-generator to produce spikes on the scope screen. This is especially easy on two-channel oscilloscopes. The spikes can be produced by a differentiator circuit (*RC* high-pass filter) following a crystal-controlled squarewave oscillator. We may then compare the known time base (the spikes) and the unknown signal.

FREQUENCY MEASUREMENT

Amateur frequency measurements tend to be a little sloppy because, in part, the FCC does not require us to operate on specific frequencies (as CBers must). All that is required is that we remain inside of band. If we do not know our exact frequency, then we can get away with sloppy technique by avoiding the band edges altogether. For years, I lost a lot of good DX because I did not trust my VFO to locate the band edge, and did not have any other means of telling where it was.

Rough Methods

Several different methods are used to make approximate measurements of frequency. These may be used when we do not have any need for great accuracy. We might, for example, want to know the approximate frequency of a spurious output from our transmitter, so that we could tell whether it was a parasitic oscillation (that is, not related to the transmitter frequency), or a harmonic of the transmitter frequency. The troubleshooting technique would be different, so we must know.

For low frequencies, we could use the oscilloscope time base to measure the period, and then take its reciprocal to find the frequency. This technique is usually limited to measurements under 1 MHz, where the accuracy of the oscilloscope time base is sufficient.

Another useful method is to use *Lissajous* patterns on an oscilloscope. One input of the oscilloscope (vertical or horizontal) is connected to the unknown signal, while the other input is connected to the output of a calibrated signal generator, or other source. When the signal generator is adjusted to an integer multiple or submultiple of the unknown frequency (2F, 3F, 4F, 0.4F, 0.2F etc.), then a stable pattern will result on the CRT screen.

Figure 23-3 shows the test set-up needed to generate the Lissajous patterns. Once the pattern is locked in, and is stable, then we know that there is an integer relationship between the horizontal and vertical frequencies. This relationship is given as

$$\frac{F_v}{F_h} = \frac{N}{N}$$

There are, however, certain practical limitations on using Lissajous patterns for frequency measurements. One is the accu-

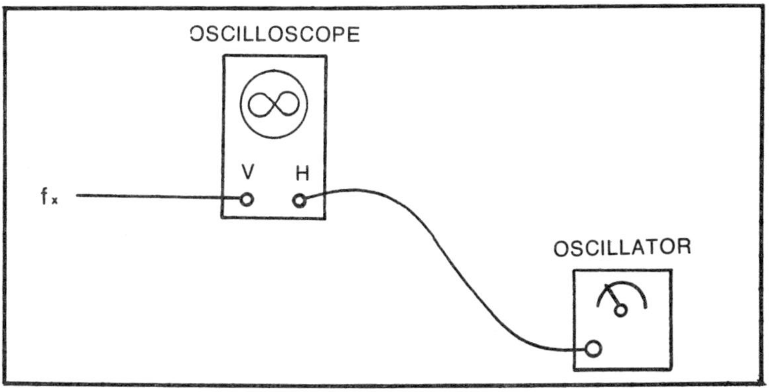

Fig. 23-3. Generating Lissajous patterns.

racy of the signal generator. If the oscillator is not well calibrated then it will be impossible to accurately use it to measure the frequency. Also, and this they don't always tell you in books, it becomes increasingly difficult to obtain a stable pattern as, a) the ratio of the two frequencies (F_v/F_h or F_h/F_v) becomes large, and, b) when the values of F_v and F_h becomes large. We might find, for example, a case where a 1:1 Lissajous pattern should result when two 14.356-MHz signals are applied to the inputs of the oscilloscope. But it is very difficult to hold either signal sufficiently stable to lock the pattern in tight.

If there is a difference in frequency the two inputs of the oscilloscope, then the pattern will rotate. The direction of the rotation tells us which frequency is higher, while the frequency of rotation tells us the amount of difference.

We can tell which direction of rotation indicates higher or lower frequency, by rocking the signal generator back and forth across the frequency at which a stable pattern is obtained. This is strictly empirical, and can be done in only a few seconds every time you make the measurement. This is better than trying to memorize a set of rules that had to take into account too many different factors!

The pattern will be locked in tight when the two frequencies bear an integer relationship to one another, but, at all other frequencies it will rotate. As the signal generator is adjusted to frequency closer and closer to the unknown, the speed of pattern rotation decreases. This is the visual equivalent of the zero beat.

An often overlooked (because it is too obvious, I suspect) technique for making rough frequency measurements is to examine the unknown signal on a communications receiver. Some modern receivers are capable of very good accuracy (100 Hz, or so), especially with a crystal calibrator and WWV. If the receiver is connected to an antenna then disconnect it, and run a separate feedline (piece of coax will do) to a "gimmick" that is coupled to the unknown signal. If there are no signals in the vicinity that will overload, or desensitize, the receiver, then it should be possible to make a descent frequency measurement using the receiver calibration.

If you are fortunate enough to have one of those high-cost, phase-locked-loop tuned communications receivers now on the market, then it might be possible for you to obtain ±10 Hz of resolution in your measurement. Most receivers, though, can only deliver 100 Hz, and some are so bad that 1 to 5 kHz is difficult to obtain.

Unless you have a professional receiver, or one that meets the same specifications as most better professional receivers, it will be best to use the receiver in conjunction with a crystal-marker oscillator that can be zero-beat to WWV or WWVH. These radio stations are operated by the National Bureau of Standards in Colorado and Hawaii, respectively. The frequencies are very accurate, and are 2.5, 5.0, 10.0, 15.0 and 20 MHz, although some changes are soon expected. The crystal marker oscillator can be zero beat to one of these frequencies. The oscillator signals can then be compared with signals on the receiver dial and the dial markings. More on this subject later in the chapter.

ABSORPTION WAVEMETERS

Figure 23-4 shows two types of *absorption wavemeter*. These instruments provide only the roughest indication of frequency, but are often useful for determining the band of operation, or whether a certain spurious signal is a harmonic or a parasitic.

Both types of absorption wavemeter use a tank circut that consists of a variable capacitor and a fixed inductor. The capacitor is equipped with a dial that is calibrated in units of frequency. The coil, in both cases, may be mounted external to the wavemeter housing, or may be internal and then coupled to the circuit under test via a short whip antenna.

It is necessary to mount the capacitor inside of a shielded metal housing so that interaction with nearby articles does not interfere with what accuracy that it might have. Even the operator's hand can detune the capacitor sufficiently to render the reading meaningless if this is not done. In most cases, the capacitor will be inside of the housing, and then interchangeable coils will be mounted on an insulated socket external to the case. The coil may then be coupled directly to the circuit under test.

The classic absorption wavemeter is shown in Fig. 23-4A. In this type of instrument, the inductor and the capacitor are connected in series with each other and an indicator lamp. The current circulating in this circuit will be greatest when the tank is tuned to the frequency of resonance, that is, when the tank-circuit frequency and the frequency being measured are the same.

Energy from the transmitter's tank can be coupled into the inductor of the wavemeter, and this will light the lamp. It takes a relatively large amount of signal to cause the lamp to light, so the technique is used mostly in transmitter work. Of course, a low-current lamp will make the instrument more sensitive, but this still does not allow it to measure small signals.

The lamp may be discharged altogether if the transmitter being tested has a plate or cathode milliammeter. The wavemeter can be coupled to the plate inductor of the stage being tested, and then tune through resonance. When the two tanks are tuned to exactly the same frequency, then the amount of rf energy absorbed by the wavemeter increases dramatically, and this will cause a brief deflection of the plate or cathode current meter.

A more sensitive version of this idea is the field-strength meter (a misnomer) of Fig. 23-4B. In this circuit, the tank of the wavemeter is used as a parallel-resonant circuit. Signal from either a coupling link, or the impedance-matching tap on the inductor (as shown) is coupled through a capacitor to the diode, where it is rectified. The pulsating DC waveform is filtered and smoothed by a second capacitor, so that a steady DC level is produced for the meter.

The milliammeter is deflected an amount that is roughly proportional to the strength of the applied signal. This wavemeter, therefore, is often used as a relative-field-strength meter (a better name). It is a good tuning aid when using certain types of radio transmitter.

No matter how well it is constructed, the absorption wavemeter and relative-field-strength meter is a crude device, capable of only very rough measurements. They are also severely limited in sensitivity.

We may use a dip oscillator to improve the sensitivity, although the accuracy may or may not be improved (depending upon model). In no instance is the accuracy of any dip instrument very good.

There are several types of dip instrument, characterized by the device used as the oscillating element, such as a vacuum tube, bipolar transistor, field-effect transistor, etc. They will be called grid-dip meters, base-dip meters, gate-dip meters, or, simply, dip meters.

All of these devices work on the same principle. If the inductor of the oscillator circuit is coupled to an inductor in another tuned circuit, then there will be an exchange of energy when the two circuits are tuned to the same frequency. When this occurs, energy from the oscillator is coupled to the external circuit being tested, and this causes a momentary dip in the grid (base, gate, etc.) current.

The dip oscillator offers the unique advantage that it can be used to test unenergized tank circuits, and this means it may be safer to use than other methods (in some cases). But, if we replace the meter in the grid (base, gate) circuit (s) with a set of earphones, then we may also use the dipper as an oscillating detector. In this

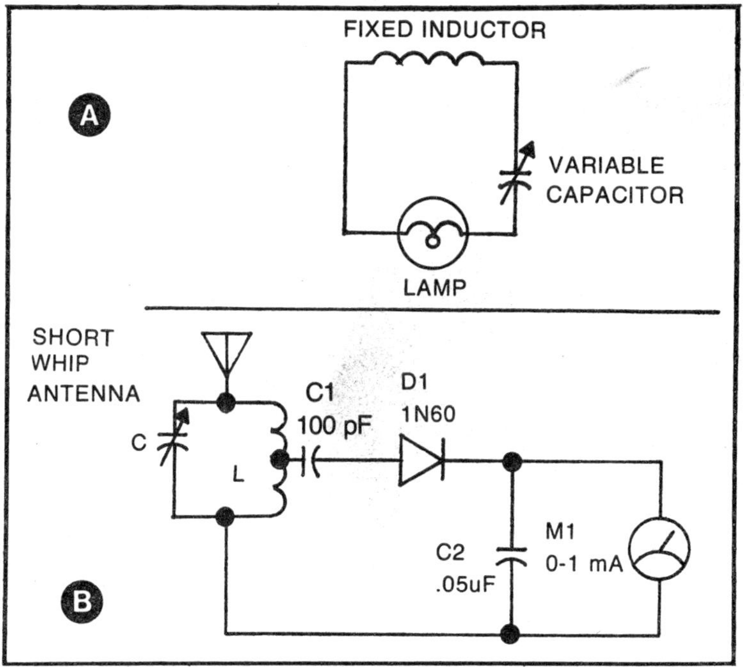

Fig. 23-4. (A), Simple wavemeter; (B), relative field strength meter.

mode, it will serve as an active absorption wavemeter that is capable of measuring the frequency of weak signals. When the dip oscillator is operating at a frequency near the frequency being measured, then a heterodyne beat is formed in the earphones. This note represents the difference in frequency between the oscillator and the unknown. When the oscillator frequency is adjusted so that this tone becomes lower and lower, and then finally disappears, then the dip oscillator frequency is the same as (zero beat with) the unknown frequency. If the dial calibration of the dipper is accurate, then we know the frequency of the unknown.

None of the techniques mentioned thus far will produce precision results. They are used only where great accuracy is not needed.

LECHER WIRES

Lecher wires are a mechanical device consisting of a section of open-wire transmission line that can be used to determine the approximate *wavelength* of signals in the 50 - 500 MHz region. At higher frequencies, a related technique called a slotted line is used to make the same kind of measurement.

A basic set of Lecher wires is made from a pair of copper wires (No. 12 or larger works best) mounted parallel to each other. The length of these wires should be at least one-half wavelength at the lowest frequency to be measured, and they should be multiples of this length, if convenient. The wires are spaced a few centimeters apart.

It is necessary to be able to measure length along the Lecher wires. So a ruler, or calibrated scale, is usually provided, and will be parallel and close to the wires. Most of these rulers are calibrated in metric units (meters, millimeters, centimeters), while some are calibrated in English units.

Some Lecher-wire sets also have an rf voltmeter consisting of a diode detector and a microammeter, as shown in Fig. 23-5. This is not strictly needed, however, if the source exciting the wires has a plate or cathode milliammeter.

A shorting bar is placed between the two conductors, and is designed so that it can be moved along the length of the conductors. This bar may be connected through a rack and pinion gear so that the operator need not touch it . . . a step that minimizes error.

This technique treats the conductors of the Lecher wires as if they were a transmission line shorted at the far end. The electrical length of the transmission line is the distance between the feedpoint and the short circuit.

Recall, from elementary transmission-line theory, that the impedance at the end of a transmission line is repeated back along the line every half wavelength. If this impedance is a short, as in the case of the Lecher wires, then the short will also appear at one-half wavelength intervals along the line.

If the Lecher wires are energized, and the bar is moved from the feedpoint (0 on the distance scale), then the voltage will be read until the short is at a distance equal to a half wavelength (i.e., 492/F). At that point, the voltage will dip, or there will be a deflection of the meter in the excitation source (such as a small transmitter).

The distance between the feedpoint, and the point where the first dip is noted, can then be used in the formulas

$$F(\text{MHz}) = 492/L(\text{ft})$$
$$F(\text{MHz}) = 300PL(\text{meters})$$

Note that the second formula can be changed to reflect the use of cm or mm by altering the constant in the numerator. In some cases, it is considered best to use a set of Lecher wires that is at least one wavelength long, and then measure the distance from zero to the

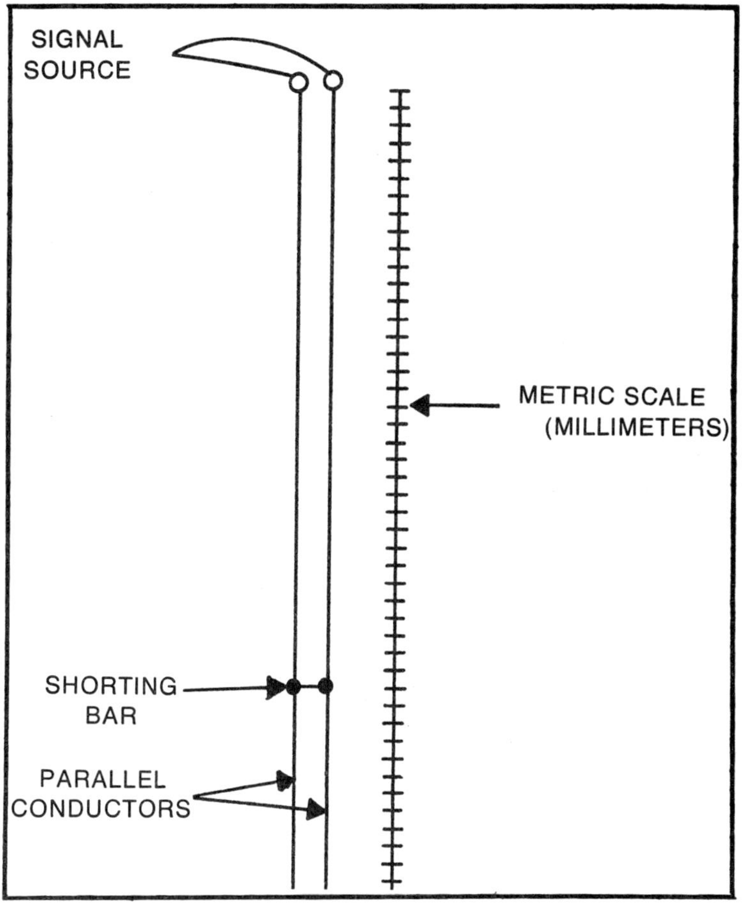

Fig. 23-5. Lecher wires.

first dip, and then average that distance with the distance from the first dip to the second dip. The averaged distance is then used in the above formulas.

The formula for use when the distances are measured in inches is

$$F(MHz) = 5906/L(in.)$$

It is reported that accuracies on the order of 1000 ppm are possible in the range 50 to 500 MHz, if good practice is followed, and the operator averages several measurements. Loose coupling to the source is also recommended for accuracy.

A technique similar to the Lecher is the slotted line of Fig. 23-6. A slotted line is another transmission-line section, but, in this case, for microwaves in the range 500-5000 MHz. The transmission line is constructed to have a slot along its length, into which an rf pick-up link and a detector can be inserted. A calibrated scale alongside the slot is also provided.

The output of the detector probe is a DC voltage that is proportional to the signal strength on the adjacent section of line. By using a ruler, or very often a micrometer dial, calibrated in metric units, we can measure the approximate wavelength of the signal applied to the line.

If the line is excited by the rf source, and the detector probe is moved along the slot, we will notice peaks and nulls at one-half-wavelength intervals along the line. If the distances at which these occur are noted, then we may use the standard formula to determine the excitation frequency.

In both Lecher wires and slotted lines, incidentally, the nulls (points of voltage minima, or nodes) are much more sharply defined than the peaks (points of voltage maxima, or antinodes).

WWV, WWVB, and WWVH

The National Bureau of Standards operates standard time and frequency radio stations of public use. One of these stations is WWV, located at Fort Collins, Colorado (near Boulder), and the other is WWVH located on the island of Kauai, Hawaii. These stations broadcast on 2.5, 5.0, 10.0, 15.0, 20.0 and 25.0 (WWV only) MHz. Check with NBS for any recent changes, reported to be imminent as of this writing.

The frequencies of these stations is maintained at ±2 parts in 10^{11} by an atomic-clock frequency standard.

Both stations also broadcast audio-range reference tones. A musical note, A-above-middle-C (440 Hz), is broadcast one minute past the hour by WWVH, and at two minutes past the hour by WWV. WWV broadcasts a 600-Hz tone during the odd minutes of every hour (i.e., 1, 3, 5, . . . 59), while WWVH broadcasts the 600-Hz tone is broadcast during the alternate minutes. Each audio tone has a duration of 45 seconds.

Voice announcements give the time in universal time (UT), and these are made every minute. The two stations are identifiable because WWV uses a male announcer, where WWVH uses a female announcer. A time-tick marker is broadcast continuously, except during the 29th and 59th seconds of each minute.

Geoalerts are broadcast during the 19th and 46th minutes of every hour, and are updated at 0400 UT every day. Voice radio-

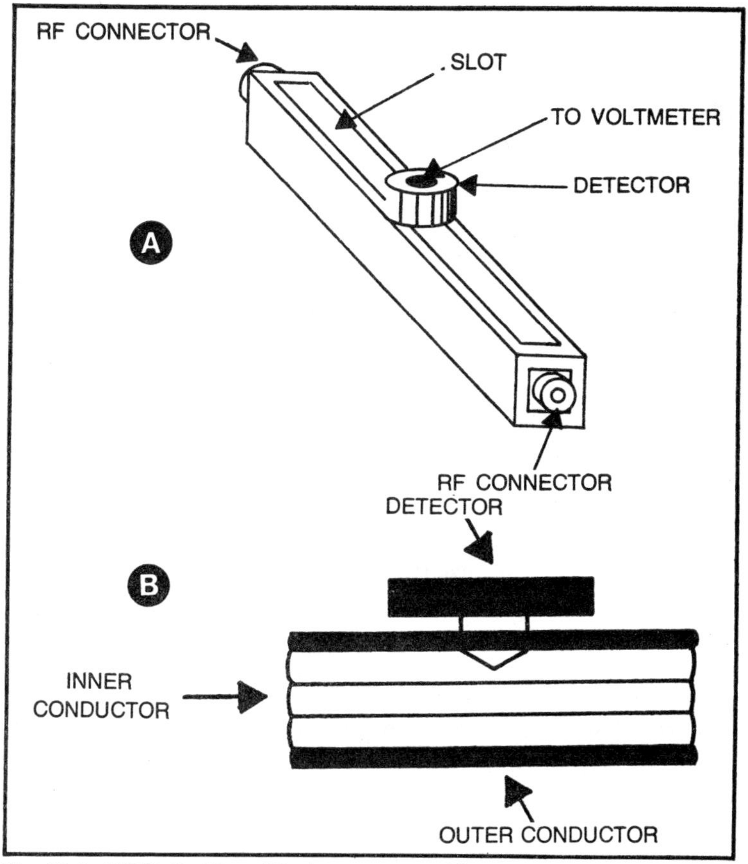

Fig. 23-6. (A), Slotted line; (B), detector assembly.

propogation predictions are broadcast during the 15th minute of every hour. These predictions are for high frequency propagation over the north-Atlantic path, considered to be one of the worst in the world.

NBS also operates a very precise 60-kHz radio station at the Fort Collins site. This station is WWVB. Besides frequency (a very accurate 60.00 KHz), the station also transmits a serially encoded binary-time signal giving UT.

Note that changes in the WWV/WWVB/WWVH services are published in *NBS Publication No. 236: NBS Frequency & Time Broadcast Services* (25 c from the superintendent of documents, Washington, D.C., 20402).

The NBS radio broadcasts may be used to compare local frequency standards, and then adjust them to best accuracy. The high-frequency WWV/WWVH transmissions are easily available to most Amateur Radio operators, and are often used to calibrate crystal-marker oscillators, counter time-base oscillators, and so forth.

Most Amateur crystal-marker-oscillators operate on 100 kHz or 1 MHz, although almost any frequency can be used if one of its harmonics falls on a WWV/WWVH frequency.

Adjustment of the oscillator is simple. It must, of course, be adjustable, which normally means a small trimmer capacitor in series or parallel with the crystal element. When both signals, the oscillator and WWV, are heard in the receiver at the same time, then an audio beat note is created equal to the difference between them. The trimmer capacitor in the crystal oscillator is then adjusted to minimize this difference, that is, create a zero-beat condition in the receiver.

The meaning of the term "zero beat" varies with different types of measurement apparatus, and betweeen different people. A normal adult can distinguish tones down to the 25 - 30 Hz range, but below that, they cannot be heard. In some people, however, hearing acuity suffers a little, and they are not able to recognize tones below 100 Hz. This limitation, and the fact that the audio sections of most receivers do not operate below 50 - 100 Hz (some –3 dB have a roll-off at 300 Hz!), keeps us from performing a zero beat accurate enough to maker frequency measurements with a high order of precision.

We may sometimes use the S-meter of a communications receiver as a zero-beat detector. The S-meter will bobble back and forth as zero beat is approached, and the frequency of this bobble is equal to the difference in frequency between the two signals. If the meter is bobbling at, say, 1 per second, then a 1-Hz difference exists . . . 100 times better than some people can detect aurally. Although it is practically impossible to obtain a true zero-beat note of 0 Hz, it is impossible to obtain a note that is less than 1 Hz.

The accuracy of the zero-beat process is improved considerably if we use the highest WWV or WWVH frequency possible. Look for the highest frequency that produces a strong signal in your area. This is so important that a precision calibration should wait until such a time that a high frequncy is able to produce a strong WWV signal in your receiver.

Let's consider an example of using the highest WWV frequency to make the zero-beat adjustment. Assume that we are going to adjust the 100-kHz calibrator in a communications receiver.

If we use the 2.5 MHz WWV signal, and the detector is capable of zero beat to within 1 Hz, then the error is divided by the same factor as harmonic is used. That is, 2500 kHz is the 25th harmonic of 100 kHz, so the error is 1 Hz/25, or 0.04 Hz. But, what is we had used the 20-MHz WWV signal? This frequency is the 200th harmonic of the 100-KHz signal, so the error is 1 Hz/200, or 0.005 Hz. Clearly, the adjustment is much closer when the higher harmonic is used.

The 60-kHz signal from WWVB is considered better for making frequency measurements than the high frequency WWV/WWVH signals. This may seem in contradiction to my earlier statement about using the highest available frequency, but it is not. The 60-kHz signal is less susceptible to fading, atmospheric gyrations, and phase-shift errors than are the errors than are the high frequency signals. The difference is so great that many professional metrology laboratories use 60-kHz WWVB comparator receivers to make measurements.

Figure 23-7 shows a sample WWVB comparator-receiver block diagram. The antenna is a shielded loop, usually in the neighborhood of 1 meter in diameter, although smaller loops have been used. In most cases, the loop antenna is resonated to 60 kHz with a variable capacitor . . . its own inductance being the *L* portion of the *LC* network. A high gain (50 - 100 dB), 60 -kHz preamplifier is usually mounted right at the feedpoint of the loop antenna, and the preamp output signal is passed to the rest of the receiver via coaxial cable. The output of this amplifier, or possibly another amplifier inside of the main receiver unit, is made available to the user via a front panel connector.

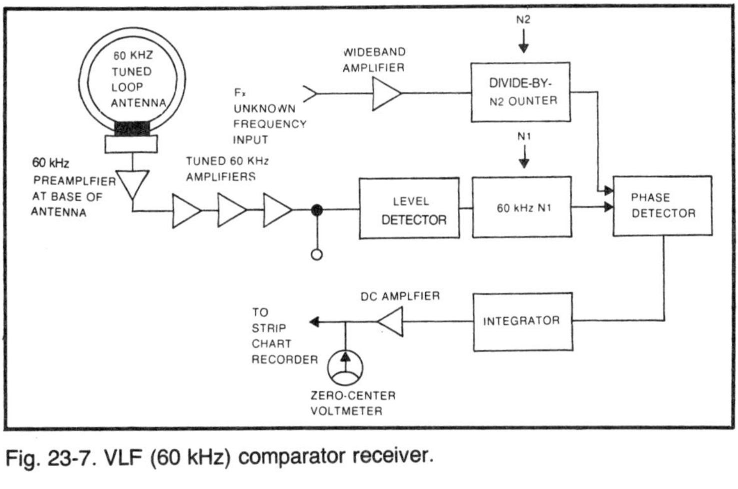

Fig. 23-7. VLF (60 kHz) comparator receiver.

Following the 60-kHz amplifier stages will be a level detector that will produce a squarewave output from the 60-kHz signal. This signal will then be used to drive a phase detector. In older model receivers, no level detector was needed because the phase detector was an analog *LC* circuit. But in modern designs, a digital phase detector is used, and these use squarewave inputs.

The frequency being measured is passed throgh a divide-by-N digital counter before being applied to the other input of the phase detector. This counter produces an output frequency of F_x/N. In some models, there is also a divide-by-N counter between the squarer circuit and the phase detector.

The phase detector is actually a coincidence detector; when the frequency and phase of the two signals are equal, then the detector output is zero. But, if there is any difference, then the output will be a string of constant-amplitude pulses whose durations vary by the frequency difference. Thse pulses can then be integrated, to be displayed on a voltmeter or strip-chart recorder.

The unknown frequency can be measured by adjusting the divide-by-N counters until there is zero voltage at the output of the integrator. The frequency of the unknown signal is found from

$$F_x \text{ (Hz)} = \frac{(N_2)\ (60{,}000)}{(N_1)}$$

Where:

F_x is the unknown frequency in hertz

N_1 is the division ratio applied to the 60 kHz signal

N_2 is the division ratio applied to the unknown signal, F_x

The use of comparator receivers has not been common among hams for several reasons. One, of course, has been expense. Until recently, when industrial surplus has become available, they were simply too costly. There are also pretty hard to build. But, today, we are a little better off. We can buy integrated circuits for most of the functional blocks in Fig. 23-7 (such as wideband amplifiers, divide-by-N counters, level detector or voltage comparators, and even tuned amplifiers. The 60-kHz amplifier may easily be a series of IC amplifiers with *RC* or *LC* 60-kHz tuning). Interestingly enough, most of the magazine-article construction projects using WWVB have used the binary-coded time signal to synchronize electronic clocks, rather than frequency-measurement applications.

ATOMIC FREQUENCY & TIME STANDARDS

Until very recently we have used the rotation of the earth on its axis as the measure of time. Frequency could then be measured by having a signal run a synchronous electrical clock, so that, by observing how much time the clock gained or lost, we could reckon the error in frequency.

However, it turns out that the earth's rotation is not the constant that we once believed it to be. In fact, it has been found that there are certain anomalies due to the waxing and waning of the polar-ice caps as the season change!

In 1964, the *12th General Conference on Weights and Measures* redefined time and frequency in terms of the atomic clock. If the clock contains a phase-locked-loop, and the reference oscillator is a rubidium-gas-cell device, or an atomic beam of thallium or cesium, then the oscillator will be locked to a very stable and very precise frequency. The atomic clock can lock a 5- or 10-MHz oscillator to ± 7×10^{-12}.

BASIC FREQUENCY METERS

There are two common forms of frequency meter in use: comparison and digital counter. Although there are several different types in each class, there are some similarities between them.

Figure 23-8 shows the elementary heterodyne frequency meter. In all heteodyne techniques, the unknown signal is mixed with a known signal from a precisely calibrated frequency source. The frequency of the beat note produced is equal to the difference in frequency between the known and unknown signals. If the known source is zero-beat with the unknown signal, then the unknown will be equal to either the known frequency or one of its harmonics (or subharmonics).

Details of frequency-counter construction are given in another chapter, so will not be repeated here.

Using Frequency Meters

In the sections to follow we will demonstrate common techniques used to actually measure frequency. Most of these

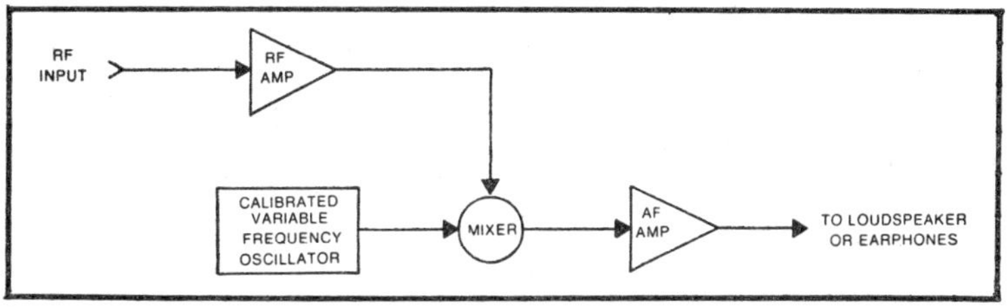

Fig. 23-8. Heterodyne frequency meter.

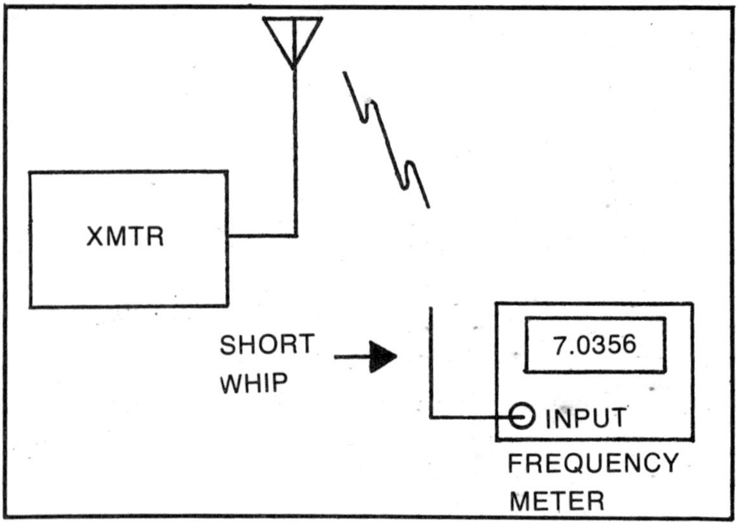

Fig. 23-9. Off-the -air use of a frequency meter.

techniques are still in common use, even though the digital frequency counter has begun to take precedence over the other methods as costs come down. In the past, actually only a decade or so ago, the cost of frequency counters was much greater. A model suitable for use by a UHF two-way radio (commercial) service shop easily cost 5000 - $8000, but today, instruments with similar specifications cost less than $1000.

The simplest method for checking the output frequency of a radio transmitter with a digital frequency counter is shown in Fig. 23-9. A short whip antenna is connected to the input of the counter through a short piece of coaxial cable, or through a coaxial connector mounted directly on the base of the whip. The transmitter is then turned on, and the radiated signal is picked up by the whip antenna. If the counter is close to the transmitter or antenna, then we may expect sufficient signal to be developed in the counter input circuit to allow counting. The transmitter's frequency will then be displayed on the counter readout.

Note that modulation on the signal will probably ruin the count, so one is cautioned against using modulated signal. Also, SSB transmitters will not produce any signal at all unless modulation is applied, so it is advisable to use the CW mode when testing SSB transmitters.

It is also worthy of note that this method may well be illegal, especially if you key up the transmitter without first listening to ascertain whether or not the frequency is in use. In commercial work, it is illegal regardless of the activity on the channel.

To prevent interference to other stations, and possible implications of illegal operation for yourself, it is best to make the above test while the transmitter is connected to a dummy load instead of the antenna. In most low-cost dummy loads, there is sufficient leakage around the rf-tight seals to allow the counter to pick up a signal, without radiating it all over the county.

Another approach is shown in Fig. 23-10. Again, the transmitter output power is applied to a dummy load instead of an antenna. A coaxial "tee" connector and a 50-ohm attenuator feeds a small sample of the transmitter output power to the counter input. It is extremely dangerous to feed the transmitter output *directly* to the counter input, and damage could (usually *will*) result. There is only one exception to this rule, and we will deal with it in a moment.

The purpose of the attenuator in Fig. 23-10 is to reduce the transmitter output power to a level that is safe for the counter. The amount of attenuation, in dB, will depend upon the output power of your transmitter. The counter specifications in the manual will tell you how much input signal the counter will tolerate, and you must provide an attenuator sufficient to reduce the output of the transmitter to this level.

Caution *Never feed the output of a transmitter, of however low a power level, directly to the input of a frequency counter unless told to do so by the manufacturer of the counter.*

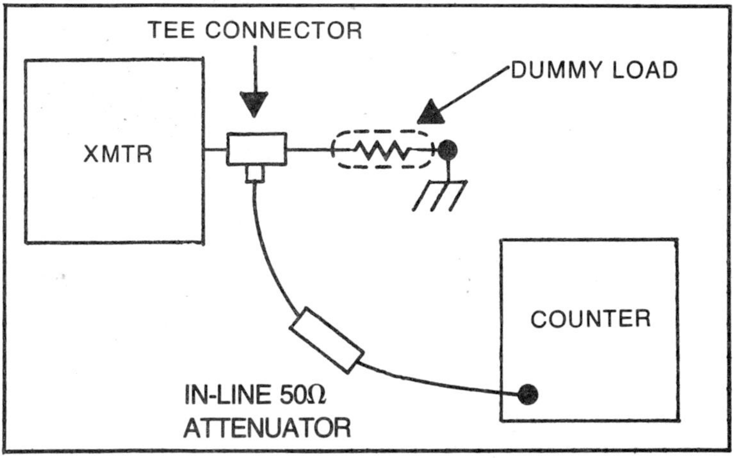

Fig. 23-10. Proper way to measure transmitter frequency causes no unnecessary interference.

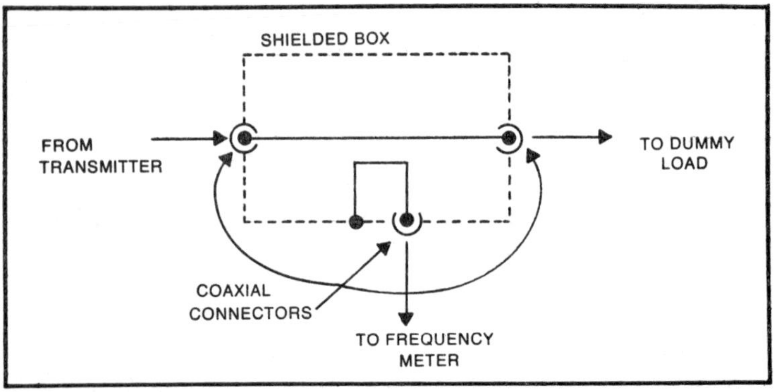

Fig. 23-11. RF pick-off box.

Those counters that allow you to connect the input directly to the output of a transmitter usually have a through-line type of rf sampler, as shown in Fig. 23-11. The rf current, passing through the center conductor of the pick-off unit, induces a small current into the loop, and this current is the actual counter input signal. Most of these types of frequency counter can be identified by the existence of two coaxial antenna connectors, one marked "transmitter" and the other marked either "dummy load" or "antenna."

Transfer Oscillator Techniques

A transfer oscillator is a stable, precisely calibrated low-frequency oscillator that can be loosely coupled to a communications receiver antenna circuit. Usually, the coupling is in the form of a small-value capacitor or a wire "gimmick." A gimmick is a few turns of wire wrapped around the antenna lead wire.

There are actually several different ways that a transfer oscillator can be used to measure the frequency of a transmitter. It is assumed in all of them, however, that we know the approximate frequency of the transmitter (i.e., that it is a 40-meter signal and is in the lower end of the CW portion of the band).

One of the most common transfer oscillator techniques is to use the oscillator on a subharmonic of the transmitter frequency. The output frequency of the transfer oscillator is adjusted until it is zero beat with the transmitter frequency, using the communications receiver as the indicator.

When using the transfer oscillator in the manner just described, it is important to know exactly which harmonic is being used. We can then multiply the know frequency of the transfer oscillator by the harmonic being used to find the transmitter output frequency.

In most cases, the transfer oscillator methods are limited to relatively low frequencies (i.e., the Amateur high-frequency bands) because, at VHF or UHF, the behavior of suitable transfer oscillators, at Amateur price levels, declines markedly.

The accuracy of the transfer oscillator in the higher end of the HF spectrum can be improved by using a method such as Fig. 23-12. In this case the output signal from the transfer oscillator is fed to the receiver through a switch. When the switch is in one position, the signal is applied to the receiver, where it is zero beat with the transmitter signal. Once the zero beat is obtained, the switch is changed, and the transfer oscillator signal is applied to the input of a digital frequency counter.

Once again, however, it is required that we know which harmonic is being heard in the receiver. After all, if the counter were capable of reading the same frequency as the transmitter, then we would use it directly and forget about that silly transfer oscillator!

In any transfer oscillator technique, if the subharmonic is too low, then we might make a serious error. In many cases, if the 20th, or higher, harmonic is used, then it may well be the case that *two* harmonics will fall close together on the receiver dial, and we may zero beat the wrong one!

One advantage of the transfer oscillator technique, even when the counter is capable of measuring the transmitter frequency

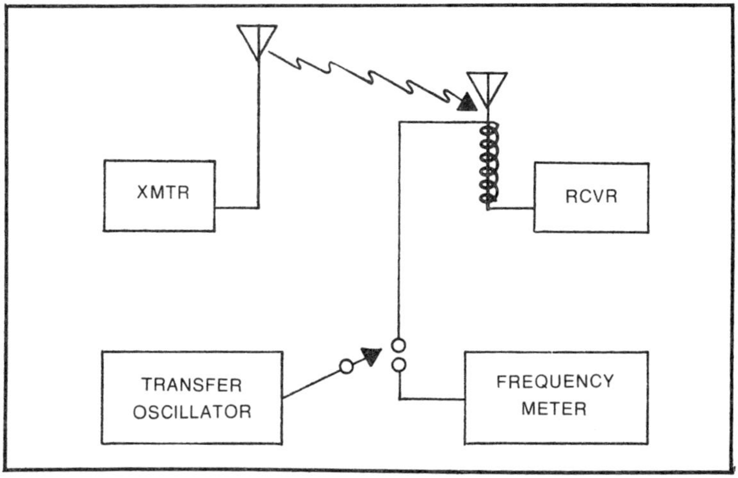

Fig. 23-12. Use of a transfer oscillator to measure frequency.

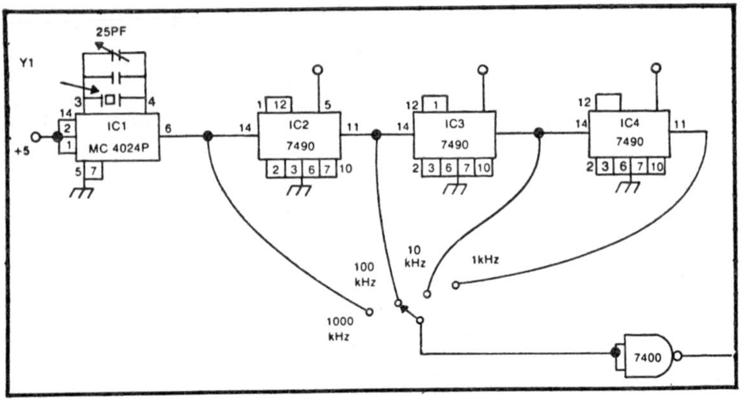

Fig. 23-13. Simple crystal marker-oscillator.

directly, is that it can be used to measure the frequency of a distant transmitter . . . not just your own transmitter, or a transmitter on the work bench.

Crystal Oscillator Markers

A technique that is related to the transfer oscillator is the use of a crystal oscillator as a marker. This allows us to calibrate specific points on the dial of a communications receiver.

The crystal oscillator used for this purpose should be a precision oscillator, and should be on a standard frequency (such as 100 kHz, 1 MHz, etc) or produce outputs on several harmonically related standard frequencies. The oscillator circuit chosen should not be a sinewave type, but nonsinusoidal to ensure that the output waveform is rich in harmonics. Examples of suitable oscillators using TTL integrated circuits shown in Figs. 23-13 and 23-14. These circuits can be used for moderately precise applications, including most Amateur Radio applications. However, where real precision is needed an oven oscillator or a temperature-compensated crystal-oscillator (TCXO) module should be used.

Figure 23-15 shows us how to use a crystal oscillator in conjunction with a communications receiver to measure the frequency of a transmitter. Note the similarity to the transfer-oscillator technique.

There are actually two different methods for using a crystal marker-oscillator in this application. The first, and least accurate, is to use the oscillator's signal to temporarily calibrate the receiver's dial. For example, suppose we want to calibrate the dial to the low end of the 20-meter band, 14.00 MHz. We would turn on a 1-MHz

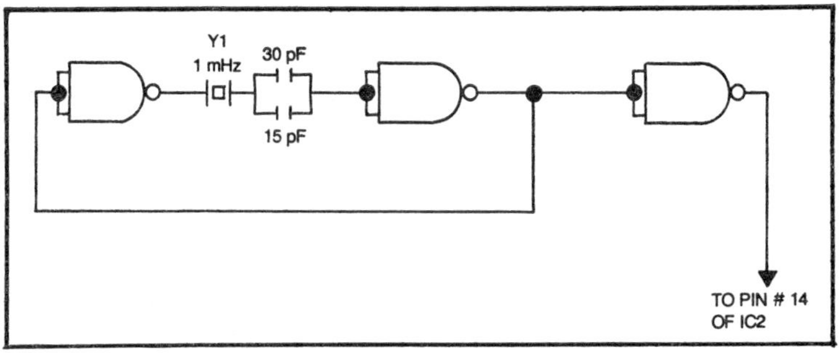

Fig. 23-14. Alternative crystal-oscillator section.

crystal calibrator, and tune the receiver in the vicinity of 14 MHz. We would then locate the signal, and tune the receiver to it (as indicated by the peak of the S-meter reading). If the receiver has a movable cursor (most good ones do), then we could move the cursor over the "14.00" mark. For frequencies close to 14.00 MHz (i.e. ± 25 kHz) we are then reasonably sure of close calibration.

This method is usually sufficient to keep us out of trouble with the FCC, ensuring that we do not operate out of band. But, it is not terribly good for accurate measurement of operating frequency. Most communications receivers have a frequency resolution of not better than 100 Hz, and many cannot resolve closer than 1 kHz. Only a few high-cost, commercial and military receivers are able to resolve 10 Hz.

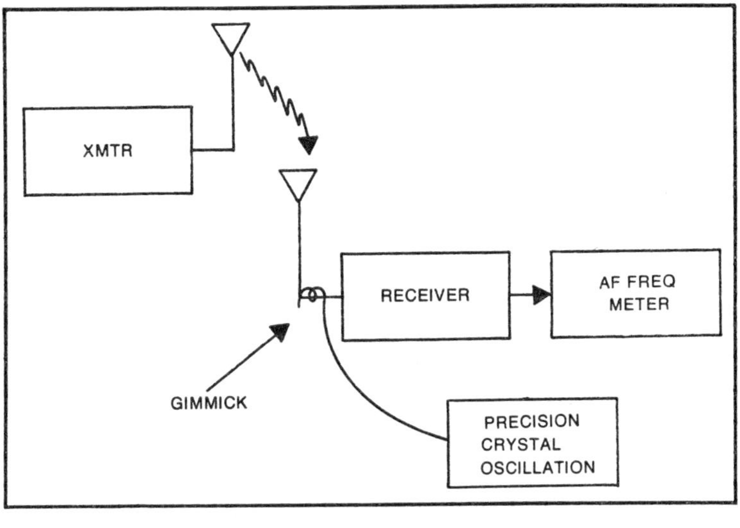

Fig. 23-15. Using a crystal marker with a receiver to measure frequency.

It is usually better to allow the crystal oscillator to heterodyne against the unknown signal, and then measure the frequency of the beat note. This beat note will be equal to the difference in frequency between the unknown and the harmonic of the calibrator. An audio frequency meter, or a low-frequency counter, can be connected to the receiver's audio output jack, or across the loudspeaker terminals. If the counter input is connected to the output of the detector, before any AC coupling capacitors are used, then a near-DC beat note can be obtained for better accuracy. First calibrate the receiver in the same manner as in the previous example, making sure that the calibrator points are close to the frequency of the unknown. Consider the setup in Fig. 23-14. It is first noted, when the 1 MHz frequency is used, that the signal falls between 7 and 8 MHz. When 100-kHz markers are used, it is then found that the unknown signal is between 7.1 and 7.2 MHz. Next, we set the crystal marker oscillator to 10 kHz, and note that an audio-frequency beat note is heard. When this tone is measured,it is found to be 3562 Hz. Since the beat note is located closest to the 7180-kHz mark on the receiver dial, we know that the actual frequency will be 7180 kHz + 3562 Hz, or 7183.562 kHz. It is considered good practice to measure the frequency three times, and then average the results. This will allow us to make a more reasonable measurement in the long run.

The calibration of the crystal marker oscillator should be checked against the N.B.S. standard frequency broadcasts of WWV and WWVH. Note that aural zone beats can be as much as 100 Hz off, even in persons who do not believe themselves to have a hearing loss. Use the S-meter to observe the zero-beat point.

Also, another point to be aware of is that we may very well have two beat notes at the same time. If, in the case above, the beat note had been in the neighborhood of 5 kHz, then we would be unsure of whether it was beating against a 7180 or a 7190 kHz signal. In those cases, go to a lower marker frequency (such as 5000 or 1000 Hz).

Chapter 24
Receiver Measurements

When we break the radio receiver down to its fundamentals, we find that all of them are pretty much alike. Very few, if any, Amateur receivers used in serious communications are of any design other than the old reliable superheterodyne. While a few construction projects, and one simple receiver offered on the commercial market, have used the direct-conversion technique, no one really takes them seriously when choosing the main station receiver. Similarly, there have been certain tuned-radio-frequency (TRF), and regenerative receivers offered in the construction articles of some magazines. But, again, no one would seriously consider these designs for their main station receiver.

The job of any radio receiver, whether for communications or broadcast reception, is to intercept the signal sent by the transmitter, and then demodulate it to recover the audio or other information impressed on the carrier at the transmitter.

The block diagram of a double-conversion superheterodyne receiver is shown in Fig. 24-1. We will consider this a "typical" Amateur receiver, even though we are mindful that single- and triple-conversion receivers are also popular.

One reason for the popularity of the superheterodyne is its ability to receive signals better than the other types. A principle reason for this is that the rf signal is converted to a lower frequency, called the intermediate frequency. In older designs, the I-F was always lower than the signal frequency. This was because lower-frequency amplifiers, especially high-gain amplifiers such as the I-F amplifier, were easier to tame. A standard intermediate frequency of 455 kHz was adopted for A-M broadcast receivers, and communications receivers up to about 10 MHz. When higher frequencies are used, then it is sometimes better to go to a double-conversion scheme, using two i-fs, one at a high frequency and the other at the lower frequency (50 to 500 kHz).

Regardless of the design of receivers, however, there are certain measurements that apply to all of them. We are, for example, concerned with parameters such as sensitivity, selectivity, image rejection (common only to superheterodynes) and noise figure.

Sensitivity is the measure of a receiver's ability to pick up, and demodulate in a usable form, signals from the antenna. Sensitivity has been specified in many different ways over the years, some of them different due to differing engineering viewpoints, and others different in order to deceive.

Two common methods for defining sensitivity seem appropriate: signal level required to produce a specified output under fixed conditions, and the "quieting" method.

In the first method, the manufacturer will specify how much signal amplitude is required to produce a given audio-output level. In most cases, the specification will be the signal level at the antenna terminals (in microvolts), with 30-percent modulation applied (400 Hz), that will develop an audio output of 500 milliwatts (some use 1 watt, but many good Amateur receivers cannot produce 1 watt of audio output under any circumstances).

The alternate form of sensitivity specification requires us to determine the amount of unmodulated signal at the antenna terminals required to produce a specified level of quieting (usually 10 or 20 dB) of the hiss-like background noise.

Selectivity is the quality of a receiver that allows the rejection of unwanted signals. A receiver should not look at the entire radio spectrum, but only a narrow window around the specific frequency containing the station of interest. Selectivity is defined in terms of bandwidth, and ideally, the selectivity of the receiver should only just match the bandwidth of the transmitted signal.

An image is a spurious response that is unique to superheterodyne receivers. Regenerative, and direct-conversion, receivers do not produce images. The cause of the image is the heterodyning process itself. The I-F amplifier is looking for a signal that is the difference between the local oscillator (LO) and the rf signal. Consequently, the I-F amplifier does not care whether the rf signal is above or below the LO—the difference will be the same. If

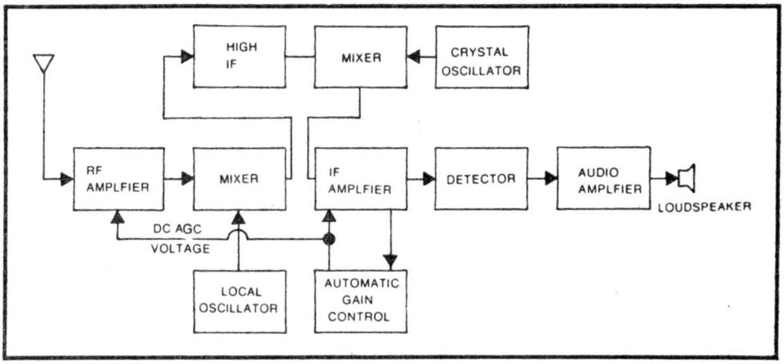

Fig. 24-1. Block diagram of a double-conversion superheterodyne receiver.

the LO is higher than the rf, then the I-F is equal to LO minus rf, and if the LO is lower than the rf, the I-F is equal to the rf minus LO.

Let us suppose that we have a receiver in which the LO is operated at frequencies above the rf, so the I-F will be equal to LO minus rf. But, we also find that any strong signal located at a frequency of LO + I-F *also* produces a difference frequency equal to the I-F!

The image rejection of the receiver is a measure of its ability to remain free from interference by image signals.

Noise, in any receiver (in the form of background hiss), is due mostly to thermal agitation within the various components of the receiver—especially in the front-end. The noise figure, in dB, is the measure of this noise.

MEASURING SENSITIVITY

The test equipment set-up for making both types of sensitivity measurement is shown in Fig. 24-2. A calibrated-output, rf-signal generator is required; uncalibrated and poorly calibrated signal sources need not apply! We also need a voltmeter across the output load (which could be a loudspeaker, if you like punishment, or a load resistor).

A dummy antenna (see Fig. 24-2B) is used between the output of the signal generator and the antenna terminals of the receiver. This network is used to simulate an infinite length transmission line.

Audio Power Method

The audio power method for measuring receiver sensitivity is used on A-M and SSB receivers. The manufacturer will specify the percentage of modulation for the input signal. Although it may be

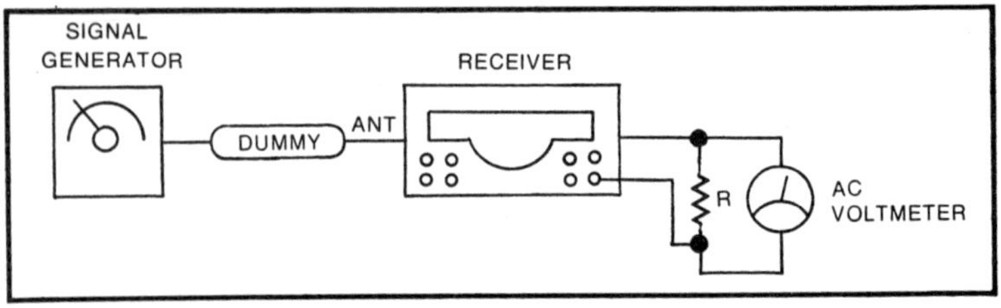

Fig. 24-2A. Alignment/measurement set-up.

anything between 25 and 100 percent, most manufacturers specify 30 percent. Also, the frequency of the modulating signal is specified, usually as 400 Hz, but in some cases 1000 Hz is used.

The audio output power is determined by measuring the rms output voltage across the load, and then applying the relationship $P = E^2/R$. Since we will usually have the manufacturers specifications as to the output power level and the load resistance, we will want to know the rms voltage required to meet those specs. This is found by algebraically rearranging the above expression to $E = (PR)^{1/2}$. In the case of the standard specification of 500 milliwatts, we find that the output voltage will be 2 volts into an 8-ohm load, and 1.4 volts into a 4-ohm load.

If the receiver has an automatic gain control (agc), and most receivers do, then this must be disabled before a proper sensitivity determination can be made. Fortunately, most Amateur receivers have a switch that allows us to turn off the agc, but, if not, the agc will have to be disabled in some other way. In many vacuum-tube receivers, it was merely necessary to ground the agc line. In a lot of transistor models, this will destroy part of the circuit, and the manufacturer requires that a voltage be used to clamp the agc off. If there is not an "agc-on-off" control on your receiver, then consult the manufacturer's service literature for instructions on how to disable the agc.

This procedure for this measurement is simple:

1. Determine the output voltage that will produce 500 mW in the loudspeaker, or in an equal value load resistance.
2. Connect the equipment according to Fig. 24-2.
3. Advance the output control of the signal generator from zero until the voltage determined in step 1 is obtained.
4. Read the sensitivity from the calibrated output dial, or the output meter, on the signal generator.

This measurement does not always tell us a great deal about the receiver, but will do two things: a) allow comparisons between receivers, and , b) allow us to tell whether the receiver has suffered any significant loss of sensitivity.

It is my belief that alignment of the receiver should not be done on a periodic basis . . . unless there is evidence that the receiver *needs* alignment. Time, especially only a year or two, does not cause alignment to shift significantly in a quality receiver! The use of the sensitivity measurement will allow you to ascertain whether alignment is needed, or whether such a job would cause unneeded wear and tear on the tuning adjustments.

Quieting Method

The quieting method for determining sensitivity also uses the same equipment set-up, as shown in Fig. 24-2A, on this case, however, the signal generator must be unmodulated.

The quieting method is used primarily on FM receivers, but has been seen on A-M, SSB and CW models as well.

1. Set the signal-generator output control to zero. As in all sensitivity measurements, turn the rf gain and audio volume controls to maximum. Turn off, or disable, the agc.
2. Measure the rms voltage across the load. The output signal at this point will be background hiss.
3. Increase the output level of the signal generator until the noise level on the output meter *drops* 10 dB (some manufacturers use 20 dB) from its zero-signal level.
4. Read the sensitivity from the calibrated output dial, or the output meter, on the signal generator.

If an AC voltmeter is used for the output meter, then the scale may well be calibrated in dB, but, if not, then use the standard dB formula:

$$dB = 20 \, Log_{10} \frac{V_i}{V_f}$$

Where:

dB denotes decibels

V_i is the initial, or no-signal, voltage

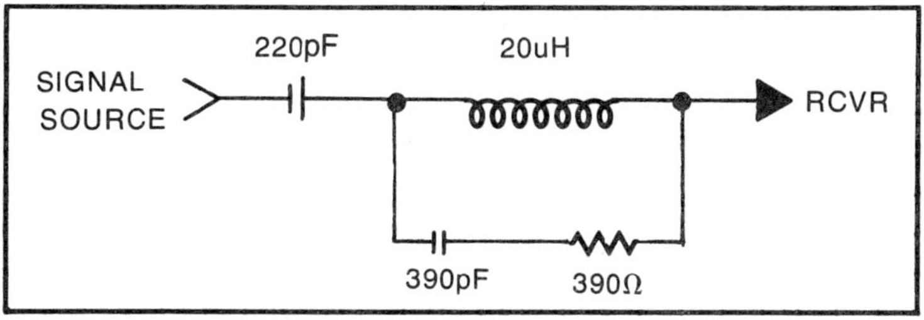

Fig. 24-2B. Dummy antenna.

V_f is the final voltage

It is necessary to ensure that only the signal from the generator is measured in these two tests. If any stations are received, then their signal is added to the signal from the generator, and that will invalidate the measurement. In most cases, if you use good shielded coaxial cables, there will be little problem. Problems are a lot more likely to occur if the receiver input uses a pair of screw terminals instead of a coaxial-cable connector. In really noisy areas, or when the receiver is poorly shielded, a true measure of sensitivity is impossible outside of a screened room.

An alternative method, used on A-M receivers, is to measure the amount of signal required to produce a signal that is 10 (or 20) dB above the background noise level:

1. Connect the equipment as in Fig. 24-2A.
2. Set the signal generator output to zero, and the receiver rf gain control to maximum. Disable, the agc.
3. Adjust the audio-gain, or volume, control for a convenient output reading (this will be a noise-level reading). If the output meter has an AC dB scale, then adjust the volume control until the meter reads 0 dB (alternatively, 0 VU).
4. Adjust the signal generator for 30 percent modulation by 400 Hz.
5. Adjust the signal-generator output level until the receiver output level rises 10 (or 20) dB. If the output meter does not have a dB scale, then calculate the level that is a 10-dB increase from the formula given earlier.
6. Read the sensitivity from the calibrated output dial, or output meter, on the signal generator.

SELECTIVITY MEASUREMENTS

Selectivity refers to the bandwidth of the receiver, and is a measure of the receiver's ability to reject unwanted signals. In the superheterodyne type of receiver, most of the selectivity is due to the I-F amplifier, and in multiple-conversion jobs, is a function of the lowest I-F used.

In most cases, therefore, we can measure the selectivity by injecting the signal into the input of the I-F amplifier chain. While some manufacturers provide a point on the I-F input transformer for connection of the signal generator, it is usually best to inject the signal into the rf port of the mixer preceding the I-F amplifier. This approach will avoid the problem of detuning the primary of the input amplifier.

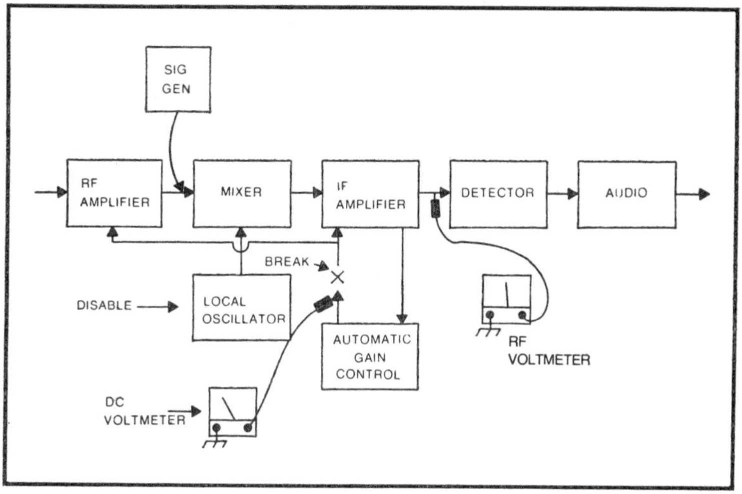

Fig. 24-3. I-F alignment scheme.

It is best, if possible, to disable the local oscillator, and to turn off, or disable, the agc circuit.

Figure 24-3 shows the equipment needed for measurement of selectivity. The output indicator will be either an AC voltmeter across the loudspeaker (or suitable load), or an rf voltmeter at the output of the I-F amplifier, immediately ahead of the detector stage. Another meter might be a DC voltmeter with an rf probe, also connected at the output of the amplifier. If the receiver is equipped with an agc, then a DC voltmeter across the agc line could be used instead. This also applies to receivers equipped with an S-meter, especially if they are well-calibrated as to dB- or μV per S-unit.

1. Adjust the signal generator for a convenient output level that overcomes background hiss, but does not overload the receiver.
2. Adjust the signal-generator frequency for the center of the I-F passband. This can be done easiest by rocking the signal generator dial back and forth to find the frequency that produces maximum output.
3. Note the output voltage from the receiver, and also note the signal-generator frequency.
4. Measure the receiver output voltage at equally spaced frequencies, above and below, but close to, the first frequency used. For Amateur communications receivers, try to obtain data at least every kHz from the center frequency, out to ± 10 kHz.

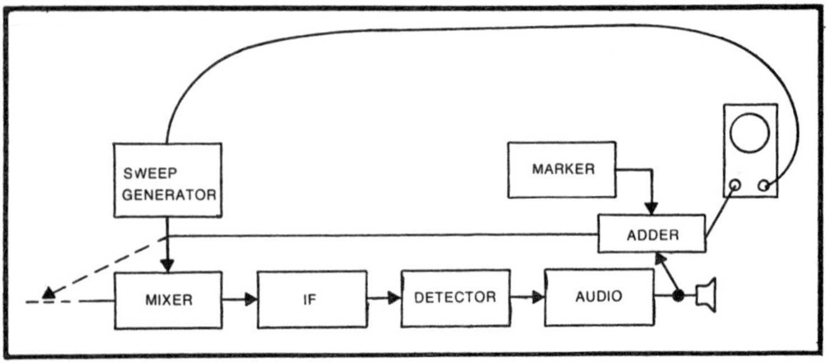

Fig. 24-4A. Sweep alignment set-up.

5. Plot the data obtained on a graph. Use a dB format, and consider the output voltage measured in step no. 3 as the 0-dB reference point.
6. The selectivity is considered to be the bandwidth between two points where the sensitivity has fallen off a specified number of dB. In most cases, the bandwidth between points where the response is down −6 dB is considered the selectivity, although some manufacturers will specify other points.

The selectivity of the typical multi-mode communications receiver will be different for different modes. The bandwidth requirement of AM, FM, SSB, and CW signals is different, so a good receiver will accommodate different reception modes with different bandwidths.

Sweep Method For Selectivity Measurement

The technique described above is too time consuming for most practical applications, so here we will consider a superior method using a sweep generator. The sweep generator will sweep back and forth across the frequency set on its dial a given number of times per second (usually 30 - 100 Hz). If we combine the output of the sweeper with the output of a crystal marker (i.e. 500 or 1000 Hz), then we can generate an output response that is amplitude-vs-frequency. The crystal calibrator, also called a marker generator, creates "pips" on the screen of the CRT to identify frequency points. If the scope has a horizontal gain control, then it is possible to calibrate the scale in kHz/div. We may also calibrate the vertical scale in dB units, but it is equally convenient to simply recognize that the −6 dB voltage points are the points where the voltage response drops to one-half of the center-band voltage. These points, one on each side of the I-F center frequency, are also called the *half-power points*.

If we do a good job of calibrating the oscilloscope, then we may read the selectivity directly from the screen. An additional advantage to the sweep technique is that it gives us the shape of the passband, and this could be critical in some cases. If the shape is poor, then an alignment may be indicated, and that is covered in Chapter 25.

IMAGE RESPONSE

We defined images in the first part of this chapter. One of the principle differences between low-cost and high-quality communications receivers is their relative abilities to reject the images.

Very-low-cost shortwave receivers, such as those ordinarily sold to not-too-serious-SWL and people who just want a shortwave receiver (but don't know why), have very poor image rejection. The problem becomes especially acute in the highest frequency band, usually 12 - 30 MHz.

There are two main tactics for reducing image response. One is to use high-Q resonant circuits in the front end of the receiver. This can produce only limited results. An even better method is to use a high i-f. When a 455-kHz I-F is used, the image in only 910 kHz away from the rf frequency, and most tuned circuits in the 12 - 30 MHz region are at least 910 kHz wide—the image sails right on into the mixer with the rf signal. On the other hand, a 9-MHz I-F would produce an image that is 18 MHz away, and that is enough difference that even the sloppiest circuit should reject it. The procedure is as follows:

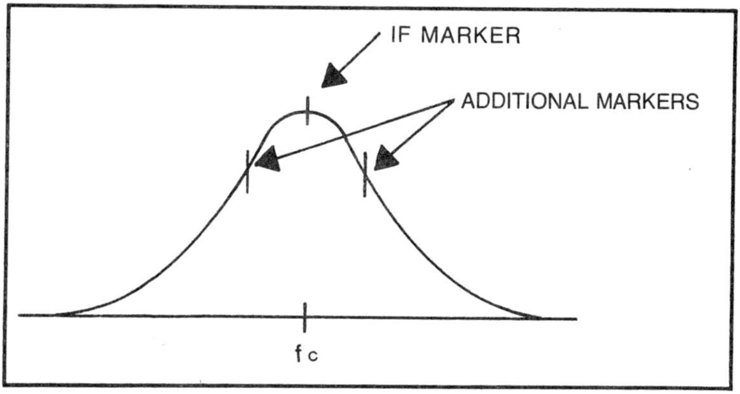

Fig. 24-4B. Response curve showing marker pips.

1. Tune the receiver and signal generator to the same frequency. If you are looking for "worst case", then use a frequency in the highest band covered.
2. Adjust the signal generator output control for 50 to 100 μV, or some output level that produces a convenient signal level in the receiver, without causing overload.
3. Record the output voltage from the receiver and the signal-generator output level.
4. Tune the signal generator to a frequency that is 2 × I-F above or below the first frequency, whichever causes the signal to be heard in the output of the receiver.
5. Adjust the signal-generator attenuator for the same receiver output level as was recorded in step no. 3.
6. Use the two signal generator output levels (in μV) in the voltage/dB formula to obtain the image rejection in dB: $dB = 20\ LOG_{10}(V2/V1)$

Chapter 25

AM and FM Receiver Alignment

It is simply amazing to note that most Amateurs seem to fall into one of two catagories with respect to alignment: alignophiles and alignophobes. The alignophile seems to be born with a "diddle stick" (i.e., alignment tool) in one hand, and will use it at every opportunity . . . regardless of whether or not there are any indications for need of alignment! The alignophile will run rampant through a troubleshooting job, testing everything by adjusting it! What they do not seem to appreciate is that they are ruining more than they are fixing. The result will be a receiver that works poorly, or not at all, because of the efforts of the alignophile.

The other extreme is the alignophobe, who is too timid to even think of doing alignment jobs! These people will automatically ship every alignment chore off to the factory, or some commercial repair shop, rather than to tackle it themselves.

The best position for the Amateur to take is one of the middle ground. Recognize that there are certain alignment jobs that are necessary, and do them if you have, or can borrow, the proper equipment. If you do not have the equipment, then do not attempt to align your receiver (unless it is an old clunker of the "I don't care" class).

Do not *ever* use alignment as a troubleshooting method. It is *not* a troubleshooting technique can only cause trouble. If the tank circuit that is "aligned" is not off, then you have only two possible outcomes, a) no change at all, or, b) you will deteriorate performance. To make matters worse, the alignment attempt will convey no useful information to the troubleshooter—with the sole possible exception of the suspected tank circuit that was found first by other means.

Second, I do not believe that a receiver or transmitter will require *annual* alignment (to "peak things up") unless it has been used in abusive situations (such as a mobile rig in rugged country). A home-station receiver, that sits sedentarily on an operating bench, will not go out of alignment in a single year.

Okay, you say, but what harm does an annual touch-up alignment do? You may be tempted to align a rig once a year just because it seems like a "good thing to do." But, is this a benign, if not helpful, process? The answer is no, it is not benign. The act of alignment causes substantial wear and tear on the alignment components, such as trimmer capacitors and coil/transformer cores. Every time you adjust one of these devices, it will wear a little bit. After a few alignments you will often find that the circuits are incapable of holding alignment for very long—and then you will need an annual alignment every few weeks.

The principle indications for alignment are, a) many years of age, b) an obviously misaligned rig from the factory or repair shop (happens more often than you care to believe), c) rugged, abusive use, and, d) replacement of tuned-circuit components or a crystal (in which case, only the affected parts need be aligned.

In this chapter, I'll give some basic, generalized instructions for the alignment of CW, FM, and AM receivers. Fortunately, the CW and AM, plus one technique for FM, alignment procedures are very similar. After all, we are dealing with radio receivers in all three cases. Note that the CW alignment instructions can be extrapolated to the SSB transmitter as well.

Before getting started with a discussion of alignment, let's review the properties of the basic forms of modulation.

In amplitude modulation, the audio information is superimposed on a radio carrier signal. The carrier is an rf signal, and is the electromagnetic signal transmitted from the antenna. Modulation is the process of altering a carrier in a manner that allows us to represent the modulating signal. Of the carrier parameters that are available for modulation, we most often alter its *existence* (on-off keying, as in CW), its *amplitude* (AM), *frequency* (FM), or *phase* (PM).

Figure 25-1 shows the elements of an amplitude-modulated carrier. The waveform in Fig. 25-1C, incidentally, can be viewed on an oscilloscope by connecting its vertical input to the output of an

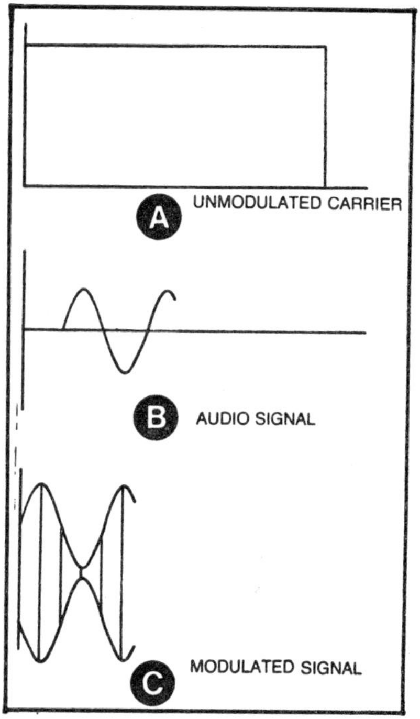

Fig. 25-1. The components of an amplitude-modulated signal.

amplitude-modulated signal generator. Note that the carrier signal is a constant-amplitude rf signal, while the modulated carrier is this same signal, but with the modulating signal superimposed on it.

An on-off signal, such as a CW transmitter would produce, would be the same as Fig. 25-1A, but will turn on and off in rhythm with the dots and dashes of the Morse code transmitted.

In FM transmitters we use one of two basic forms of angular modulation. One is pure FM, while the other is phase modulation, or PM. Note that, in practical situations, many PM transmitters are designated FM. Most 2-meter FM transmitters, for example, are actually PM transmitters. The receiver cannot tell the difference, and both behave pretty much the same as far as the operator is concerned. The PM circuit, however, has traditionally been a lot easier to implement in real transmitters.

FM terminology - deviation and swing. In any given FM modulator an audio frequency signal of one polarity will cause the carrier frequency to increase, while the opposite polarity signal causes the frequency to decrease.

Many people become bogged down at this point, with unfamiliar terminology and false impressions. One point is confusion over the sweep width and deviation. The distinction between these is shown in Fig. 25-2.

The maximum frequency change from the unmodulated (carrier), frequency to either high- or low-frequency limits is called the *deviation*. A certain FM transmitter might, for example, shift from 52.000000 MHz to 52. 000005 MHz, a change of 5 kHz, during positive modulation peaks. The deviation of this transmitter is, therefore, 5 kHz.

The entire width of the FM signal, from its lowest extremity to its highest, is called *frequency swing*. If a perfectly linear FM modulator is given a perfectly symmetrical sinewave audio signal, then the positive and negative deviations are equal, so the swing will be 2X deviation.

The only time that frequency swing and deviation would be the same is in the unlikely event that the transmitter was modulated by unipolar waveforms.

Another term, ***modulation index***, often is used in connection with frequency modulation. It is defined as the ratio between total deviation and the modulating frequency. If a 2-meter FM transmitter can transmit audio signals up to 3 kHz, and deviate up to 5 kHz, then the MI at maximum conditions will be 5 kHz/3 kHz, or 1.666.

Some Amateurs believe that the total frequency swing and the deviation are determined by the frequency of the modulating audio frequency. This is true (after a fashion) in AM systems, where we talk about bandwidth. In pure FM, however, deviation (and thus frequency swing) is determined solely by the amplitude of the applied audio signal. The term "100-percent modulation" in FM systems is a relative term that is *defined*. In AM systems, 100-percent modulation is determined by the physical properties of the modulated carrier. The FCC has set 100-percent modulation at 75 kHz deviation for FM broadcasters and certain radio telemetry applications, 25 kHz of deviation for broadcast-television audio signals, and 5 kHz for landmobile and marine FM two-way radios. Amateur equipment usually follows the landmobile standard of 5 kHz being defined as 100-modulation.

PHASE MODULATION

To prevent confusion, it might be well to explain a thing or two about the two major related forms of angular modulation. Phase modulation (PM) is a type of modulation that is often called FM, but is not. It is the type of modulation used in most VHF and UHF two-way radios, including Amateur equipment. The deviation in PM

Fig. 25-2. Deviation vs audio-modulating frequency.

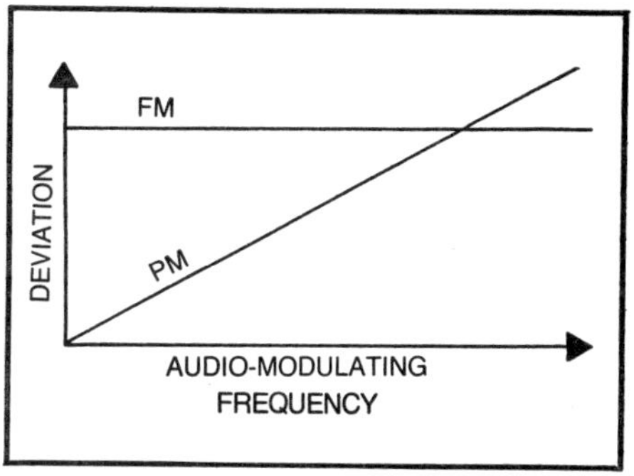

is directly related to the modulating frequency (in distinct contrast to true FM). The process of phase modulating an rf carrier with an audio signal causes a 6 dB/octave rising, or preemphasis, characteristic. Because of this, it is necessary to deemphasize the audio recovered at the receiver to restore the proper audio balance. The few true FM transmitters available use artificial preemphasis in order to achieve a superior signal-to-noise ratio, as well as ensure compatibility with PM transmitters in the system.

FM Fidelity and Noiseless Reception

Many people erroneously believe that FM is inherently hi-fi, and noise free. The first of these suppositions is probably because FM broadcasters are allowed to transmit audio frequencies up to 15 kHz, and this (or at least was once) deemed "high" fidelity.

The second supposition is probably due to the efforts of sales people to sell FM broadcast receivers and mobile communications equipment. FM systems are *capable* of noise-free operation, but it is not necessarily quaranteed. The FM detector is sensitive only to phase or frequency variations, and *rejects* amplitude variations. However, impulse noise, the most common type of interference, tends to amplitude-modulate the radio carrier. In the receiver, the noise is seen as an AM component on an FM signal. If the signal has an amplitude capable of driving the limiter stage into hard limiting, then most of the noise spike will be clipped off. This gives us the impression of noise free operation . . . until the time comes when we tune in a station that is not strong enough to quiet the receiver! In mobile use, the FM transceiver will give noise-free operation only when the station being received is relatively strong. As the signal strength goes down, at the fringes of the service area, the noise level will come up. There will be a remarkable increase of noise interference if the signal merely drops below the hard-limiting level.

CW ALIGNMENT

In this section we will discuss the generalized instructions for alignment of the CW receiver. Note that a particular manufacturer will give you their own set of instructions. If those are available to you, then use them. My instructions are highly generalized, and are intended to guide you in alignment of a receiver for which no instructions are available, or one that is being homebrewed.

Adjust the rf gain control for maximum, and set the BFO at the "zero" (usually center of range) position. Tune the receiver to the high end of the band.

I am assuming that the receiver has an agc system. It is necessary to disable the agc, even though it kills the best level indicator that we have—the S-meter. The agc will cause you to make mistakes because of its behavior at high signal strengths.

Apply an I-F signal to either the input of the i-f amplifier chain, or better yet, the input of the mixer circuit. The latter point will provide some isolation between the rf signal generator and the first tuned I-F circuit.

We do not want a signal that is too strong, yet we must have at least enough signal strength to work with. The best advice seems to be to advance the rf-output control on the signal generator until the noise (hiss) disappears, and then back off just enough to bring in a little noise. A barely perceptible background noise indicates a signal of optimum amplitude.

Use an rf voltmeter, or demodulator probe on a DC voltmeter, to measure the I-F amplifier output at the detector input. Adjust the tuned circuits (usually coil slugs inside the i-f transformers, but sometimes a capacitor) for maximum output indication. Once you have peaked up all the I-F amplifier adjustments, or when you notice that subsequent adjustments are seemingly futile, then reduce the output level of the signal generator and do it all again. Repeat the process several times, or until it is apparent that no further improvement can be obtained.

Techniques for BFO adjustment vary somewhat. In the case where the BFO frequency is variable, there is usually a null point at which the signal in the center of the passband should be zero beat. Adjust the signal generator for the correct I-F frequency (precisely), and then adjust the BFO tuning slug (not the knob! the slug inside the transformer can!) for zero beat.

In the case where the BFO is crystal controlled, as in most SSB receivers, set the signal generator for an on-channel, center-of-passband signal, and then adjust the BFO tuning slug for the correct

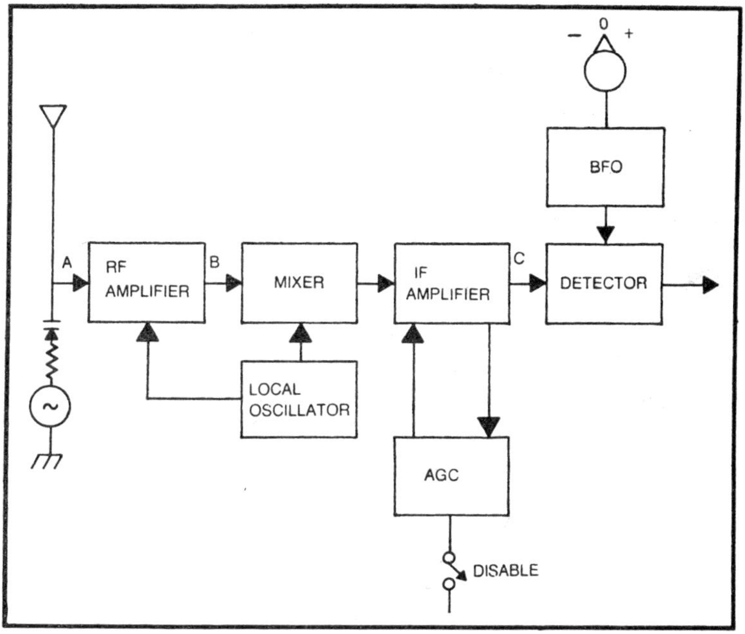

Fig. 25-3. SSB/CW receiver.

audio frequency beatnote in the output. This will have to be done independently for USB and LSB in SSB receivers.

If a fixed-frequency crystal filter is used anywhere in the amplifier, and in most Amateur receivers this is the case, then the correct I-F frequency will be a little different from the frequency published in the service manual or spec sheet. To find the true frequency, rock the signal-generator dial back and forth to find the point at which maximum output occurs. This will then be used as your on-channel signal.

Next, move the signal generator to the antenna terminals of the receiver, and adjust it to produce a signal equal to the top band edge. Most rf signal generators are not sufficiently accurate to permit us to believe the dial calibration, so use a crystal calibrator that has been zero-beat with WWV, or a digital frequency counter to measure the exact frequency.

Adjust the local-oscillator trimmer *capacitors* for correct alignment of the high-end frequency with the dial marking. If the receiver has a movable dial cursor, then it should be replaced in the middle of its movement range.

Next, move the signal generator frequency to the low end of the dial, and retune the receiver dial to the low end. This time,

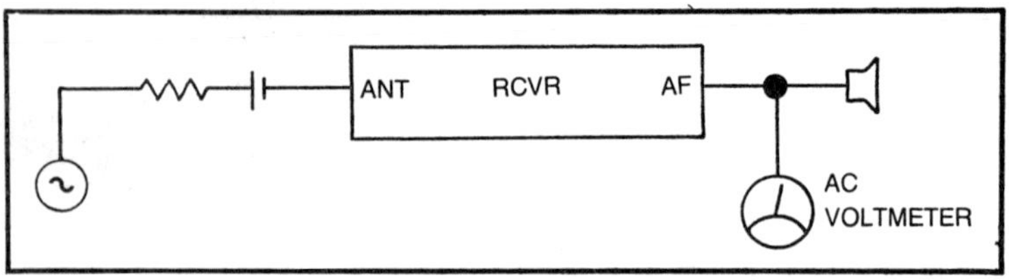

Fig. 25-4. An AC voltmeter can be used to measure output.

adjust the local oscillator *inductor* (slug or core) for proper alignment with the dial marking.

Note well that these are highly interactive adjustments. Proper local oscillator tracking across the entire band depends heavily on performing the above procedure several times, back and forth, to insure that both high and low ends are on frequency.

In double- and triple-conversion receivers, we use pretty much the same technique, except that we have to account for the "high I-F" and the first mixer/LO. If we apply a precisely calibrated rf signal to the antenna, then we can adjust the first LO (usually a crystal oscillator) and the high I-F transformers by peaking the output, using their tuning adjustments. It is not usually necessary to generate a separate signal for the high I-F and first local oscillator, although this is sometimes recommended by receiver manufacturers.

AM ALIGNMENT

Figure 25-4 shows a typical set-up for the alignment of an AM receiver. We follow pretty much the same procedure as for the CW receiver (which will not be repeated here), so only the points of difference must be noted. In the AM receiver, we may use the recovered modulation, that is, the audio output, as an indicator of signal strength. Although some people use their hearing as the output indicator, and this is a reasonable practice on low-cost broadcast receivers (and for people with good hearing), I prefer to use an AC voltmeter or oscilloscope as the output indicator. The signal generator can be connected as before, or be connected directly to the antenna input terminals for the entire alignment. I personally prefer to use a separate connection at the mixer input for I-F alignment, and then move the signal-generator connection to perform the front-end alignment. This seems to produce more predictable results.

FM ALIGNMENT

There are two approaches to FM alignment: swept and unswept. The use of a sweep generator is, technically, the best, but there are many service people who are able to properly align FM receivers without the use of sweep generators or special FM signal generators. In fact, two of the major producers of commercial FM two-way radios used to publish the instructions for unswept alignment only!

Sweep alignment. A typical sweep-alignment equipment set-up is shown in Fig. 25-5. The sweep generator is able to produce a simulated FM signal, and can be adjusted to produce deviations compatible with the receiver. Most will have a fixed modulating frequency, either 1000 or 400 hertz.

The marker generator provides small "pips" on the frequency-response trace, which help determine the frequency at a particular point on the oscilloscope curve. The marker and the sweep outputs are fed to an adder, where they are mixed together. The adder combines all fo the required signals into one composite that can be applied directly to the FM receiver. Direct connection, without the adder, cna cause interaction between the marker and sweep signals, which can distort the response curve.

Many test equipment manufacturers now offer alignment generators that include marker, sweeper, and adder functions inside of one cabinet. To the commercial servicer, this means less clutter, and quicker work, and so is useful. The use of individual devices, however, is in no way inferior.

I-F. The alignment instructions for most FM receivers will specify where the signal is to be injected into the receiver. Some will

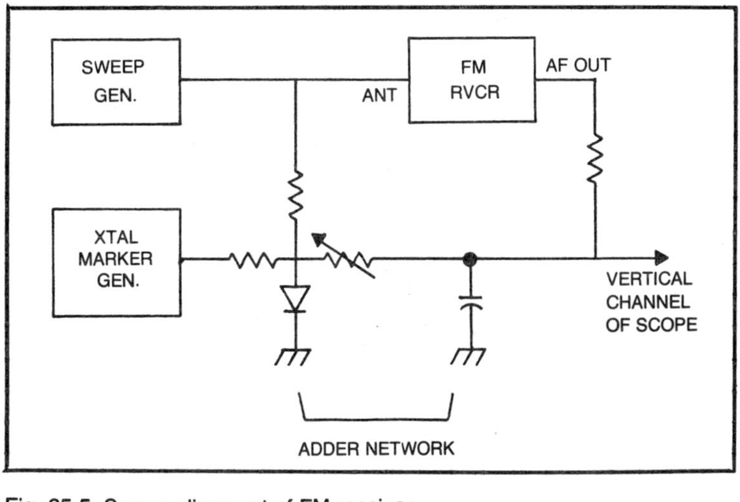

Fig. 25-5. Sweep alignment of FM receiver.

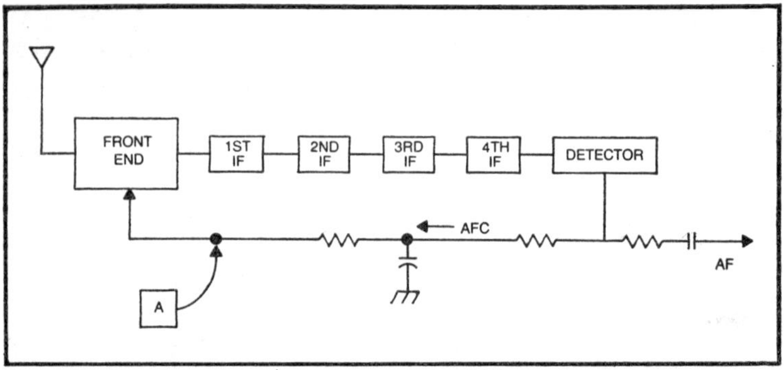

Fig. 25-6. Location of null-meter probe, paint "A".

use the antenna terminals, while others will use some other point, usually in the area of the mixer. Follow these instructions to the letter for best results.

If such instructions are not given, however, or are unavailable to you, then try squirting a signal from the output of the adder into the mixer circuit, using a short piece of hookup wire. This is called a "gimmick." The gimmick may, for example, be carefully dropped inside of the first I-F transformer (this is why it should be *insulated* hook-up wire!). If the first transformer is not amenable to dropping things inside, then connect the output of the adder to the grid, base, or gate (as needed) of the mixer stage. The hot side of the adder output cable can be connected to this point through a 0.001 - 0.01 μF capacitor.

Most instructions also specify the sweep width, or deviation, of the input signal. If this is not given, then set it to approximately one-third to one-half of the normal deviation seen by the receiver for peaking the I-F stages, and greater than the normal deviation for tuning the FM demodulator. Most manufacturers also specify that the modulating audio be 400 or 1000 Hz.

A simplified block diagram of an FM receiver is shown in Fig. 25-6. Connect a high impedance, zero-center, voltmeter to point "A." Most analog VTVMs (or their solid-state counterpart) can be adjusted for the pointer to be at center-scale when the input voltage is zero. In fact, most VTVMs have a zero-center scale, along with the other scales, just for this purpose.

With a proper signal (as described above) applied to the input of the receiver, or I-F amplifier chain, adjust the secondary of the demodulator transformer for zero volts. The meter will read a positive voltage on one side of the correct point, and a negative voltage on the other side of the correct point.

Next, connect an AC voltmeter across the audio output and adjust the primary of the demodulator transformer for maximum. Then peak all of the other i-f tuned circuits for a response curve as shown in Fig. 25-7.

Front end. To align the front end, it is necessary to apply a proper signal to the antenna terminals of the set. The signal, perferably, should have a frequency that corresponds to one of the points on the dial if a slide-rule, or other "analog," dial if used, or a channel frequency if the receiver is crystal controlled.

After injecting the proper signal, zero the receiver local oscillator by turning the associated trimmer capacitor (or, occassionally, a slug-tuned-coil adjustment) until the voltage at point A is again zero.

If a receiver is crystal controlled and has more than one channel, this last step must be performed for each channel. An alternative procedure for Amateurs is to "net" the receiver local oscillator to the local repeater. Ideally , all repeaters operate precisely (±0) on their chosen or assigned channel frequency, but we all know that real repeaters are sometimes off frequency a little (depending upon whose frequency counter was used to set it up). This latter procedure might work better for some local situations, because even well-equipped professionals sometimes make it a habit to use a base station, or a single frequency meter to net in a system.

Next, peak the rf amplifier and antenna adjustments for maximum output. Monitor the response curve during this procedure. The marker pips will help you determine whether or not the

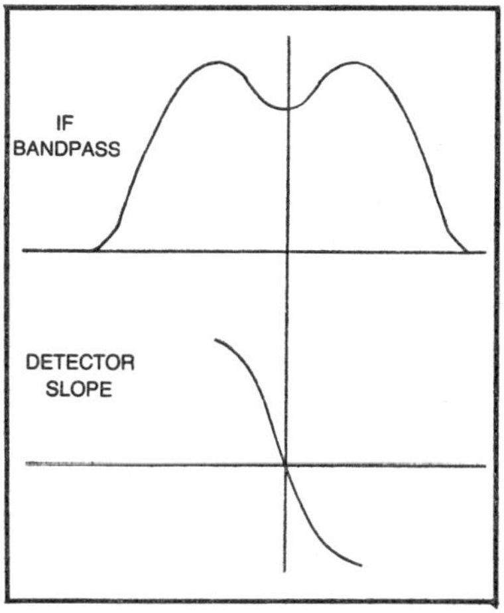

Fig. 25-7. I-F response and detector slope.

response is correct at the significant frequencies. We don't want a lopsided response! Incidentally, it is best to sacrifice a little gain for a correctly shaped response curve.

What is a "correctly shaped response curve?" The curve in Fig. 25-7A shows a slight dip in the center of the passband. This dip should be not more than 10 percent of the overall trace amplitude. Furthermore, the two humps of the trace should be within 10 percent of each other. If the curve that you obtain on your set does not have a dip, yet has the correct bandwidth, then for crying-out-loud don't work too hard to *obtain* the dip. The dip is from an imperfection in a tuned circuit . . . and the 10-percent figure is a compromise ("trade-off" in engineer's jargon). If the response that you obtain has less than 10 percent, or is perfect (flat-topped), then be happy, and go on to greater things. I remember one commercial servicer who spent a lot of time one afternoon trying to get that 10 percent dip, when the trace most easily obtained was almost ideal and perfect!

NONSWEPT ALIGNMENT

Although sweep alignment is the best technique for FM communications receivers, the use of a nonswept "CW" signal (unmodulated) can provide as good a job, if used properly.

Equipment for nonswept alignment varies greatly, and almost any decent signal source can be used. The types of equipment needed can be broken down into two categories: a level indicator and a stable signal source. The level indicator may be a DC coupled oscilloscope, or a VTVM capable of zero-center operation. No really fancy indicators are needed. The signal source can be service type of signal generator (but only if the modulation can be turned off or disabled), or a crystal-controlled oscillator that you build. The only real criterion is that it be stable and reasonably precise. In the case of receivers that use fixed-frequency bandpass filters, a tunable signal source is suggested. The crystal oscillator can be used, if a trimmer capacitor is connected into the oscillator circuit to vary the actual operating frequency over a small range.

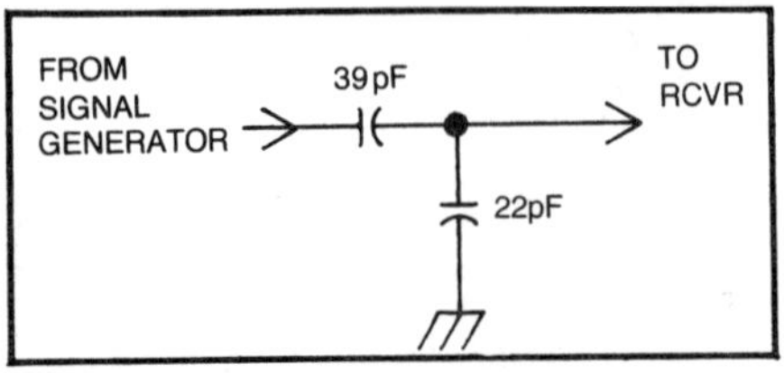

Fig. 25-8. Dummy antenna for VHF receivers.

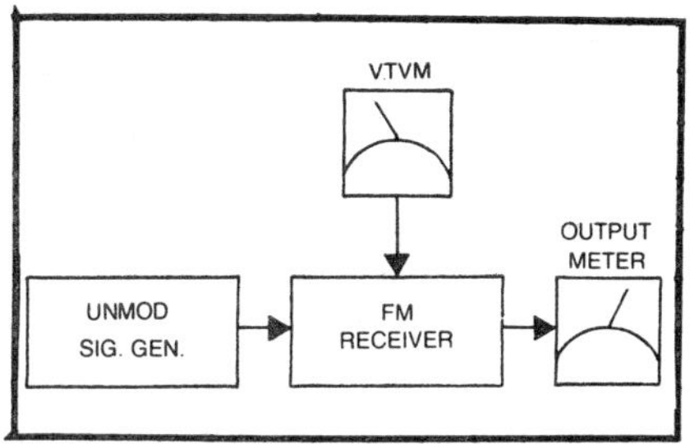

Fig. 25-9. Nonswept alignment set-up.

One other requirement that is not often met by low cost "service-grade" instruments, and most homebrew projects, is low rf leakage. The signal generator must be double or triple shielded to prevent the rf signal to escape by any path other than the output attenuator and connector. Many poor-quality signal generators produce enough rf when the attenuator is at minimum to drive an FM receiver into hard limiting—and *that* we don't need when aligning the receiver!

I once built a fairly decent alignment oscillator for the FM broadcast band, using International Crystal (10 No. Lee St., Oklahoma City, Oklahoma) OX-crystals, and the oscillator and buffer kits offered by International. These were housed inside a double-shielded cabinet. The inner shield was a Bud die-cast aluminum box, and the outer case was a tightly sealed aluminum chassis (most chassis, incidentally, will not do. . . they are not rf tight!). The output frequencies were 10.7 MHz for FM I-F alignment, and 9 MHz to produce markers at 90, 99, and 108 MHz. Of course, for Amateur work, different markers are needed.

Most sets require some sort of dummy load between the signal source and the antenna connector. This is to simulate a transmission line. A dummy antenna which I have used on 150-MHz FM receivers is shown in Fig. 25-8. It can be built inside of a small shielded housing, or if you are clever, inside of an antenna connector. It should be connected directly to the receiver antenna terminals for proper operation. The cable from the rf signal generator may then be connected to the input of the dummy antenna.

The test set-up block diagrammed in Fig. 25-9 will produce a satisfactory alignment job for most receivers. Connect the output of a signal generator tuned to the first I-F (10.7 MHz in most sets, but possibly anything in the 9-15 MHz range) to the input of the FM mixer via a 0.001-0.005 μF capacitor, or via a "gimmick." Do not connect the generator directly to the I-F transformer, because its cable capacitance will seriously detune the stage. Connect a high-impedance DC voltmeter to the point marked "X" in Fig. 25-10, or a similar point if another type of demodulator is used.

Adjust the secondary of the demodulator transformer for a null (zero voltage). This is why the zero-center meter is specified. The voltage at point "X" will be positive if the transformer is tuned on one side of the correct point, and negative if tuned to the other side of the correct point.

There are two major factors which, if overlooked, can make identification of the null difficult, if not impossible. One is overloading of the I-F amplifier chain, which, in many receivers can be caused by the agc circuit. Another problem is when the meter is set on too high a scale. The visual effect of adjusting the secondary of the transformer to zero is obscured by too small a movement of the meter pointer.

In many receivers, it is possible to disable the agc circuit during alignment, and if fact, some manufacturers provide for this to be done easily.

In other cases, however, we must be sure to keep the signal generator output level so low that agc action cannot effect the sharpness of the null. For the alignment of the detector secondary, most authorities tell us to adjust the signal generator to an output level that barely "quiets" the receiver. As the circuits are peaked up, it might prove necessary to readjust the signal generator output *downwards*. The correct level for peaking the I-F and RF trimmers is slightly below quieting. This will both make the agc action noneffective, and provide you with a more "contrary" audio output, allowing you to hear the peaking take place.

BASIC I-F ALIGNMENT

To peak the I-F transformers, connect the meter to a point that corresponds to "Z" in Fig. 25-10. An alternative point in receivers that use a discriminator instead of a ratio detector, is the input of the limiter. The conduction voltage across the emitter resistor of the limiter might also be used. In fact, if you do not have an rf probe for the voltmeter, or if the DC present at the limiter input does not vary linearity (or at all) with signal strength, then you might have to use the emitter-resistor voltage drop.

Connecting the output indicator to the limiter, however, does not allow you to adjust the limiter output transformer (usually the primary of the demodulator/detector transformer). This may have to be peaked up by using point "Z" and a very weak signal, and then go to the input of the limiter to align the rest of the amplifier.

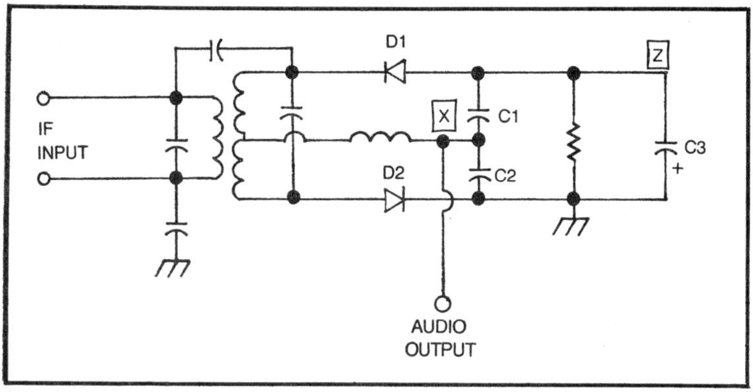

Fig. 25-10. Ratio-detector circuit.

It is extremely important to use a *high impedance* voltmeter in these alignments, unless the manufacturer has specifically provided you with features that allow the use of a VOM.

During I-F transformer peaking, the signal source may be left connected in the same manner as it was in the detector alignment. Again, make sure the signal level is not too great, and does not cause overloading of the receiver's circuits. Poor alignment will surely result.

Do not attempt to use a station on the air as a signal source. It is very difficult to find a station that will maintain exactly the correct strength for alignment purposes. The only exception to this rule is when you are netting the local oscillator to a local repeater frequency.

Try to avoid using your ears as a level detector. While it is usually a good idea to listen to the output while alignment progresses, depend upon the meter, not your ears, as the proper output indicator. The receiver will be quieted over several turns of the alignment slugs; a good reason why your ears are not reliable as an output indicator.

Work from the detector end of the cascade chain towards the front-end. Adjust the secondaries of each transformer *first* and then adjust the primaries. When you have finished, go back and do it again. The alignment should be done several times, or until no further improvement is noted. This is necessary because these adjustments are highly interactive.

Front-end Alignment

The alignment of the front end of an FM receiver, using an unswept oscillator, is pretty much the same as for a swept system.

Use a signal generator that produces an output signal on the channel frequency, or on a subharmonic of the channel frequency. It is necessary that this signal be stable (preferably crystal controlled and precise), and that its amplitude be easily adjusted.

The LO can be set in one of two ways. One is to adjust the LO trimmer for maximum signal at the output of the I-F amplifier. The other is to adjust it for a zero condition at point "X" in the demodulator, or on the afc (automatic frequency control) line. The rf and antenna trimmers may be peaked up in the usual manner.

Note well that we are dealing with very high frequencies in this circuit. Make sure that your alignment tool is *nonmetallic*. These circuits are especially sensitive to stray capacitances, and changes in inductance, due to local magnetic fields. Metallic tools, whether ferrous or nonferrous, will affect both L and C components. It is also wise to leave all metallic covers in place. The stray capacitance of these shields can cause the alignment to change when you put the rig back together. Fortunately, in cases where this is a big problem, the manufacturer usually provides access holes in the cover for alignment purposes.

Digital detectors. Digital detectors do not work on the same principle as tuned detectors. They do not, in general, require any alignment. The sole exception *might* be a case where the duration of the one-shot pulse used in these circuits needed adjustment. I have not, however, ever seen that to be required, so view it as purely theoretical.

Quadrature detectors. The methods previously detailed work on sets that use either a ratio detector or a Foster-Seeley discriminator. In years past, this was all that was needed, because all FM detectors in two-way radio equipment was one or the other. But, today, we might also find an integrated-circuit quadrature detector in use. These ICs have become increasingly popular, with the MC1357P and ULN2111 devices being common.

The circuit in Fig. 25-11 uses a quadrature-detector IC for FM demodulation. Note that the input transformer is an ordinary i-f transformer, and that no special demodulator transformer is used. There is, however, a single LC circuit used to provide a 90-degree phase shift at the center of the I-F passband (usually 10.7 MHz).

For alignment, we connect an rf demodulator probe to point "M" in Fig. 25-11. The 10.7-MHz signal is injected into the base or gate of the mixer stage through a dummy load consisting of a 270-ohm resistor in series with a 0.0047 μF capacitor.

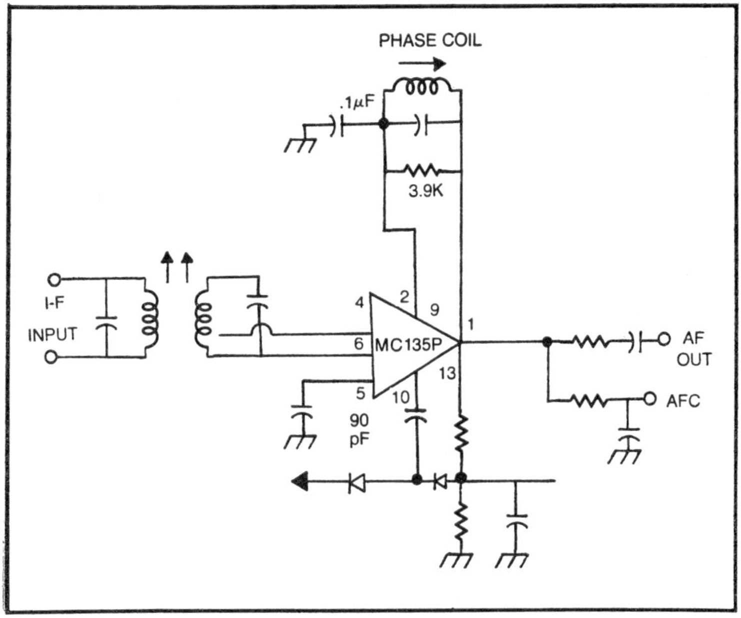

Fig. 25-11. IC Quadrature-detector based on the MC1357P.

Adjust all the transformers for maximum signal at point "M," keeping the signal-generator output low enough to not obscure the peak readings.

Alignment of the phase coil is a little different. Use the same 10.7-MHz, unmodulated, signal, but the output indicator should be an AC voltmeter across the speaker terminals. Alternatively, an oscilloscope may be used. We will be looking for the background noise (hiss).

As the phase coil is adjusted through its entire range, you will note that there are two peaks in the background noise. The proper adjustment point is the *null* between these two peaks, which is very nearly the mechanical midway point.

Note that an even better method is to use a sweep generator modulated by a sinewave, or an FM signal generator (also modulated by a sinewave). A total-harmonic-distortion analyzer on the output will then tell you when the phase coil is properly adjusted. Many professional hi-fi servicers have used the THD analyzer to align all forms of FM detector transformers.

Chapter 26

Antenna and Transmission Line Measurements

Antennas and transmission lines are very much a part of Amateur Radio communications, but are unfortunately sometimes overlooked as a cause of problems in the system. The antenna and its associated transmission line are so important that a small defect will render your station ineffective. Oddly enough, Amateurs have developed most of the practically oriented antenna literature available. Many professional journal articles and textbooks either evade the issue of how to get a real antenna working, or specify procedures that require very expensive test equipment.

The amateur is, by and large, left without such fancy equipment, and must fend for himself. We have been able to develop reasonably good techniques for testing and tuning both VHF and HF antennas using very low cost, easy to use, test equipment. In this chapter, we will take a look at the antenna system and some of the tests and test equipment that might be required.

ANTENNAS

A radio antenna is a structure or device that radiates an electromagnetic signal into space when it is excited by an electrical current. The radiating effect will take place to some degree at all frequencies, but the efficiency is poor except at certain *resonant* frequencies. At these frequencies, the efficiency increases tremendously.

There are many different types of radio antenna, as witnessed by the fact that several dozen complete books are available on the

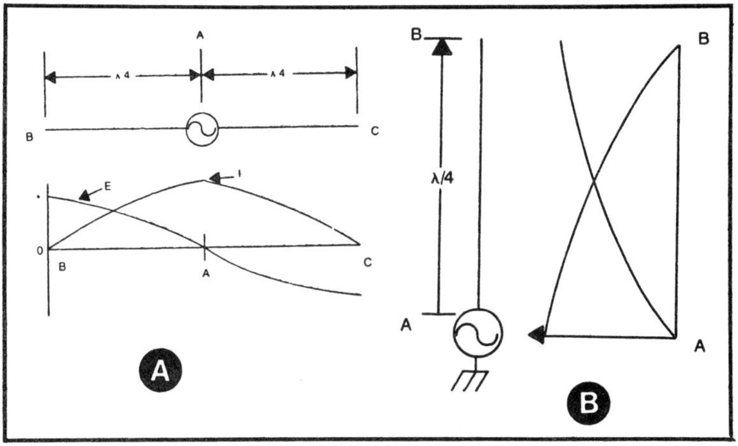

Fig. 26-1. Half-wavelength horizontal antenna (A), and quarter-wavelength vertical antenna (B).

subject. In this discussion, however, we are going to remain general, presenting the theory and practices in such a way that you will be able to extend it to the specific problem at hand. Although some mention will be made of directive arrays, we will use as our examples the high-frequency, horizontal half-wavelength dipole, and the vertical, quarter-wavelength radiator.

Figure 26-1 shows our two fundamental antennas. The half-wavelength dipole is shown in Fig. 26-1A, while the quarter-wavelength vertical radiator is shown in Fig. 26-1B. These antennas seem similar in many ways (they are!), but a principal difference does exist: the polarity of the electromagnetic field that has its electrical component horizontal to the earth's surface, while the vertical antenna radiates its electrical field perpendicular to the earth's surface. Both, of course, radiate both electrical and magnetic fields, but they are at right angles to each other. The polarity of the *electrical* field determines the polarity of the antenna.

The length of the dipole shown in Fig. 26-1A is one-half of a wavelength. In free-space (that is, many wavelengths from any object or surface that could deform the field) the length of the dipole is given by the formula:

$$L_{(ft)} = \frac{492}{F_{MHz}}$$

Nearer the earth's surface, the required length is a little shorter, on the order of

$$L_{(ft)} = \frac{468}{F_{MHz}}$$

The first equation, using 492 as the velocity factor, is only valid when the antenna is located in so-called free-space, many wavelengths from any conducting surface. The second equation, using 468 as the factor, is not based on theoretical considerations as much as on observation of practical antenna systems. We find that the 468 factor *usually* works reasonably well in most cases.

Equations for the quarter wavelength antenna are essentially the same as above, except that we must substitute 246 and 234 for 492 and 468, respectively.

Also shown in Fig. 26-1 are the voltage and current distribution patterns along the lengths of the radiators. In both antennas, the feedpoint is the point of lowest impedance, which is the point were the current is the maximum. The feedpoint impedance is a complex impedance of the form $Z = R_0 \pm jX$. But, we find that the reactive component, $\pm jX$, goes to zero, and the antenna is resistive.

Resonance is the condition where the impedance is purely resistive, $R_0 = Z$. At frequencies above resonance, the antenna becomes reactive, and the phase angle between I and E show it to be inductive reactance. At frequencies higher than resonance, the antenna is too long, so some capacitance reactance in series with the antenna, to exactly cancel the $+jX_L$ component of impedance, is needed to resonate the antenna.

At frequencies lower than resonance, the situation reverses and the antenna becomes too short. This antenna is capacitive, meaning that the reactive component of impedance will be $-jX_C$. This antenna may be resonated by placing an inductor in series with the radiator to exactly cancel the capactive reactance.

TRANSMISSION LINES

It is rarely practical to locate the radio transmitter at the center of the antenna, or even at the base, as shown in Fig. 26-1. This is simply not practical—the transmitter must be located, at best, a few feet or a few dozen yards from the antenna. A transmission line is the cable that carries the radio signal from the transmitter to the antenna.

A transmission line may be a pair of parallel conductors, as in Fig. 26-2, or it might be a coaxial cable, as is most commonly used in Amateur stations. A true transmission line has properties that are far more complex than a simple "shielded cable" or "twisted pair." We will leave the discussion of transmission line theory to more advanced texts, and apply ourselves here only to matters such as the surge impedance and the velocity factor of the line—matters deemed important in troubleshooting and adjusting antennas.

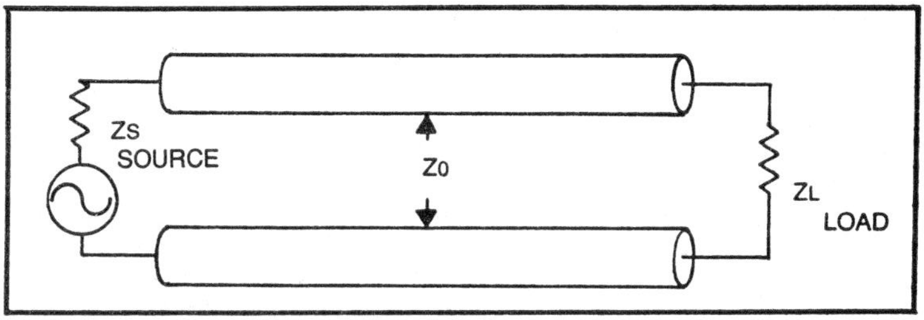

Fig. 26-2. Transmission-line analogy.

The velocity factor arises from the fact that radio waves do not propagate in transmission lines as rapidly as they do in free space. The free-space velocity of radio waves is the speed of light, on the order of 300,000,000 meters/second. But, in a radio transmission line the velocity of the wave is a lot less than this speed. It will be a fraction of the speed of light, and this fraction is usually expressed in terms of a decimal number between 0 and 1. The velocity factors of three commonly employed transmission lines are:

Foam-filled coaxial cable	0.80
Polyethylene dielectric coaxial cable	0.66
Foam twin-lead (TV antennas)	0.82

The velocity factor alludes to the fraction of the speed of light at which signals will propagate along a normal piece of the transmission line. For example, the foam dielectric coaxial cable has a velocity factor (V) of 0.80. This means that the signal will propagate at 0.8 X the speed of light, or about 240,000,000 m/s.

The velocity factor of a piece of coaxial cable must be used when trying find an *electrical* length for a piece of transmission line. For example, if we want an electrical half-wavelength of cable (it is unlikely that we would want a physical half wavelength, because this is an electrical system), then we would modify the half-wavelength formula presented earlier to account for the velocity factor of the cable:

$$L_{(ft)} = \frac{492\ V}{F_{MHz}}$$

and, for a quarter-wavelength:

$$L_{(ft)}\ \frac{246\ V}{F_{MHz}}$$

The *characteristic impedance*, also called *surge impedance*, of a transmission line is associated with the distributed inductance and capacitance of the cable per unit of length. Practical values for Z_0

vary from 100 to 1000 ohms for twin lead and open-wire pairs, and 10 to 100 ohms for coaxial cable.

We can also observe that Z_0 is the value of the load resistance that will permit maximum transfer of electrical power between the transmitter and the load. If any other value of load resistance is used, then not all of the power applied by the transmitter will be abosrbed by the load. Some of the power will be reflected back down the line towards the transmitter, and this results in standing waves.

ANTENNA SYSTEM MEASUREMENTS

The antenna system is potentially very complex, and its measurement is also sometimes complex. Yet, at the same time, it is also possible for us to get an antenna working properly (or very nearly so) using very simple equipment and almost naively simple techniques. True, our measurements will not always have the quantitative accuracy that professionals need, but we can get it working!

The key parameters which interest us are *impedance, standing-wave ratio, resonance, radiation pattern*, and *gain*.

Antenna Impedance

A number of different factors affect the impedance at the feedpoint of any practical antenna that you might build. Some of them are under your precise control, while others are at the mercy of "fate." In our simplified world, in which we will admit only the horizontal half-wave and vertical quarter-wave antennas, we find that "ideal" impedances are on the order of 73 ohms for the horizontal and 35 ohms for the vertical. In reality, we usually find that antennas have impedances that vary markedly from these textbook values. The horizontal dipole, for example, will exhibit radiation resistances between 5 and 125 ohms as the antenna height varies from 1/25 to 1 wavelength above the earth's surface. The 73-ohm figure in the textbooks is merely a theoretical consideration for free-space situations. Similarly, the 35-ohm impedance of the vertical antenna has a nominal value of 35 ohms, but will range from 5 to 80 ohms in real-life antennas. Mobile installations are particularly bad, ranging from 5 to 30 ohms.

STANDING WAVES AND SWR

Standing waves on antennas and transmission lines are simple phenomena, but are very often misunderstood. A certain mystique

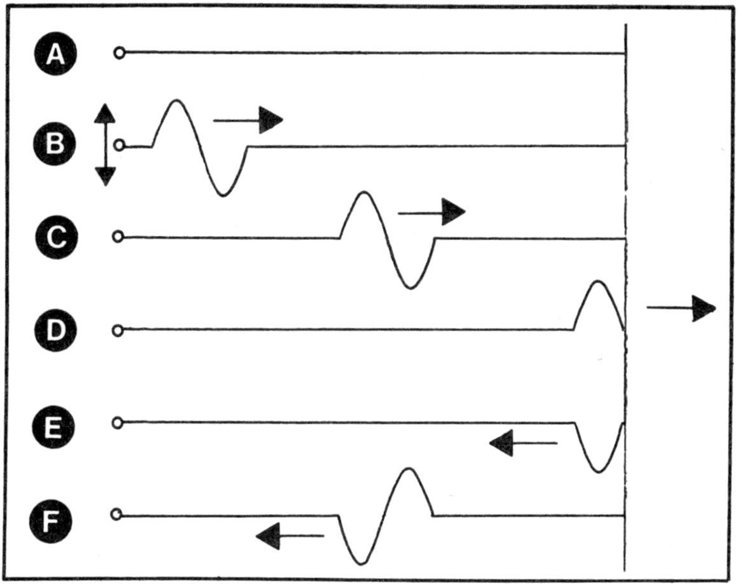

Fig. 26-3. Analogy for standing waves.

has built up around this topic, especially among Citizens Band radio operators. Many people spend a large amount of effort and money to optimize the VSMR of an antenna system, which is good unless it passes the point of diminishing returns.

Before discussing SWR, let's first determine the nature of standing waves. We may do this best by mechanical analogy using a model consisting of a rope anchored at one end. The phenomenon is the same, only the medium is different.

The rope model is shown in Fig. 26-3. The rope is held taut by the operator at one end, and is anchored firmly to the wall at the other.

When the operator imparts a single up and down motion to the free end of the rope, a wave is created (Fig. 26-3B). The wave will travel at a given velocity down the rope until it hits the wall, at which point it is *reflected* back toward the source end of the rope. A phase reversal takes place on reflection.

If the rope were ideal, and the wall was perfectly rigid, then the wave would reflect back towards the source completely undiminished. But like real electrical circuits, there is *loss* in the system, so the reflection is of slightly less amplitude than the incident wave. The wave will reflect back and forth down the line until it is totally diminished.

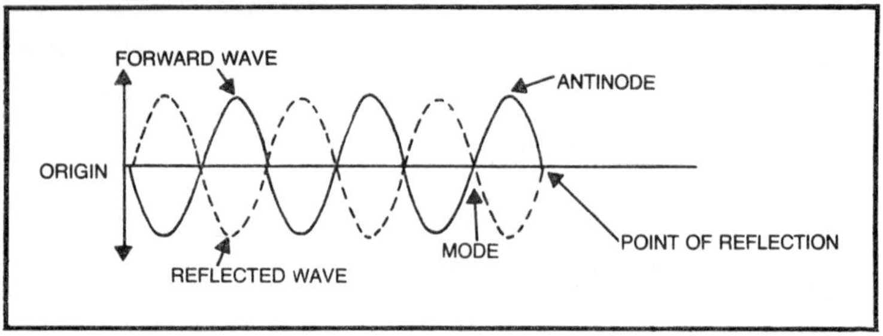

Fig. 26-4. Standing waves.

The situation in Fig. 26-3 is analogous to applying a pulse to a radio transmission line. But what happens when a *continuous* oscillation (from a radio transmitter) is applied to the line? The answer to the question can be seen in Fig. 26-4. In the mechanical rope analogy, the source end of the rope is moved up and down in a rhythmic motion, simulating connecting a radio transmission line to an electrical oscillator. The operator's hand in this case, is in fact, a mechanical oscillator.

The wave applied to the rope is a sinewave. After the first wave has reflected from the wall, we will actually have two waves present on the line; an incident, or forward, wave and a reflected wave. The forward wave travels from left to right in Fig. 26-4, while the reflected wave travels from right to left.

The wave amplitude at any given point will be the algebraic sum of the forward and reflected waves. When viewed from the side, the rope will take on a blurred appearance, showing the nodes and antinodes of Fig. 26-4. These are known as *standing waves* because they do not move with respect to the length of the rope.

Figure 26-5 shows three situations of standing waves on a transmission line. Figure 26-5A shows the situation when the forward wave is totally reflected back towards the transmitter. This type of situation can be caused by either of two things: a shorted load (zero ohms), or an open load (infinite resistance, so that it cannot absorb any power).

Interestingly enough, the pattern for both conditions is essentially the same except for the location of the nodes and antinodes. If the line is shorted, then the voltage at the load end is zero, so the nodes (points of minimum voltage, in this case, zero volts), are spaced at half-wavelength intervals along the line back from the short. But if the line were open, the voltage at the load end is maximum, so the antinodes will be located at half wavelengths from the load end of the cable. In other words, the two patterns are identical, except for a 90-degree phase shift in the relative positions of the nodes and antinodes.

Most antenna systems, however, do not totally reflect the energy from the transmitter back down the line from the antenna.

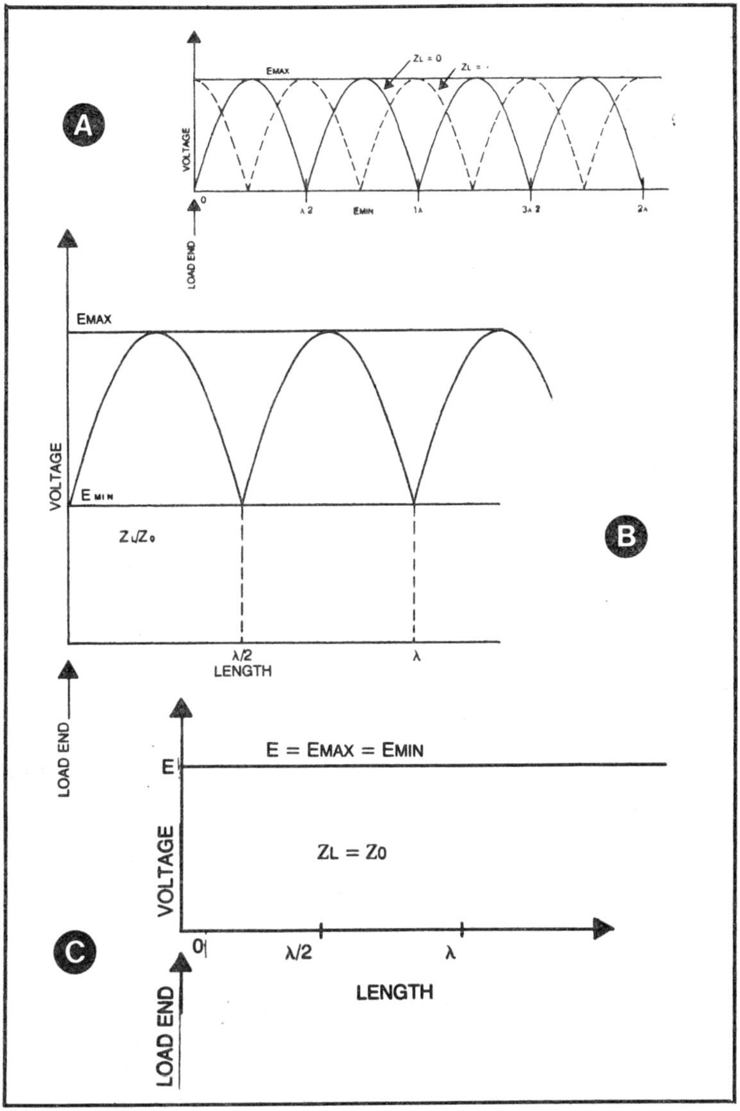

Fig. 26-5. Standing wave conditions on transmission lines: (A), when load is open or shorted; (B), when load and surge impedances are unequal but nonzero/noninfinite; (C), when load and surge impedances are equal.

Some of the energy is radiated into free space. This means that the reflected wave will have a lower amplitude than the incident wave, by an amount that accounts for the energy radiated. This means that the cancellation of the forward wave by the reflected wave is far from total. The graph in Fig. 26-5B shows this situation; note that the voltage at the nodes is not zero. The *standing wave ratio* (SWR) is given by:

$$SWR = E_{max}/E_{min}$$

or, if current instead of voltage is measured:

$$SWR = I_{max}/I_{min}$$

Some textbooks designate these as VSMR (*voltage* standing wave ratio) and ISWR (current standing wave ratio). We may also calculate the SWR of an antenna system from the following:

$$SWR = Z_0 / Z_L$$

Where:

Z_0 is the surge impedance of the transmission line and
Z_L is the antenna feedpoint impedance.

and, in the case were Z_0 is less than Z_L

$$SWR = Z_L / Z_0$$

or, if power measurements are made instead of voltage

$$SWR = \frac{1 + \sqrt{P_r / P_f}}{1 - \sqrt{P_r / P_f}}$$

and, finally,

$$SWR = \frac{E_f + E_r}{E_f - E_r}$$

Where:

E_f is the forward voltage
E_r is the reflected voltage
P_f is the forward power
P_r is the reflected power

One last case is shown in Fig. 26-5C. This case represents the situation where the load impedance (Z_L) is exactly equal to the transmission line surge impdeance (Z_0).

When the source and load impedances of any electrical system are equal, then we have a maximum transfer of power between the two. This is the case when the VSMR is 1:1, and means that all of the power applied to the transmission line is radiated. The line is said to be "flat." This situation never exists in real systems, but the approximations that are possible are quite good.

Measuring SWR

Any of the equations presented above for the determination of SWR can be used as the basis for an instrument to measure SWR. We can use the antenna impedance, the power in the forward and reflected waves, or the voltages and currents along the line.

All of the methods work, and we can all find numerous instruments on the market that use each of them. In most low-cost SWR instruments, either the line current or the line voltage will be used as the basis for measurement.

The use of low cost instruments, however, has led to one of the most persistent myths regarding SWR measurement, and it is especially prevalent amongest CBers. The erroneous belief is that you can "adjust" the SWR by trimming the coaxial cable between the antenna and the transmitter. The notion is totally false, but is given credibility by the fact that voltage- or current-based (low cost) "SWR meters" make it *appear* to be true. It is true that the VSMR numbers obtained from the measurement will reduce to 1:1 as you trim the coaxial cable. But that fails to take into account the fact that those numbers are valid *only* at the base (or feedpoint) of the antenna, and at electrical half wavelengths back down the transmission line—where the antenna impedance will be reflected. The transmission line must, therefore, be an integer multiple of an electrical half wavelength, or the readings obtained are invalid. The relative reading, however, will be accurate. When you use a cheapie SWR meter to reduce the SWR to a minimum, it will be at a minimum regardless of the coaxial cable length. The correct length for the coaxial cable will be

$$L = \frac{492\ N\ V}{F}$$

Where:

L is the length in feet

V is the velocity factor of the transmission line

N is an integrator (1, 2, 3 . . .)

F is the frequency of operation in megahertz (MHz)

Another limitation on the use of the cheapie SWR meter is in systems that are grossly improper. If the antenna length is too far off, the SWR will be high, and in that area the errors are terrific. Such antennas have a large reactive component to their impedance, and that will interfere with the correct measurement of SWR.

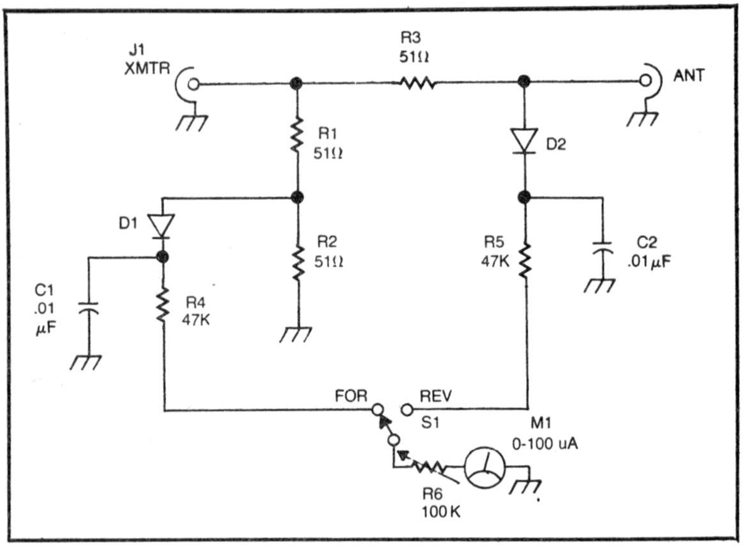

Fig. 26-6A. Resistance-bridge type SWR meter.

But, do not get the idea that the low-cost SWR meter is a hopeless case! These meters are very valid if we understand their limitations, and what the readings they give mean (or don't mean). If you use the SWR meter as a relative indicator to adjust your antenna, then the results will be just as good as if you had used a more costly instrument. The numbers obtained will not be the same, but the working of the antenna will be equally good. The SWR minimum will be located at the same points on the adjustment, regardless of the type of meter used.

Understand that the cheapie SWR meter is a good tool, but do not take the numbers seriously if the cable electrical length is unknown, or if the antenna impedance is grossly reactive.

If a true rf watt-meter is used to measure the forward and reflected *power*, then we may plug the resultant readings into the power SWR formula to obtain VSMR. Alternatively, if you use a Bird Electronics Model 43 Thru-line rf watt-meter, then use the nomograph based on that formula, as supplied by Bird with the instrument.

SWR Meters

In addition to the rf watt-meter, there are several different types of low cost SWR meter. The circuits for two of these are shown in Fig. 26-6. These instruments are voltage- and current-based, respectively.

The *SWR bridge* (a voltage-based instrument) is shown in Fig. 26-6A. It is basically a Wheatstone-bridge circuit. The bridge consists of four arms: resistances R1 through R3, and the radiation resistance of the antenna being measured. The transmitter, or a low-power rf source such as a signal generator, is used to produce the exitation potential for the bridge. Note that this circuit is a *resistance* bridge, so is incapable of measuring the reactive component.

The circuit in Fig. 26-6A is set up to measure the radiation resistance of a 50-ohm antenna because resistor R3 is set at 51 ohms. If this resistor were changed to 75 ohms, then the bridge will also work accurately in 75 ohm systems (such as dipoles). In practice, many amateur SWR meters of this type are designed to

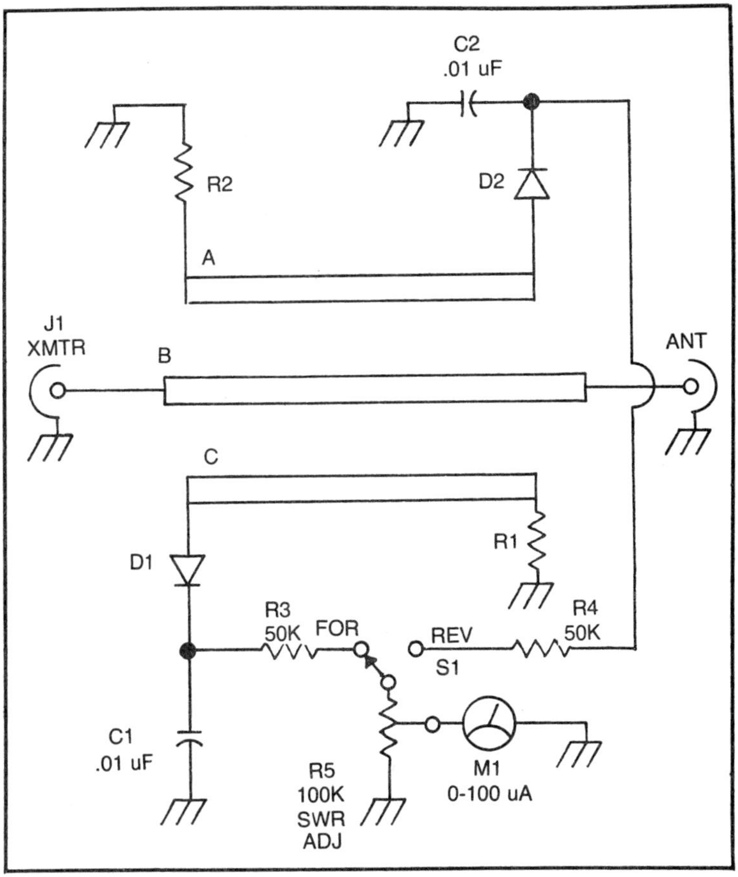

Fig. 26-6B. Current-transformer, or "monimatch" instrument.

work on 50-75 ohms systems by setting the value of R3 to 68 ohms. This results in an error, but a negligible error in most cases, at the gain of a wider-range instrument.

The operation of this circuit, and all similar circuits, is as follows:

1. Set S1 to *forward*
2. Adjust R6 for a full-scale deflection
3. Set S1 to *reflected*
4. Read VSMR from the special meter scale.

An alternate form of SWR meter is shown in Fig. 26-6B. This instrument is based on the current transformer, and can be left in the line while transmitting. The loss through the instrument is small enough to be ignored. The heart of this instrument is the pick-up unit, labeled A through C.

Conductor B is the center conductor of the transmission line, while A and C are identical to each other, and are spaced equidistant from B. In older units conductor B was a piece of coaxial cable, while the other two conductors were small diameter enameled wires threaded beneath the shield braid of the cable. In most modern instruments, however, the pick-up unit is made of foil patterns on a printed-circuit board.

Resistors R1 and R2 are selected to have a resistance equal to the surge impedance of the transmission line. Again, in some units, the manufacturer offers a trade-off situation by using a 68-ohm resistor to accommodate both 50- and 75-ohm antennas with equal (but negligible) error.

The operation of this unit is the same as for the voltage bridge previously discussed, so the instructions will not be repeated here.

ANTENNA IMPEDANCE MEASUREMENTS

There are several different types of antenna impedance bridge on the market today, but almost all of them are based on one or more variations of the Wheatstone bridge. The basic form of the Wheatstone bridge is shown in Fig. 26-7A. In most instruments, Z4 is the antenna radiation resistance, Z3 is fixed and both Z1 and Z2 are variable.

A simple resistance bridge of the form shown in Fig. 26-6A will often be used, but this type of circuit measures only the radiation resistance component of the antenna impedance. The reactive component is totally ignored. This is not too much of a problem, because most Amateur antenna systems are operated near reso-

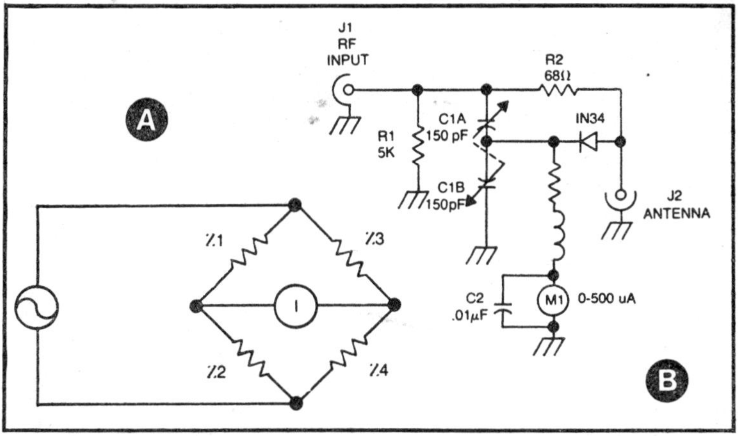

Fig. 26-7. Basic Wheatstone bridge (A), implementation with dual capacitor (B).

nance where the reactive component is low, or even zero. Bringing an antenna to resonance can be done without the use of a complex impedance bridge.

Figure 26-7B shows another form of antenna impedance bridge in which Z1/ Z2 is replaced with a differential capacitor, C1. This instrument is also limited to measuring the resistive component of impedance.

A low-cost, commercial radiation-resistance bridge is shown in Fig. 26-8. This is the Leader Model LIM-870A, and is especially designed for Amateur radio use. It will measure impedance from near-DC to 150- MHz. The range of resistances that it will measure is 0 to 1000 ohms.

Although none of the bridges offered thus far can measure the reactive component of the antenna impedance, they can give an experienced operator a subjective appraisal of the reactance by the depth and broadness of the null when measuring the radiation resistance. If the reactance is zero, or very low, then the null obtained will be deep and very narrow. But, as the reactive component increases, we now find that the null becomes less deep and becomes much broader. While this is a subjective technique, probably disdained by measurement purists, it offers the low-budget user some advantage.

We can modify the basic bridge circuit to obtain a relative indication of reactive component. In some cases, a balancing network (see Fig. 26-9) consisting of a series inductor and a variable capacitor is used. This forms a series-resonant circuit tuned to the antenna's frequency of operation. The inductor (L1) forms a *balance*

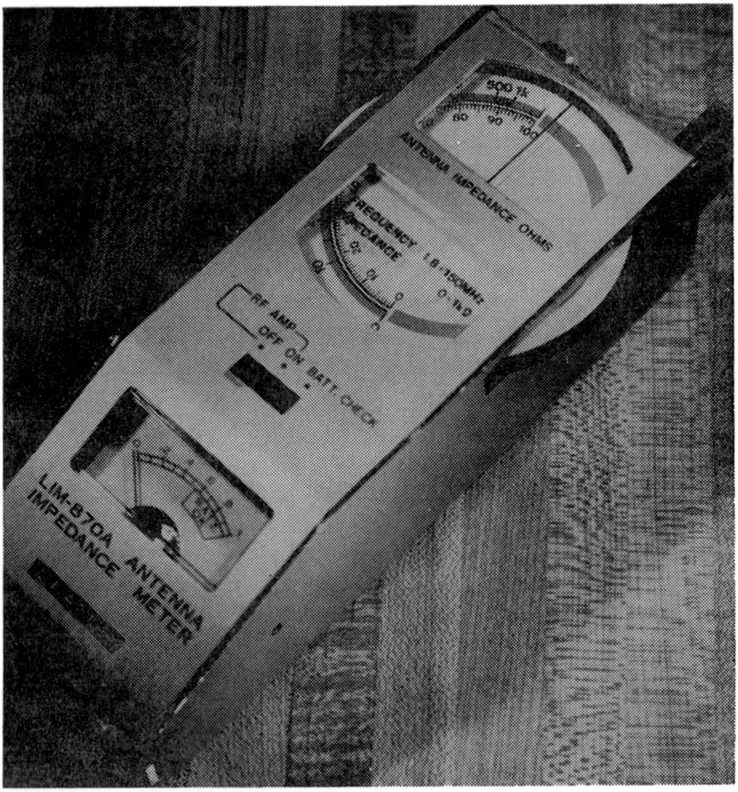

Fig. 26-8. Typical amateur antenna-impedance meter.

control, and is set to resonate with C2 when C2 is exactly in the center of its range. The impedance of L1/C1 under this condition is resistive, and will have a very low value.

Another technique that is often employed is shown in Fig. 26-9B. In this circuit, two capacitors are used, C2 and C3. Capacitor C2 is calibrated in nondimensional units, and is variable. This type of circuit requires the capacitor to be of the "straightline capacitance" construction. Capacitor C3 is a fixed, precision, type that has a value equal to that of C2 at midrange. The two capacitances, then, will cancel each other out in the balanced bridge circuit.

The dial connected to C2 reads zero at midscale, and in nondimensional units either side of midscale. If the antenna is purely resistive, then C2 will null the bridge at midscale. But if it is reactive, then the null will occur on either side of midscale. If the antenna has a capacitive reactive component, then C2 will read lower in scale, and if it is inductive, then C2 will read higher.

Note well that these measurements are not absolute in any sense of the word. We must obey the same constraints here as for the simple SWR meters discussed earlier. The bridge must be connected to the antenna at the feedpoint, or through a coaxial cable that is an integer multiple of one half wavelength. Note also, that, since we are looking for numerical results, this constraint is even greater than in the SWR meter, where relative readings were useful . . . there is no such thing as "relative impedance."

Noise Bridges

A noise bridge is a type of Wheatstone-like bridge circuit that uses a "white-noise" (well, would you believe off-white?) generator as the signal excitation source. This type of bridge has certain well-known advantages over other types. Two examples of commercial noise bridges intended for the Amateur Radio market are shown in Fig. 26-10. The Palomar Engineers RX bridge is shown in Fig. 26-10, while the Omega-T noise bridge is shown in Fig. 26-10B.

A representative circuit for a noise bridge is shown in Fig. 26-11. The wideband noise signal is generated by a reverse biased Zener diode, and this produces a wide band of white noise. This noise is then amplified in a wideband (150 MHz) amplifier, and is

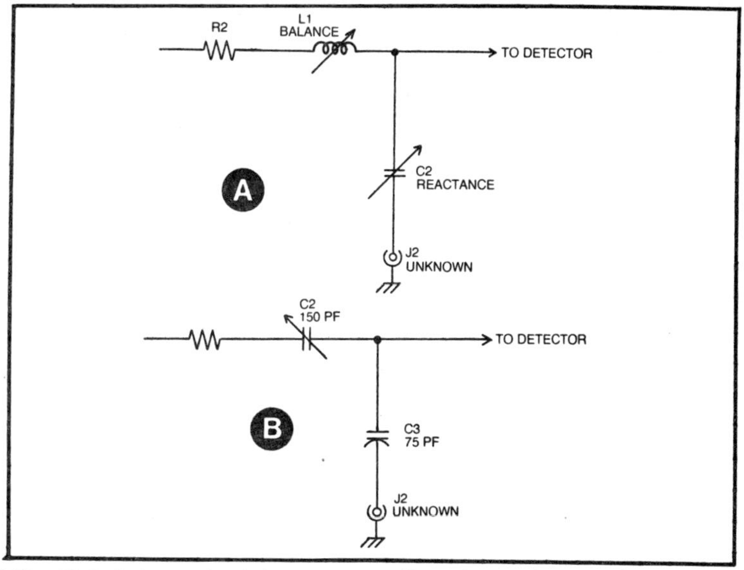

Fig. 26-9. Balance circuit for reactance (A, and capacitor compensation for reactance measurement (B).

Fig. 26-10. Antenna noise bridges: (A), Palomar Engineers, (B), Omega-T.

then applied to the primary of a special trifilar wound transformer, T1. The ratio of T1 is 1:1:1, and all three windings are trifilar wound on a toroid core. The dots shown indicate the same ends of L1, L2, and L3.

One end of L2 is connected to a noninductive potentiometer that has a dial calibrated in ohms, from zero to the full-scale range of the potentiometer. In most cases, this will be 250 ohms, but some have been published that used 1000 ohms. Also connected to L2 is a variable capacitor that has a zero-center dial in the same manner as discussed previously. Capacitance C1 is exactly balanced by C2 so that C1 = C2 when X = 0.

The detector used with a noise bridge is a general-coverage communications receiver that is equipped with an S-meter (actually the S-meter is a highly desirable feature). The receiver will show a great noise level at all frequencies except the antenna's resonant frequency.

There are several principle uses of the noise bridge in Amateur Radio work. One could, for example, find the resonant frequency of an antenna, the correct physical length of a piece of coaxial cable that is to be a certain electrical length, and certain other applications. Let us now examine how some of these are done.

Antenna's resonant frequency. There are two basic forms of noise bridge; one uses C1/C2, while other does not. There are, then, two separate procedures for using these two different bridges. For those bridges that lack C1/C2:

1. Set R1 to the anticipated radiation resistance of the antenna being tested (50 ohms or 75 ohms).
2. Tune the communications receiver to the anticipated resonant frequency for the antenna.
3. Adjust the receiver tuning *very slowly* above and below the design frequency until a null, or dip, is noted in noise level. The S-meter is especially helpful in this measurment.
4. Adjust the receiver dial and R1 for the deepest null obtainable. The receiver dial is tuned to the antenna's resonant frequency, while R1 is set to a value equal to the radiation resistance. Note that, in most low-cost instruments, the

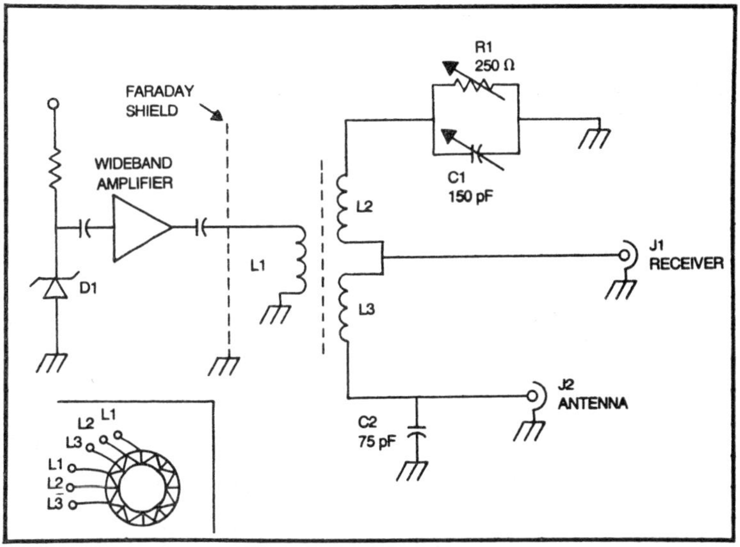

Fig. 26-11. Simplified noise-bridge schematic.

receiver tuning and R1 are slightly interactive, so repeat the measurement several times to find the optimum result.

If the noise bridge uses C1/C2, then use the modified procedure below:

1. Set R1 to the anticipated resistance value
2. Set C1 to midscale (zero)
3. Tune the receiver to the design frequency of the antenna
4. Adjust C1 for a null, or dip, in the noise level. The receiver S-meter is especially useful in this respect.
5. If the null occurs when C1 is in the X_c region, then the antenna is too long. But if the null occurs in the X_L region, then the antenna is too short.
   a) If C1 indicates in the X_c region, tune the receiver downband to the resonant frequency, as indicated by a deepening of the null.
   b) If C1 indicates in the X_L region, then tune the receiver upband to the resonant point.
6. As in the previous example, these adjustments are all interactive. Adjust R1, C1 and the receiver dial for the best null.

Coaxial Cable Length

The physical length of a piece of coaxial cable will vary from the electrical length because of the velocity factor of the line. The velocity factor will be expressed as a decimal fraction, such as 0.66, 0.80, etc. If the velocity factor is known, then we may multiply the half-wavelength, obtained from the formula 492/F, by the factor to find the correct physical length of a half wavelength of cable. We do not always know the velocity factor of any given piece of transmission line. How do we find an electrical half wavelength in that case? We must find a technique for locating half-wavelength points along the line. A noise bridge can be used, in conjunction with a receiver, to make such a determination.

1. Use the equation $492/F_{MHz}$ to find the half wavelength in *feet*.
2. Set the *X* control on the noise bridge for zero, and the *R* control to some resistance just slightly above zero (such as 2 - 5 ohms).
3. Connect one end of the cable to the *antenna* terminal on the noise bridge, and short-circuit the other end.

4. Tune the receiver to the frequency at which the cable should be one-half wavelength, and then tune slowly downband until a null is heard. A general-coverage receiver is best for this, because the null point will probably be located below the Amateur band.
5. Trim the cable length, approximately 1 - 3 inches at a time.
6. Each time the cable is trimmed, short out the end and repeat the above steps. You should notice that the null frequency moves upband each time.
7. Continue steps 1 - 6 until the null frequency is the same as the required frequency.
8. This is only approximate, because the noise-bridge *R* control was set near zero, not at zero. Now check the result by using a noninductive, carbon composition resistor in the 10 - 250 ohm range in place of the short. The *R* controll should null at the value selected.

Finding the Velocity Factor of Coaxial Cable. We can also use the noise bridge to find the velocity factor of a piece of coaxial cable. We may rearrange the standard formula to the form below:

$$V = (L \times F)/(492N)$$

Where:

V is the velocity factor
L is the length in feet
F is the frequency in megahertz (MHz)
N is an integer (1, 2, 3, . . . N)

(Convenience usually dictates setting N=1 at low frequencies.)

1. Cut a test length of coaxial cable. The length selected should be a physical half-wavelength at some frequency well within the tuning range of your receiver. The frequency (or length) selected should not be near a band edge.
2. Calculate the free-space frequency of the cable from $F = 492/L_{(ft)}$
3. Connect one end of the cable to the *antenna* terminal on the noise bridge, and short-circuit the other end.
4. Set the *R* control on the noise bridge in the range 2-5 ohms.
5. Tune the receiver until a null in the noise is heard.
6. Using the length *L* found in Step no. 1, and the frequency from step no. 5, use the equation above to calculate the velocity factor.

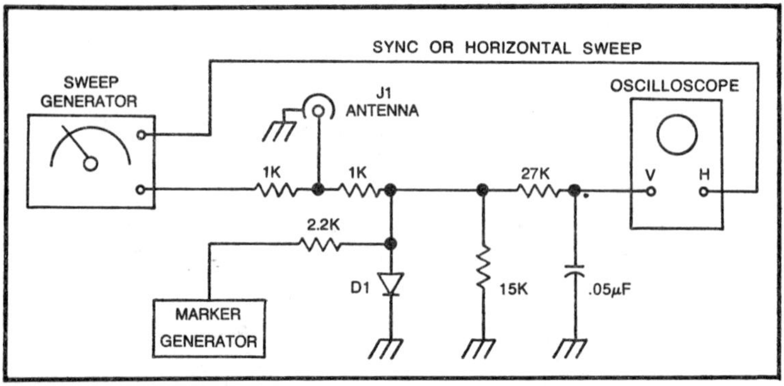

Fig. 26-12. Test equipment for swept measurements of antennas.

Using a Sweep Generator

A frequency-modulated (FM), or sawtooth-modulated, sweep generator can also be pressed into service to "wring out" antennas and transmission lines. The equipment needed to make these measurements is shown in Fig. 26-12. The signals from the sweeper and a crystal-controlled marker generator, which is used to help identify points on the curve displayed on the oscilloscope, are combined in a nonlinear resistor/diode adder circuit. The output of the adder is a DC voltage, and remains relatively constant at frequencies near resonance, a null or dip is noted. A sample oscilloscope trace is shown in Fig. 26-13. This particular trace is common to transmission lines that are shorted at one end. By noting the frequencies at which the marker "pips" occur we can determine at which frequency this cable is one-half wavelength.

ANTENNA GAIN MEASUREMENTS

One of the favorite bugaboos in advertisements aimed at users of two-way radio equipment is the gain of the antennas offered. There has been so much "creative spec writing" in the matter of antenna gains, that at least one Amateur Radio publication refuses to permit the gain of an antenna to be stated in the advertisements. The situation among CB operators is even worse than among Amateurs (who presumedly have *some* insight into the matter). Some CB antennas are claimed to have +40 dB of gain, which, when exicted by a kilowatt linear, should be able to light cigarettes in Europe!

We cannot, in this text, get too deeply into the different ways to "measure" antenna gain. We will limit the discussion to ways in which you can make valid *comparisons* between antennas. The numbers obtained will be valid, in many cases, but you are advised that the art of making these measurements is such that you should take the results with a grain of salt.

Two different types of figures are bandied about in talking about directional antennas: *forward gain*, and *front-to-back ratio* (or *front-to-side ratio* in some antenna designs).

The antenna gain is not really "gain" at all. A power amplifier produces gain by drawing power from a DC power supply, and converting it to rf energy. But the "gain" we obtain in an antenna system does not increase the *total* energy one bit. The gain is the result of *redirecting* the energy delivered by the transmitter in one, or two, directions. In a dipole, for example, we direct the energy in two directions, forming a "figure-8" pattern. In a Yagi antenna, which is said to have "gain" over a dipole, we force all of the available energy into one direction. We will at least "gain" a two-fold increase in the energy content of the signal in that direction, and more if the beam width is narrower.

How do we measure gain of antennas in the absence of an antenna test range or indoor test chamber? Figure 26-14 shows a test configuration for measuring the front-to-back and front-to-side ratios, while Fig. 26-15 shows a method for determining forward gain.

In both cases, the antenna is installed in a way that allows it to be rotated. A constant-output test signal is generated by a transmitter. This test signal should be located at least "many wavelengths" from the antenna, and must have a constant level. This is necessary to ensure that the results obtained are due to antenna-gain differences, and not changes in the level of the test signal. Beware of using distant stations, especially those that are "skip." These stations tend to fade a lot, and could easily fade enough during the course of the measurement to invalidate results.

To find the front-to-back or front-to-side ratios, we must determine the relative difference between the signal strengths

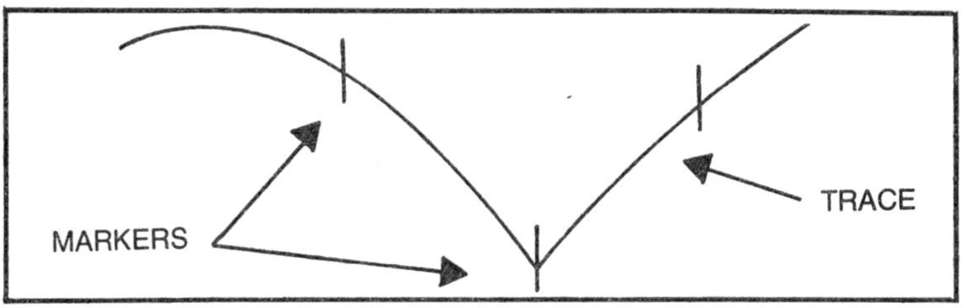

Fig. 26-13. Trace from swept measurement.

when the antenna is pointed *toward* the signal source, and when the antenna is pointed away from (F-to-B) or at 90 degrees to, (F-to-S) the signal source. The signal-strength difference, in decibels, is the measurement sought.

The receiver's S-meter will give a rough indication of the decibel difference, but only if you know that its S-unit/dB conversion factor is accurate (most are not!). A better method is to use a calibrated rf step attenuator (see the ARRL *Radio Amateur's Handbook* for designs) in conjunction with the receiver's S-meter. The S-meter is then used as a relative indicator for an "equal-deflection" measurement. In this method, we first aim the antenna away from the signal source, and set the attenuator to zero (it is out of the circuit). Note the exact position of the S-meter pointer, and mark it on the S-meter face if necessary. Now, rotate the antenna so that it faces the signal source. The S-meter reading should have increased markedly. Now, adjust the attenuator, in small steps, until the S-meter reads exactly the same as it did before. The amount of attenuation needed to cause an equal deflection of the S-meter pointer is the front-to-back, or front-to-side, ratio of that antenna.

If the signal source was a distant station, then it may have faded enough when you made the rotation to invalidate your measurements. The sure for this is to make the measurements on a local station, no more than a few miles distant. If that is not possible, then try making the measurement at a time when the band in use seems stable. Even then, make the measurement eight or ten times, with a view towards averaging them together. Foreign broadcast stations generally make better signal sources than Amateur DX stations because of their higher power.

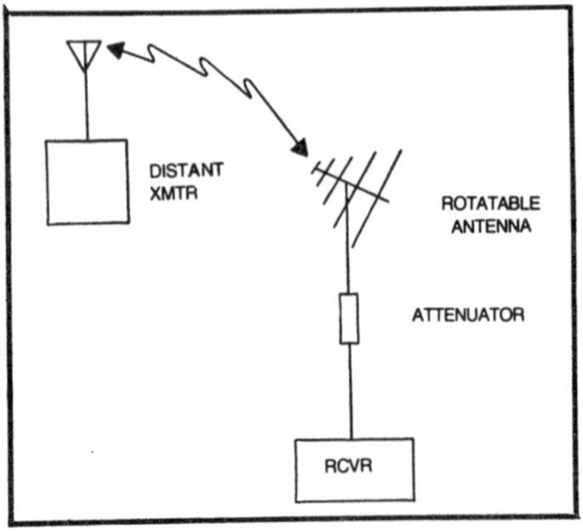

Fig. 26-14. Front-to-back ratio measurement.

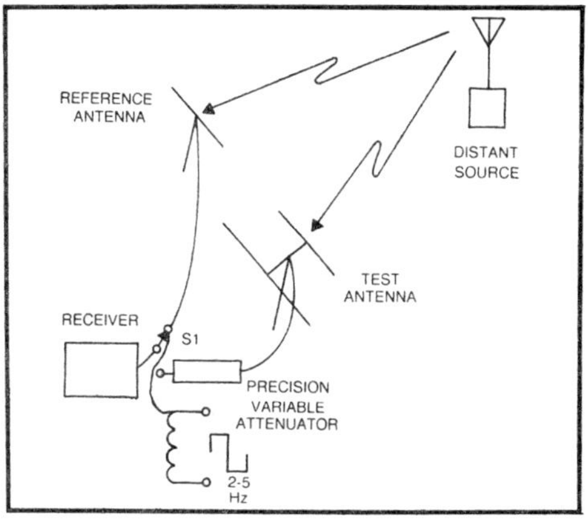

Fig. 26-15. Forward-gain measurement.

The measurement of antenna forward gain is done similarly, and with a similar test set-up. As amateurs, we do not have an easy way to measure antenna gain relative to isotropic radiators, so must use a comparison with a dipole reference antenna. Once again, we use a distant station (cross town or, if absolutely necessary distant in the "DX" sense of the word).

Switch S1 in Fig. 26-15 is an electrical coaxial relay, and is pulsed by a squarewave to switch between the two antennas at a rate of 1 to 5 Hz. A precision step-attenuator is inserted in series with the transmission line of the higher gain antenna.

We can expect the S-meter on the receiver to slowly oscillate back and forth between two points because we are connecting the antenna input of the receiver alternately to two signals of different strengths.

The step attenuator is adjusted so that the difference between the two signals alternately applied to the receiver input becomes zero (or at least negligible). The S-meter oscillations will become less and less, until they cease altogether when the signals from the two difference antennas have equal strengths. The amount of attenuation needed to stop the oscillation of the S-meter is the gain of the test antenna over the reference antenna (usually a dipole cut to the same frequency). If the dipole reference antenna has more gain than the test antenna, then it is usually the practice to write the number with a minus sign. For example, a mobile antenna may be found to have a gain of −0.7 dBd (dB gain compared to a dipole). This means that the test antenna has a gain that is 0.7 dB less than the gain of the dipole (whatever that is). A three-element beam, on the other hand, may be found to have a gain of 7.6 dB greater than a dipole, so this would be written 7.6 dBd.

Chapter 27

Testing Radio Transmitters

Radio-transmitter measurements are relatively few in number, but they must be made to verify that the transmitter is operating in accordance with the FCC rules and regs, with good engineering practice, with best performance. We might also have to make certain measurements when troubleshooting the transmitter, or when we first buy, or build, the beastie in order to verify the manufacturer or designer's specifications.

RF POWER MEASUREMENT

The measurement of rf power is a little more difficult than the measurement of audio or DC power because the instruments become more difficult to accurately calibrate as frequency increases. However, even homebuilt, crudely designed, rf-power meters are acceptable when accuracies of only ± 20 percent or so are required. In fact, there was a time when ± 20 percent accuracy was deemed sufficient. Certain low-cost professional instruments will measure rf power to within ±5 percent, and even better accuracy is possible if you are willing to spend the money.

Fig. 27-1 shows a method often used in A-M broadcasting, and in the old days when marine pleasure boat radios were A-M and operated in the 2 - 3 MHz band. Here we have a thermocouple rf ammeter connected in series with the antenna or dummy load, so as to measure the line current supplied to the load by the transmitter.

Thermocouple meters are inherently rms-reading devices, so will yield a good measurement of the rf power present in the circuit.

If the load is restrictive, and it should be, then we can calculate the power from the current using the expression $P = I^2R$, where P is in watts, I is in amperes, and R is in ohms.
Example:

Calculate the rf power dissipated in a 50 ohm resistive load if a line current of 2 amperes is flowing.
Solution:

$$P = I^2R$$
$$P = (2)^2 (50)$$
$$P = (4)(50) = 200 \text{ watts}$$

Most commercial rf ammeters currently being sold are usually valid up to 50 or 60 MHz without requiring any frequency compensation to the reading. Beware of older types, such as might be bought at a hamfest or surplus junk-dealers shop, because some of the older models were very sensitive to changes in frequency. Those are marked (most of the time) "calibrated on a ⅛-steel panel," or something similar.

Rf voltmeters can also be used to measure power although there are some problems here. Fig. 27-2 shows the use of a simple rf voltmeter to measure rf power. Again we depend upon a simple, well known, basic relationship to form the measurement. In this case we are looking at $P = E^2/R$. But, the problem we find is that the current is a peak-reading device, and the peak voltage can be related to the power only when the rf waveform is a sinewave. This is not often a problem in Amateur applications, because the key-down. CW, output signal signal is supposed to be a sinewave—or else someone will complain about four severe harmonic radiation.

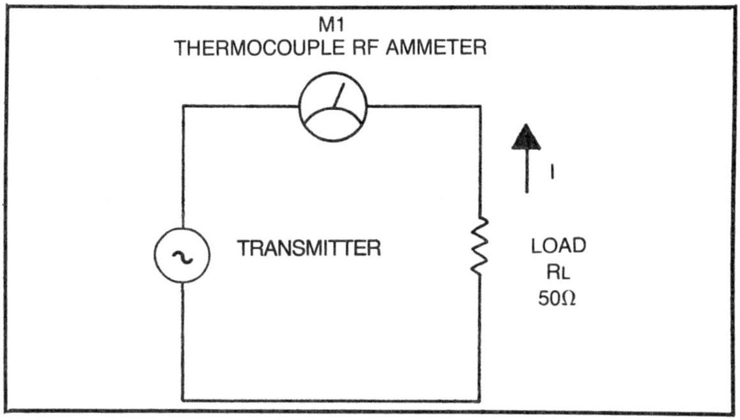

Fig. 27-1. Thermocouple RF ammeter used to measure rf power.

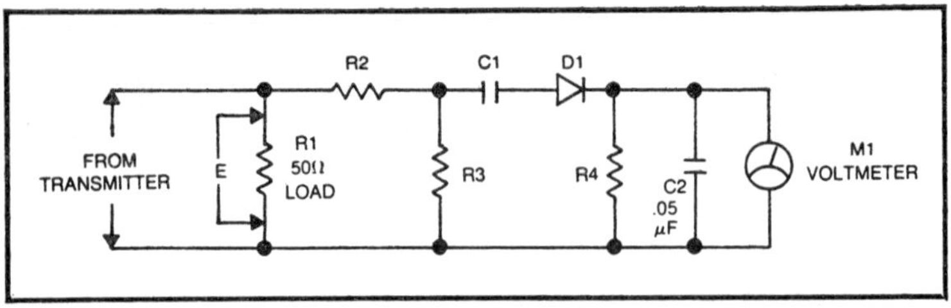

Fig. 27-2. RF voltmeter to measure power.

The voltage read on the meter in Fig. 27-2 will be proportional to the rf power dissipated in the load. If the power waveform is sinusoidal, then we may safely calibrate the face of M1 in units of power.

The simple rf-power meter in Fig. 27-3 is now very popular, and this circuit forms the basis of many of the lowcost amateur rf wattmeters and VSWR meters now on the market.

The heart of this type of rf wattmeter is the toroidal current transformer that delivers an output voltage proportional to the line current flowing in the transmission line. This rf voltage is rectified, filtered, and then used to drive a DC-current meter.

Capacitors C1 and C2 are used in this circuit to compensate for frequency, so the instrument will operate reasonably well over the range 2-30 MHz.

Note that the thermocouple meter, being inherently rms-reading, provides the most valid measure of power. However, it is not too linear, and much scale-crowding is noted at low-current levels. The main use of thermocouple meters, then, is to calibrate the other types.

Most of the really accurate power-measuring instruments are based on the measurement of the heat created by the power. Whenever any electrical current flows through a resistance, heat is generated. In fact, some textbooks use a heat analogy with DC to explain the meaning of the term rms power. The same idea gives us a means for accurately measuring rf power.

One wag has suggested that we measure the power output of Amateur power amplifiers in the full-gallon (1-kW) range by using a thermometer mounted on the wall of the shack, and a dummy load. We would theoretically divert all of the power from the antenna into the dummy load, and then measure the temperature rise in the room! While this was intended as a joke, it hits very near the mark.

One early thermal rf wattmeter placed the dummyload inside of a container of oil, and a mercury thermometer, visible from the outside, was used to measure the rise in temperature of the oil. A nomograph, or simple formula, could then be used to calculate the rf power from knowledge of the temperature change and the time required to make the change.

This method is too cumbersome for day-to-day work. Modern thermal r-f wattmeters, however use something similar. A dummy-load resistor is placed inside of a thermally isolated chamber (see Fig. 27-4A), along with a thermistor, or other fast-acting thermal-sensor device. The external connections of the sensor go to an electronic thermometer circuit that has on output-display device calibrated in units of power (watts) instead of degrees.

Thermal sensors have the advantage of being inherently rms reading, and will operate over a wide frequency range. While the frequency response of the thermal sensor might be very slow (milliseconds to seconds), the frequency response of the instrument is the frequency response of the dummy load. VHF and UHF dummy loads are well within the range of Amateur builders.

One limitation of most basic thermal wattmeters is that they are limited by the amount of power that they will handle. Some

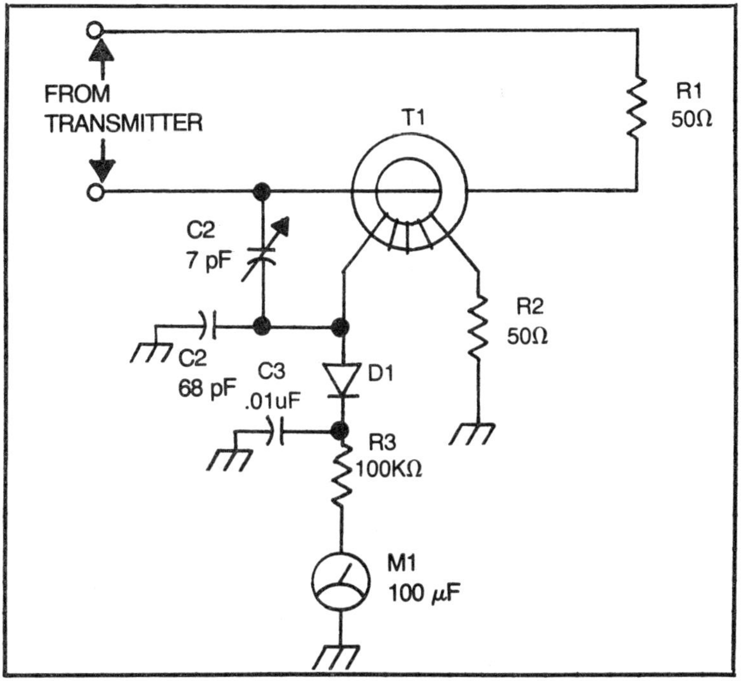

Fig. 27-3. Current transformer RF watt-meter.

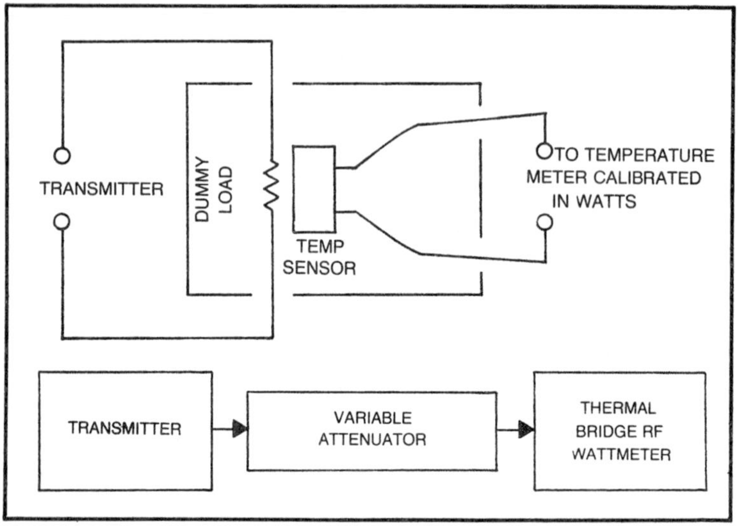

Fig. 27-4. (A), thermal method for measuring rf power; (B), method for increasing full scale range.

models, for example, will safely dissipate only 100 milliwatts, and that is one of the most accurate models on the market! Figure 27-4B shows the proper method for extending the range of the thermal wattmeter. Here we see an attenuator in series with the thermal wattmeter. This is a precision device that will reduce the amount of power to a level within the range of the wattmeter. The actual attenuator used may be a precision step atteuator, with the decibel steps marked on the switch, or individual fixed attenuators with the attenuation factor marked on the barrel of the device. Attenuation factors are marked in decibels, so we can find the actual power level produced by the transmitter by solving the following dB equation for P1:

$$dB = 10 \, Log_{10} \, (P1/P2)$$

Where: dB is the attenuation in decibels
P1 is the transmitter output power
P2 is the reading on the thermal wattmeter
Log_{10} refers to the common, or "base-10" logarithms

DUMMY LOADS

Dummy loads are important to most transmitter measurements for two reasons: 1) it is rude, and probably illegal, to measure the rf power on the air, if that is your sole reason for turning on the transmitter, and, 2) it is less accurate to use the antenna instead of a known load.

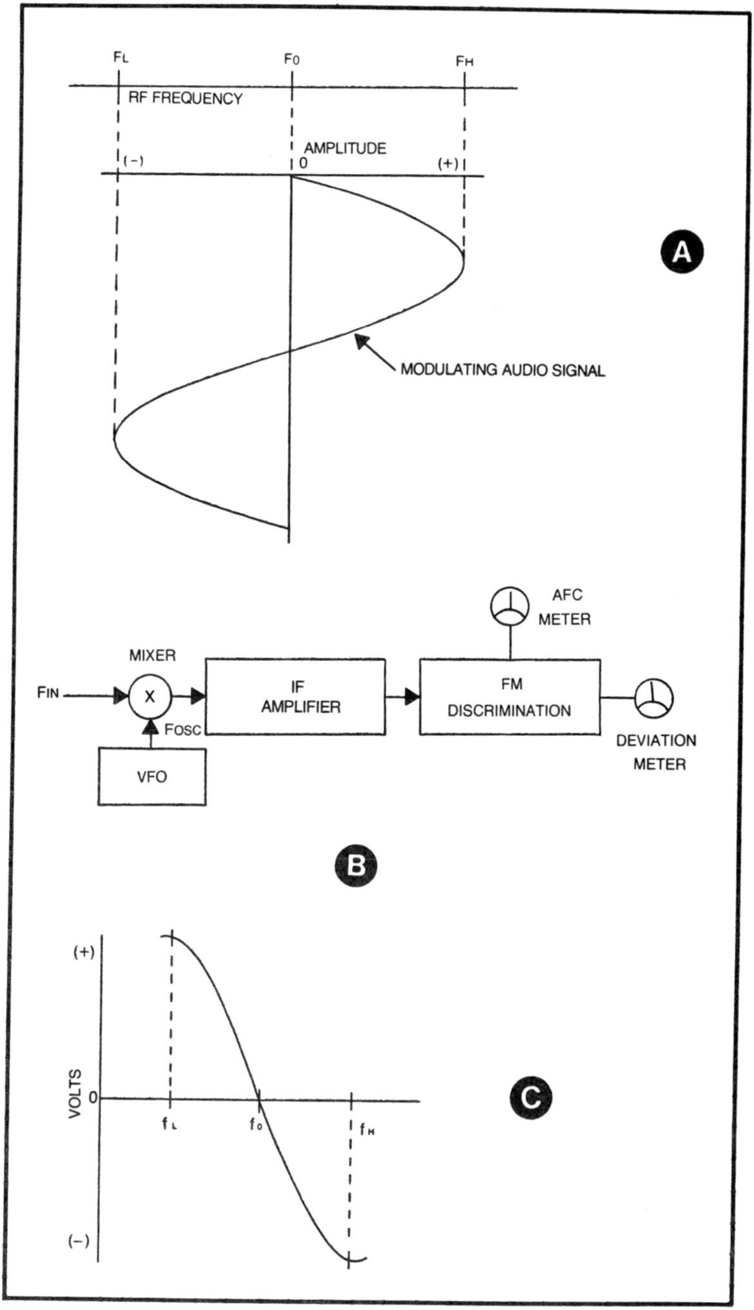

Fig. 27-5. (A), FM deviation; (B), FM deviation meter; (C), detector curve.

A dummy load must be capable of withstanding the rf power of your transmitter for more than a few minutes, although, in practice, underpowered loads are used if the time-on (duty factor) is kept low. Furthermore, it must have the correct resistance, usually 50 or 75 ohms in Amateur Radio work. Above all, the load must be nonreactive. This means that the resistors used to make the load must be noninductive types. Ordinary wire-wound resistors, usually used as power resistors, will not do! Carbon, or reverse-pitch wire-wound resistors (good to 30 MHz only) are needed.

The dummy load must be shielded, so that rf radiation does not escape. Sometimes, when adjusting transmitters, the harmonics and parasitics might not be fully suppressed, so would cause interference with other stations if allowed to escape. This is one prime reason why a well-shielded dummy load is superior to an active antenna for servicing transmitters.

The old-fashioned use of a light bulb as a dummy load is not a good idea. For one thing, the resistance changes as the bulb heats up, and for another, it is not shielded. Some people tried to solve the radiation problem by painting the light bulb with one of the metallic paints now available. They would leave a small "peep hole" in the paint so that they could get a rough (*very* rough) idea of output by observing brightness. That sounded like a good idea to me, until one day, on 40-meter CW, I worked a guy over 10 miles away who was loading his 200-watt transmitter into such a load! He was topping S-9!

MODULATION MEASUREMENTS

It is sometimes necessary to measure the percentage of modulation of both a-m and FM transmitters (we lump PM transmitters with FM in this case).

FM/PM Deviation. The relationship between the amplitude of the audio signal and the transmitter's output frequency is shown in Fig. 27-5A. When the audio amplitude is zero, then the transmitter frequency is the so-called "carrier frequency," or the frequency is the you think that you are operating on, F_o. When the audio-signal amplitude increases in a positive direction, then the carrier frequency will increase along with it, to a maximum value of F_h at the audio peak. Similarily, on negative excursions of the audio signal, the transmitter frequency will decrease, reaching a value of F_L on the negative audio peak.

A simple form of FM deviation meter is shown in block form in Fig. 27-5B. It is merely a receiver with an FM detector that has an output meter, and an afc meter to ensure that it is correctly tuned.

The output-voltage curve of a discriminator type FM detector is shown in Fig. 27-5C. The voltage on one side of the discriminator circuit, when the receiver is tuned directly on-channel, will be positive and is proportional to the negative deviation, while on the other side it is negative and proportional to the positive deviation.

If the deviation meter is equipped with a variable-frequency oscillator for the LO, as most are, then a meter on the automatic frequency control (afc) line is needed to insure correct adjustment of the VFO. The afc voltage will be zero when the unit is tuned to the exact center of the received signal. The output voltmeter, which is calibrated in units of frequency (for deviation) can be switched between the two sides of the detector for positive and negative deviation.

AM Measurements. In an amplitude-modulated system, the audio signal adds to, or subtracts from, the unmodulated rf carrier signal. This is shown in Fig. 27-6. The region marked "a" is the unmodulated carrier signal. Zones "b" and "c" are the modulated signal. In this case, a sinewave was used to modulate the carrier, producing the characteristic peaks and valleys shown. We can generate this modulation waveform on an oscilloscope, using a wire gimmick around the transmission line, as shown in Fig. 27-7, which loosely couples the transmitter signal to the scope input. The waveform will be as shown only when the audio signal applied to the transmitter microphone jack is a sinewave, and the scope timebase is adjusted to show just one or two cycles of the modulating frequency. The percentage of modulation is given by

$$\% \text{ mod} = (c - b)(100)/(c + b)$$

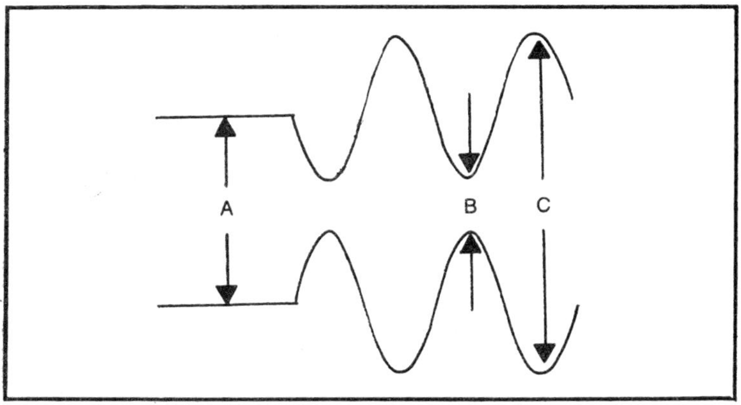

Fig. 27-6. Amplitude-modulated carrier wave.

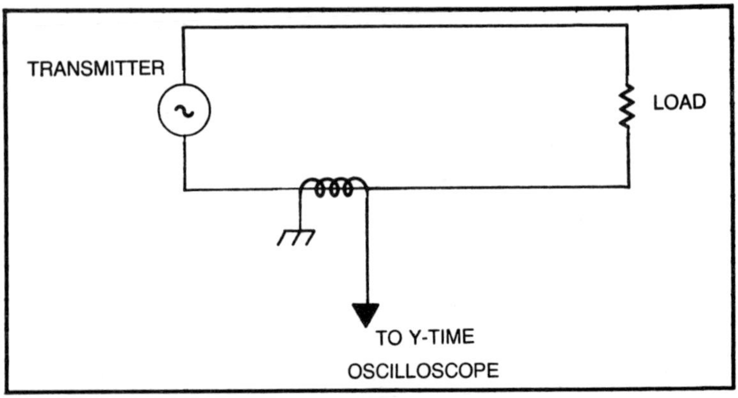

Fig. 27-7. RF pick-off to oscilloscope.

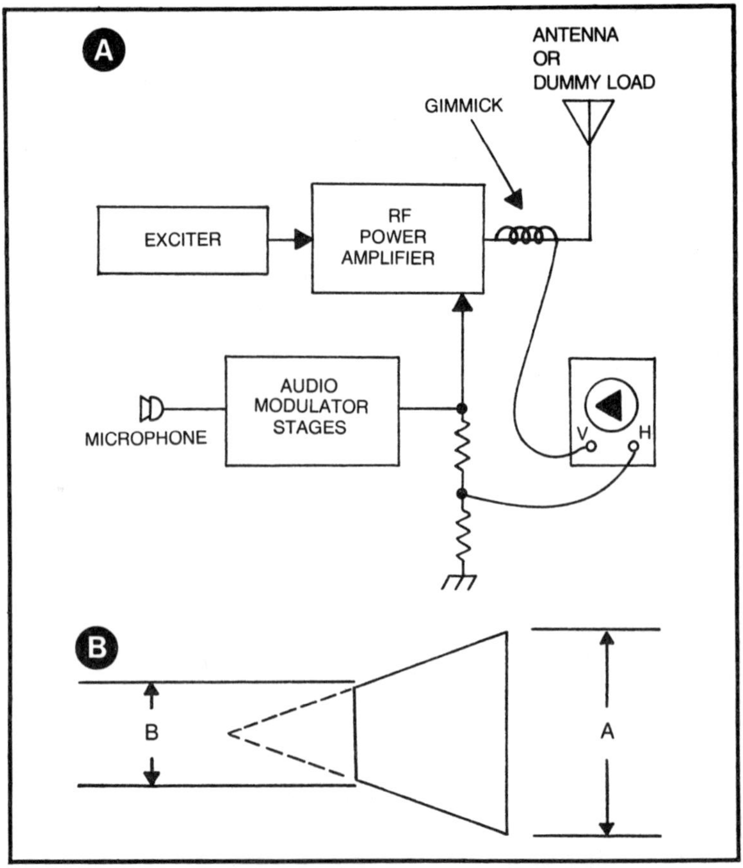

Fig. 27-8A. Generating the trapezoidal pattern. B. Trapezoidal pattern.

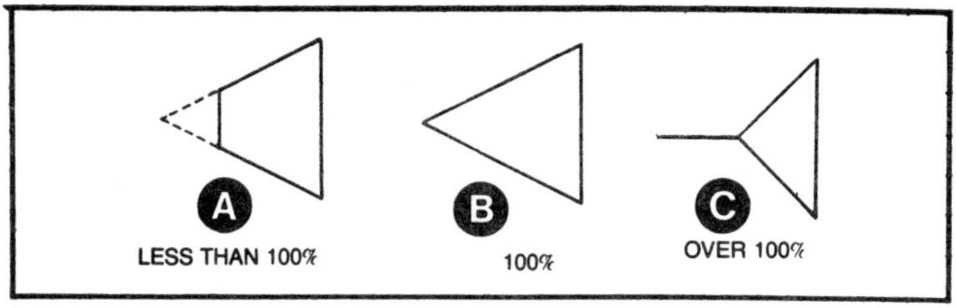

Fig. 27-9. Trapezodial pattern for (A), less than 100% modulation; (B), 100%; (C), over 100%.

Another method for measurement of sinewave percentage of modulation is the *trapazoidal pattern* on an oscilloscope connected as in Fig. 27-8A. In this case, a sample of the rf signal, containing the modulation, is applied to the vertical input of the oscilloscope, and a sample of the audio vertical input of the oscilloscope, and a sample of the audio modulating signal is applied to the horizontal input. Note that an oscilloscope with an "external horizontal input" is needed for this test.

An example of the trapazoidal pattern is shown in Fig. 27-8B. Using the notation of this figure, we can calculate the percentage of modulation from

$$\% \text{ Mod.} = (A-B)\ (100)/(A+B)$$

Figure 27-9 shows three typical trapazoidal patterns for less than 100-percent modulation (Fig. 27-9A), exactly 100-percent modulation (Fig. 27-9B), and overmodulation, i.e., greater than 100 percent (Fig. 27-9C).

SSB Measurements. A single-sideband, suppressed-carrier (SSSSC or, simply, SSB) transmitter is a special case of the AM

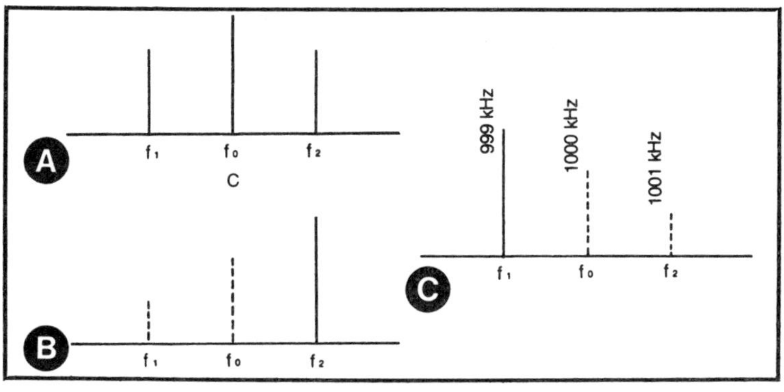

Fig. 27-10. (A) AM (DSB); (B) USB; (C) LSB.

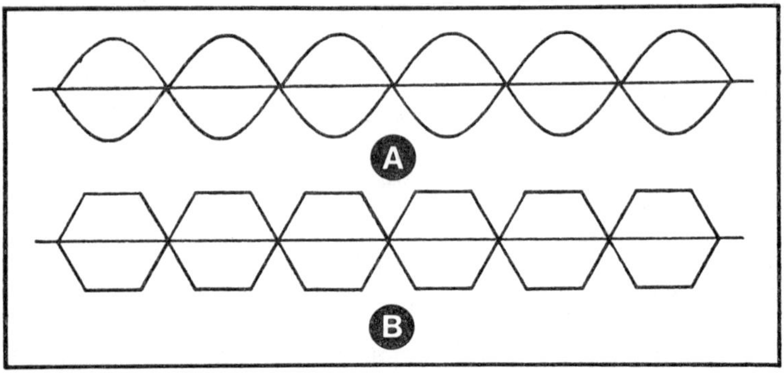

Fig. 27-11. Normal AM/SSB envelope. Flat-topping at B indicates overmodulation.

transmitter, in which the carrier and one of the two possible sidebands are suppressed. Figure 27-10 shows the frequency spectrum of a 1000-kHz transmitter, amplitude modulated by a 1-kHz sinewave (this set of frequencies was selected for the easy arithmetic!)

A straight AM transmitter will produce the spectrum shown in Fig. 27-10A. Here, the 1000-HKHz carrier is shown alongside of the sum (1001) and difference (999) frequencies created by the modulation process. These are called sidebands.

In Fig. 27-10B, the lower sideband (999 kHz) and the 1000-kHz carrier have been eliminated, and only the upper sideband (1001 kHz) remains. Similarily, in Fig. 27-10C, the upper sideband and the carrier are eliminated, and only the lower sideband remains.

The advantages of SSB transmission are, a) narrower bandwidth of the transmitted signal, and b) all of the power goes to transmitting the audio modulation, rather than splitting it three ways between two sidebands (only one of which is needed for communication) and the useless carrier.

If the SSB transmitter is connected to show the modulation envelope (Fig. 27-7), then we can observe the patterns as in Fig.

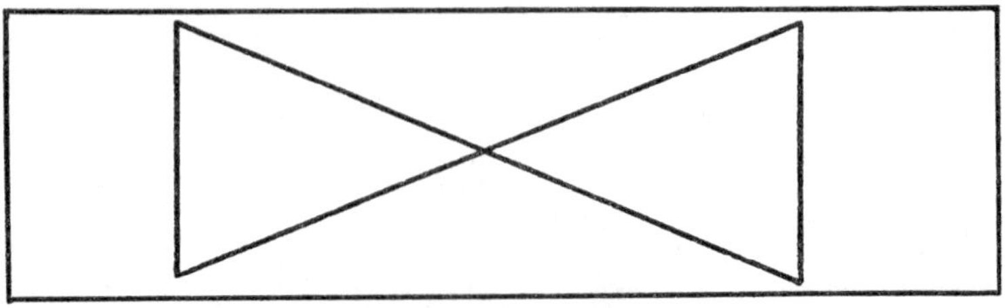

Fig. 27-12. Two-tone bow-tie pattern.

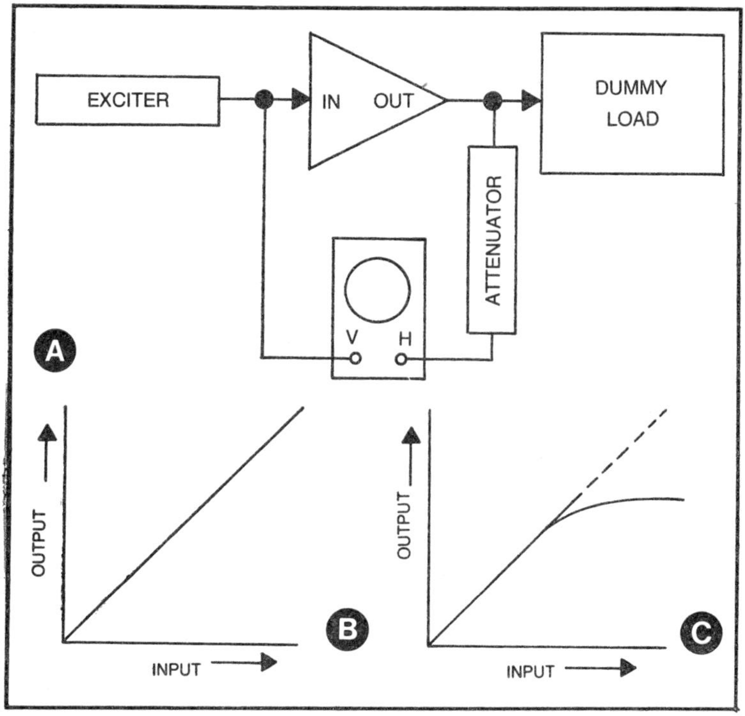

Fig. 27-13. Linearity testing circuit (A); linear pattern (B); nonlinear pattern (C).

27-11. Figure 27-11A shows the normal sinewave pattern, when the SSB transmitter is operating properly. But, if the audio level is too high, and the transmitter is overmodulating, we will see the flat-topping of Fig. 27-11. Many monitor scopes for SSB transmitters are nothing but formalized versions of Fig. 27-7.

If we connect the SSB transmitter to form a trapezoidal pattern, and then modulate it with a two-tone sinewave test signal, then we will generate the bow-tie pattern of Fig. 27-12.

The linearity of an amplifier can be checked on an oscilloscope. It is not possible to use the previously shown connections for this purpose, because they are not sensitive enough to show levels of distortion that are bad news. But, we can use an oscilloscope with an external horizontal input, as in Fig. 27-13, to make this determination. Here, we are making a Lissajous pattern, using the input and output signals of the linear amplifier as our scope inputs. A perfectly linear amplifier will produce a straight line (as in Fig. 27-13B) on the CRT screen, while non-linearity shows up as a curved line (Fig. 27-13C).

Chapter 28

Troubleshooting FM Transceivers

For many years, the use of frequency modulation (FM) was limited to a few experimenters. The mode never achieved a large following. In the early 1950s, Hallicrafters offered an FM transceiver for 10-meters, but only a few were sold or used on the ham bands. It was not until the advent of the VHF-FM repeater, a move orginally begun in California, that FM became a popular mode for amateurs.

Commercial two-way-radio users have been on FM for many years. The use of FM is standard practice in the land-mobile service. In fact, early Amateur FM stations depended upon used land-mobile commercial equipment (often *very* used!)

In the late 1960s, the FCC forced small-boat owners to switch to VHF-FM for interboat and ship-to-shore communications. The new marine frequencies were VHF, located just above the 2-meter Amateur band. The old type of boat radio was a 2-3 MHz A-M rig, and was subject to more types of interference than you can imagine . . . unless you have been there! The switch to VHF-FM made better communications for most small boat owners. Those who ventured off-shore any great distance could also use 2-3 MHz, but only on single sideband. The HF license was contingent on also having a VHF-FM license.

REPEATERS

Starting in the late 1960s, Amateurs took to 2-meter FM in a big way. Now, most Amateurs have at least something on 2-meters, so that they can take advantage of local repeaters.

A repeater greatly expands the distance that a low powered mobile can cover. Usually situated on a high location, with good quality equipment, a repeater can hear most mobiles, even those with a power-level too low to be heard by other mobiles in the area. Two frequencies are used for repeaters. The repeater station transmits on A and receives on B, while the users rigs are equipped just the opposite; they transmit on B and receive on A. The receiver in the repeater station will pick up the transmission of the mobile on frequency B, and demodulate it to recover the audio signal. The recovered audio is then used to modulate the repeater transmitter on frequency A. If the mobile rigs are equipped to receive on A, then communications between them, through the repeater, is possible.

The widespread installation of Amateur repeaters on 146 MHz, and the increasing number of repeaters active on the 220 and 430-MHz bands, has led to the offering of large numbers of FM transceivers, both mobile and portable. Some recent base stations seem to do everything but wash the dishes!

In the recent past, most commercial FM transceivers were crystal controlled, and could accommodate 6, 12, or 23 (I wonder where those crystal selector switches came from? channels. This was adequate for most local situations (there are some cities with more than 12 repeaters!), but could not be of much use when traveling from city to city. Today, many transceivers are offered that use a phase-locked-loop synthesizer which has up to 600 or 800 channels!

RECEIVER FRONT ENDS

The block diagram of a typical FM receiver, or the receiver section of an FM transceiver, is shown in Fig. 28-1. Most of these receivers are at least superheterodyne (the day of the 2-meter superreggerative mess is gone forever!), and most are double-conversion superhets.

The signals from the antenna are in the 144 to 148-MHz band, and are amplified by an rf amplifier stage. Note that the rf amplifier in most receivers serves three purposes: gain, selectivity, and isolation. The gain is actually minimal, because most of the receiver's overall gain is provided by the I-F amplifier stage later on. The selectivity is provided by the tuned circuits in the rf amplifier, and freedom from certain types of interference depend heavily upon the quality of the rf amplifier circuits. The last purpose served by the rf amplifier is to isolate the receiver. The local oscillator signal feeding the next stage, the mixer, can radiate from the antenna if there is no rf amplifier. This could create interference to other stations, or to other services outside of the Amateur bands (a real no-no).

The local oscillator in VHF-FM rigs is seldom on the injection frequency required by the mixer. The frequency of the crystal LO usually falls in the 5 to 18 MHz region, and is then multiplied up to the VHF region. The multiplication factor for that receiver must be known before proper crystals can be ordered. Factors such as X12, X24, X16, etc., are common.

The mixer stage combines the local-oscillator signal and the rf-amplifier signal to produce four output frequencies, RF, LO, RF+LO, andRF-LO. In most, perhaps all, cases, a tuned circuit or crystal filter at the outpt of the mixer selects the *difference* frequency (RF-LO(, and this is called the intermediate frequency, or I-F.

Most 2-meter FM rigs use an I-F in the 9 to 15-MHz range, which is in obediance to the "approximately 1/10" rule for I-F systems. The most popular I-F for 2-meter FM rigs is 10.7 MHz, the same as used in almost all U.S. made (and Japanese) FM broadcast receivers. It seems wise, and cheaper, to use existing parts, rather than to have "specials" made up for one manufacturer.

If the receiver is a single-conversion job, then the entire i-f amplifier, usually 3 or 4 stages in cascade, will be operated at 10.7 MHz (or whatever), and the demodulator is at 10.7 MHz.

But, if the receiver is double conversion, as is our example in Fig. 18-1, there will be a second mixer stage, so that the high i-f will be heterodyned down to some low frequency (usually 455 kHz). In the example shown, the 10.7-MHz high i-f is beat against the signal from a 10.245-MHz crystal oscillator to form a 455-KHz signal.

The low I-F in our example, provides the bulk of the amplification for the receiver. Also, the best selectivity is provided in the low I-F chain. Crystal filters, or high-Q tuned circuits, are used to restrict the bandwidth of the receiver to the required amount.

It is not practical to use a low frequency for the first I-F, because of image rejection problems. The purpose of selecting a frequency approximately 1/10 of the RF frequency is to ensure good image rejection.

I-F SYSTEMS

Many ingenious circuits and devices have been used to improve the performance of i-f amplifier stages. Simple tuned circuits are relatively poor compared with complex *LC* filter circuits, crystal-lattice filters, and so forth. Until recently, the price of the crystal-lattice filter was excessive, so only the best rigs used

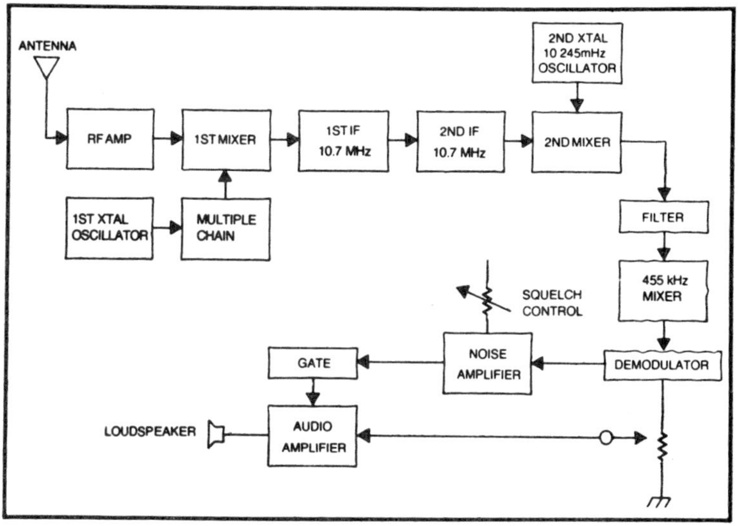

Fig. 28-1. VHF-FM receiver block diagram.

them—often as an option at extra cost. Today, advances in the construction of the ceramic filter has made it possible to use these filters in almost all rigs.

Many FM receivers use a limiter-amplifier as the last stage in the I-F section. A limiter is a special stage that is easily driven into saturation by positive-going excursions of the signal, and into cut-off on negative-going excursions. The purpose of the limiter is to clip off the *amplitude* peaks. The modulation of the FM signal is, after all, carried in frequency variations. Yet, we find that impulse noise, lightning, auto ignition, etc., tends to amplitude-modulate the signal. By clipping off the amplitude peaks, we severely attenuate the noise, thereby giving the FM receiver its much-heralded freedom from noise. The operation of an FM receiver is not guaranteed noise free, however, despite the claims of the car-radio salesman!. FM is noise free *only* when there is sufficient signal to drive the receiver into hard limiting. When the signal is too weak to overdrive the limiter, then there may be some noise interference.

FM DEMODULATORS

The details of FM demodulators have been covered in Chapter 12 (detectors), and channel 25 (alignment), so will not be repeated here. Most Amateur FM receivers use either a ratio detector, Foster-Seeley discriminator, or an IC quadrature detector. I am not aware of Amateur equipment using the pulse counting, or digital, detector, although there is no reason why such could not be done. At least one high-priced, high-fidelity stereo FM tuner uses the pulse-counting detector.

Following the demodulator, of whatever type used, will be the audio amplifiers. Typically, there will be a one-or two-stage preamplifier, followed by a power amplifier that develops 250 to 2000 milliwatts (depending upon model) for the loudspeaker. In most recent designs, an integrated circuit is used which contains the entire audio section.

Squelch is the circuit that cuts off the receiver output during periods when no signal is being received. If the squelch is not used, or if it misadjusted, then everytime there is no station on-channel there would be a loud hiss for us to "enjoy". Most squelch circuits operate by using a noise amplifier that detects this hiss level. If the noise level decreases markedly, indicating that a station is present, then the squelch circuit will open a gate that allows audio to pass through the audio-amplifier stages. Most of the audio ICs intended for use as the entire audio section of communications equipment will have a gate terminal that is triggered by the squelch, or will have the squelch function built into the IC. Only a potentiometer is needed to add squelch to a circuit in the latter case.

TRANSMITTER CIRCUITS

In a typical FM transceiver, internal relays, or diode switching networks, will select the transmit or receive function, as dictated by the position of a push-to-talk switch on the microphone.

The block diagram of a typical FM transmitter, or the transmitter section of a transceiver, is shown in Fig. 28-2. Once again, as was true of the receiver local oscillator, the transmitter oscillator is at a lower frequency, a subharmonic of the operating frequency.

The crystal oscillator usually operates in the 6-to- 18-MHz region for most 2-meter FM rigs, depending upon the multiplication factor used in the design of the transmitter. In the case shown, the crystal oscillator operates in the 12-MHz region. The modulator (a PM circuit) operates on the 12-MHz signal, and, once the modulation is added, the frequency is multiplied to the operating frequency in a chain of multipliers. The first multiplier is a doubler, which has an output frequency of (2 × 12 MHz), or 24 MHz. This 24-MHz signal is fed to the input of another multiplier, in this case a tripler, which produces an output frequency of (3 × 24 MHz), or 73 MHz. The final multiplier is another doubler, and it takes the 73-MHz signal and doubles it to 146 MHz. The multiplication factor for this

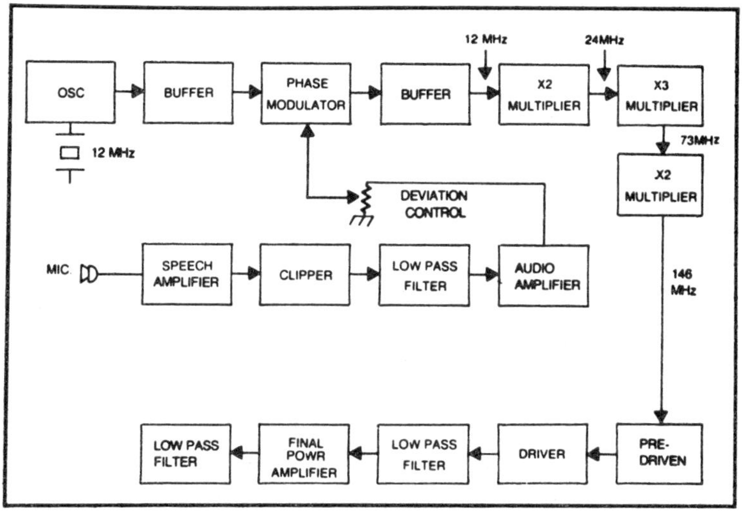

Fig. 28-2. 2-meter FM transmitter block diagram.

transmitter, then, is 2 × 3 × 2, or 12. This means that we must divide the channel frequency by 12 in order to find the correct transmitter oscillator-crystal frequency. For example, suppose that we wanted to operate on a frequency of 146.31 MHz. We would need a crystal on a frequency of 146.31/12 = 12.1925 MHz.

Some transmitters use the final amplifier as another multiplier, in an attempt to save money in the manufacture of the rig. However, the multiplier is always a nonlinear stage, and lots of extra frequencies are created. Hopefully, these would be filtered out by the final amplifier tank circuits, but this is rarely the case. A hallmark of such rigs is complaints from others about interference. The complaints will increase as the "tune" of the final stage becomes worse.

The first element in the transmit audio stages is the microphone. It will produce a small audio signal that is applied to an input preamplifier. The audio signal is first amplified, and is then limited in a clipper circuit. But, clipping, while it does tend to make the average signal level higher, also produces a lot of spurious signals in the audio rang (clipping is always a nonlinear process). To reduce this "hash", a low-pass filter crucuit follows the clipper circuit.

A typical audio clipper circuit from an FM rig is shown in Fig. 28-3. The paralleled "front-to-back" diodes clip off both the positive and negative peaks of the audio signal, making the signal a near-square wave. At first glance, it would seem that these diodes would

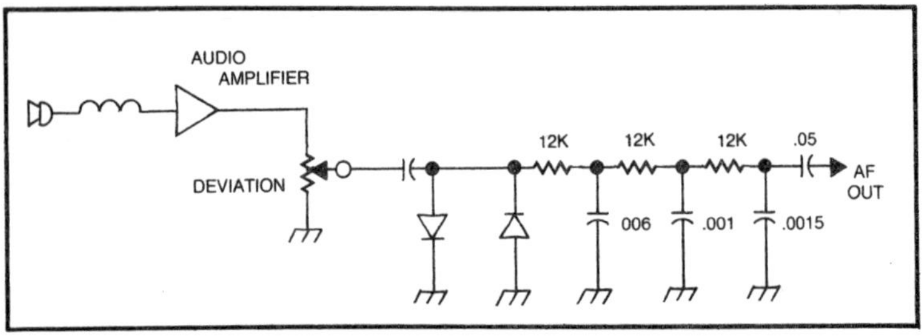

Fig. 28-3. Audio clipper and low-pass filter.

eliminate all signals because one shorts out the positive peaks and the other the negative peaks. But the silicon diodes do not conduct current until the signal is greater than 0.6 to 0.7 volts, so we will see a 1.2 - 1.4 volt signal across these diodes.

The low-pass filter is an *RC* network following the diode clippers. This section serves two purposes. One, of course, is to eliminate the harmonic distortion in the audio signal created by the clipper action. The other is to limit the audio bandwidth of the transmitter modulator. Most communications equipment is bandwidth-limited to something on the order of 300 to 3000 hertz.

The low audio frequencies are attenuated by the 0.05 μF capacitor at the output of the low-pass filter. Such a low-value capacitor will have this effect due to the impedance of the circuit.

High frequencies, of course, are reduced by the low-pass filter circuit itself. Some improvement of the signal-to-noise ratio results from the filtering. However, the main reason for the filter is to reduce splatter and distortion created by the clipper. Clipping adds a type of raspy-sounding distortion because the process of clipping adds harmonics of the audio signal which were not originally present. The roll-off of the filter rounds the corners of the waveforms, and this returns the sound to a more natural balance.

Deviation Control

If the deviation control shown in Fig. 28-3 looks suspiciously like an audio volume control, then congratulations . . . you recognized a basic truth. That's exactly what it is! The deviation of an FM transmitter is proportional to the amplitude of the modulating audio signal. So to control deviation, we must insert a "columе control."

In the circuits of Fig. 28-2 we see the modulator is located after the oscillator and an oscillator buffer amplifier. This is a phase-modulator circuit. The phase modulator will use something like a reactance tube or transistor (often a JFET) to vary the phase of the rf signal.

If the transmitter is true FM, however, a means must be provided to modulate the oscillator directly. In some older equipment (very little, actually, for most simply used PM), they used a reactance tube across the crystal to vary the apparent capacitance seen by the crystal. But this proved to be not terribly good. A more recent approach, shown in Fig. 28-4, is to connect a variable-capacitance diode (Varactor) into the crystal circuit. A Varactor is a special diode whose junction capacitance is varied by the reverse-bias voltage across the diode. When the audio signal adds to, or subtracts from, the reverse bias, the capacitance of the diode PN junction will vary with the modulation.

The change made possible through the use of a Varactor is not very large. This is not important, because the deviation is multiplied by the same factor as the crystal frequency. In the previous case, where a factor of X12 was used, we would only have to deviate the crystal 417 Hz in order to achieve a transmitter deviation of 5 kHz.

The reactance modulator is shown in Fig. 28-5. This circuit uses a JFET transistor, a very common arrangement. At the drain of the JFET, two signals are combined. One is the direct signal coming through the gate/drain capacitor from the rf input (i.e., the oscillator signal). The other is the phase-inverted signal from the same source. The phase inversion is caused by the amplification action of the JFET.

The amplitude of the direct signal will remain constant, but that of the phase-inverted signal is varied by the modulating audio signal. The length of the vector (see Fig. 28-5B) that represents this signal will vary proportionally to the modulation. The resultants, therefore, also vary.

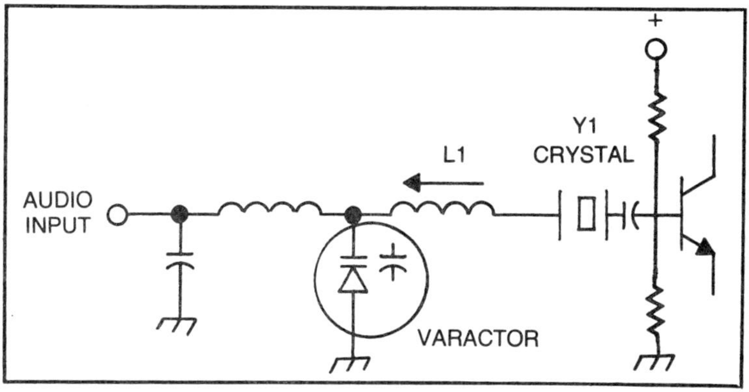

Fig. 28-4. Varactor FM modulator in crystal oscillator.

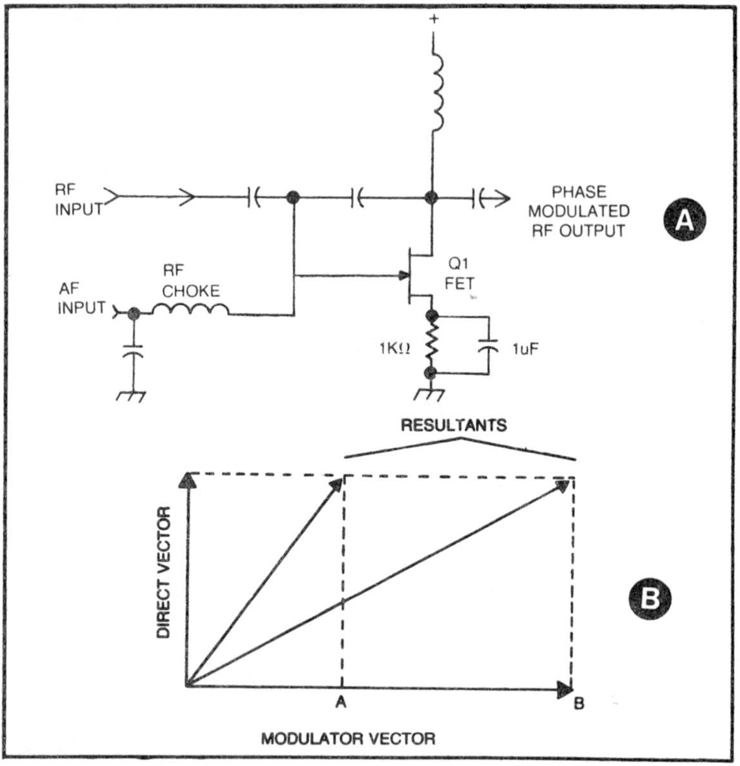

Fig. 28-5. Reactance modulator (A), and the vector relationships (B).

Reduced Power Circuits

Some transceivers are equipped with a reduced-power switch. One popular rig puts out approximately 25 watts in the high-power mode, and about 1 watt in the reduced-power mode. In almost all cases, a riduculously simple method is used to achieve this capability, and it is easily added to existing rigs. They reduce the drive level to the final amplifier, or to the last two amplifier stages. They frequently do this by changing the V+ level applied to one of the earlier rf amplifier stages following the multiplier (see Fig.28-6). When the Hi-Lo switch is in the "Lo" position, the 100-ohm resistor is in series with the collector supply to one of the predriver stages. But, in the "Hi" position, the resistor is shorted out, and the stage receives the full power-supply V+ voltage. Although reduced drive in some vacuum tube circuits could cause the destruction of the tube, in transistor circuits it merely cuts down the current drain of the transistor.

Power Output Stages

Most of the advances in solid-state technology, that make mobile FM possible for VHFers, is the invention of VHF power transistors. Until the last few years, VHF power transistors suffered from low power levels, low power gains, and very high cost. The transistors once used in popular landmobile VHF-FM rig cost the customer $165 each. Fortunately, at typical Amateur power levels, all of these problems have been overcome by modern designs. Even 150-watt mobile power amplifiers are offered for less than $300, while 25 watt amplifiers are had for a "song."

Two features makes the VHF power-amplifier circuit of Fig. 28-7 interesting, and both are in the base circuit. There is no apparent source of positive voltage, as required to forward bias the transistor (supposedly a precondition for operation!). If there were no driving rf signal present, then there would be no bias and no transistor collector current! The rf signal from the driver is sufficient to cause base current to flow during about half of each cycle. In other words, this stage can act as a class-B or class-C amplifier ((actually, most class-C stages use a small-value series resistor in the base circuit).

Older transmitters used an rf choke between the grid of the vacuum tube and ground (in lieu of a tank circuit). In transistor circuits, however, a different approach is used. Rf chokes typically have too much stray capacitance, and that tends to resonate with the inductance of the choke at some frequency. This can lead to spurious output frequencies, caused by parisitic oscillations.

In transistor circuits we use ferrite beads in place of the rf choke, and this is the second thing that I find interesting about the circuit in Fig. 28-7. The beads have a hole through their centers, and are slipped over a piece of wire. The beads cause an inductance

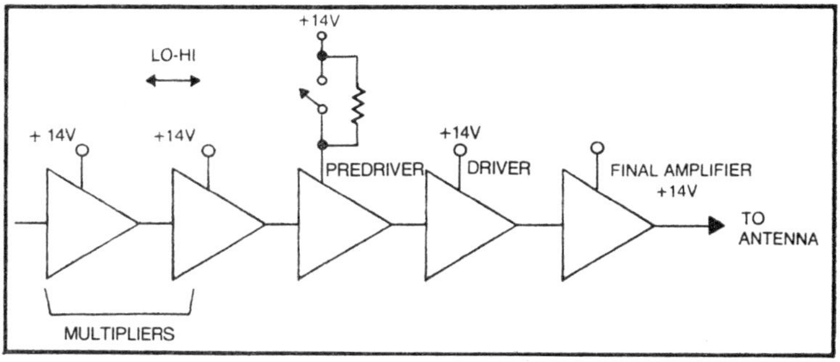

Fig. 28-6. High-low power switch in RF power amplifier.

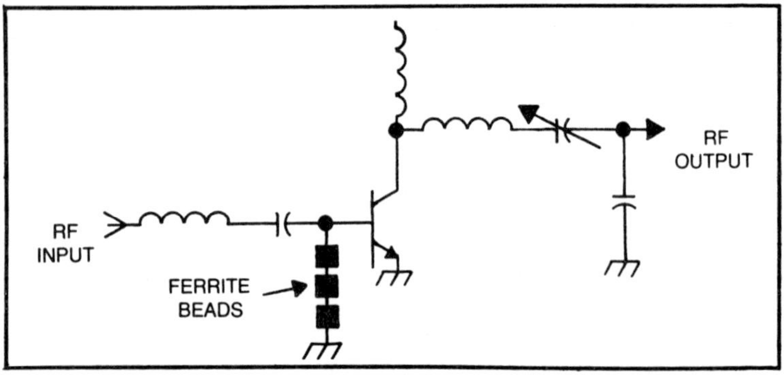

Fig. 28-7. RF power amplifier in VHF range uses ferrite beads over a short piece of wire as an RF choke.

to be seen by the rf signal, and a short circuit is seen by the DC levels.

TEST EQUIPMENT

Some years ago, I worked as an apprentice technician in a shop that did a lot of 2 - 3 MHz marine service. We used nothing more than a Simpson VOM (an old 260), a surplus WW-II BC-221 (or was it an LM?), a homebrew crystal oscillator for checking receivers, and a beat-up Harmmarlund Super-pro receiver. All of that equipment cost less than $500. Today, a well-equipped communications shop to service VHF-FM equipment can spend many kilobucks on fancy equipment. Fortunately, not all of that equipment is always necessary. To be sure, some jobs cannot be done without it. But for the most part, the professional servicers uses it because it makes the repairs go faster, and time is money. Also, some of it is needed to meet FCC specs, which are tighter on most commercial equipment than on Amateur gear.

Servicing VHF-FM receivers is pretty much like servicing any other receiver (see earlier chapters). The transmitter section can often be serviced with a voltmeter. Many of the stages in the transmitter use a base resistor, and a DC voltage is developed across this base resistor when signal is present. So, using the DC voltmeter as a signal tracer, we can isolate the defective stage.

Converted commercial surplus equipment often is equipped with a service jack (Motorola is famous for this type of design). A test set, consisting of a 50 μA DC meter and some switching, is used to isolate the stage. The technician in the two-way shop knows what reading to expect at each setting of the switch, so can quickly isolate the defective stage (so says the advertisements for the test set!).

Chapter 29

Scrounging For Parts

There is a certain mystique about elaborate electronic test equipment, and this causes many people to think that a large array of fancy test equipment is needed. Even those who can obtain older, but good, items still want the latest and fanciest test equipment on the market . . . or will consider themselves unable to cope with their troubleshooting chores. These attitudes are not always valid. Much troubleshooting can be done with very simple and low-cost equipment, or hardly any equipment at all!

Signal tracing or signal injection procedures can be done with almost any signal generator. You can use one of those high cost, synthesized lab jobs, a lab-grade generator of an older time, a so-called "service-grade" generator, (which no service tech in a right frame of mind would use), or a 1-Khz squarewave "noise generator."

Basic troubleshooting is not complicated, and the signal generator need not be a high-quality instrument. However, do not try to perform an alignment job on any equipment that you value unless a proper signal generator is used! Poor signal generators will louse up an alignment job faster than almost any other cause! I remember fouling up a Hammarlund HQ-110 alignment job by (blush) 100 kHz (!) by using a so-called "service grade" instrument. Proper alignment requires a proper generator.

But what *is* a "proper signal generator?" How can Amateurs acquire proper equipment? Where can you buy good used test equipment?

First, learn to recognize "good stuff" when you see it. A signal generator with a cheap, friction-drive frequency-control dial is not a good instrument. Nor is the instrument that uses an output level control that resembles an audio gain control, i.e., a simple potentiometer. A good rf signal generator will have a frequency-control knob that feels like the control on a good receiver! The output-level control will be calibrated in terms of microvolts or dBm (dB referenced to 1 mW into a 50-ohm resistive load).

If the output connector is a microphone jack, then don't count on the signal generator being of too high quality. An exception to that rule were certain older Hickock instruments made for the radio-TV trade many years ago. Most good rf-signal generators used a regular coaxial connector (SO-239, BNC, type-N, etc.) as the output connector.

Also look at the cabinetry and, especially, the seams. A good rf-signal generator must be rf-tight; the cabinet must be tight enough to prevent rf leakage. In some "instruments," and I use the term loosely, the leakage signal was greater than the signal coming through the attenuator at low settings. Hardly a situation conducive to accurate measurements or alignment.

Finally, heft the best. A good signal generator of older design was heavy . . . all of that shielding weighs a lot! If the generator is old, and does not seem heavy, then it probably lacks much . . . and should be shunned. Keep in mind, though, most of the best-quality, modern signal-synthesizer generators weight less than the junk of a decade ago.

Good oscilloscopes can be recognized by their brand names. Look for used Hewlett-Packard (H-P) and Tektronix models. These were the leaders for many years. Shun most of the kit-built oscilloscopes unless you are hard up . . . and can get the beastie for a song. I prefer plug-in types, in which the oscilloscope mainframe will accept a number of different vertical and horizontal plug-in units. Most Tektronix, and many H-P models, in the junk market have this feature.

WHERE TO FIND?

Oscilloscopes and signal generators, as well as other prime pieces of test equipment, that were an engineer's dream only a decade ago are now on the surplus and junk markets. Modern instruments do so much more (at a high price, of course,) and are a lot more satisfying to the professional who must work with the instrument day in and day out. But, one man's junk is another man's treasure . . . we Amateurs can easily overlook the fact that someone else has a new, shiny, super-duper equipment that does all but wash windows. That old clunker will do a lot more for us than for the junk man!

Electronic-surplus outlets can often supply high-quality test equipment that is ten to twenty years old. On some items, the demand is so good that you will have to pay whatever is the going price. But, on many of these instruments, only a few Amateurs are interested, and all the professionals consider them to be electronic boat-anchors. This is a situation ready-made for a little horse trading.

At hamfests, for three years prior to this writing, I have seen prices on once-prime pieces of test equipment drop from $500 - $1000 range to the $100 - $300 range. I saw a Tektronix 545 go for $150 after it was proved to be in working order!

If you have some money to spend, then don't overlook the possibility of buying new stuff. Some of the instruments meant for the radio-TV repairman are good enough for many Amateur jobs. Oscilloscopes, especially, have come down in price. Look into models by B&K and Leader, for ready-built, and Heath for kit-form instruments. Note that Tektronix has several lines lower in cost than their premier stuff, namely the TM-900 series, and that Telequipment line they import from England.

If you are unable to buy all of the test equipment needed for properly aligning and troubleshooting equipment, then it might be possible to arrange for several local hams to go together on a test-equipment bank. In some cases, especially small towns or clubs, one person will buy the signal generator and another an oscilloscope. In still other cases, hams have been known to buy test equipment as a club project, using club-treasury funds. The equipment would then be stored at the home of a (trusted) member, or, if a club station is available, at "the ol' club house."

In short, there is now less reason than in past decades for not owning good test equipment. The equipment may be old, and somebody else's junk, but it will serve *you* well.

Chapter 30

Rejuvenating Old Equipment

Over the years, I have been a teenager without funds, a college student without funds, and a working husband and father who had funds . . . that were committed to mortgages, diapers, and so forth. As a consequence, I have owned a rather lengthy list of older communications receivers, transmitters, and the like. I have owned some of the finest equipment on the Amateur market . . . of twenty five years ago. In many cases, it is possible to equip a functional ham-radio station with other peoples glittering cast offs! Let them go for the new, Super Bandbuster II, complete with digital readouts and fancy controls. Give me an old Hallicrafters or Hammarlund or . . . Collins 75A4!

There are actually quite a few good, old, receivers to choose from. A tour through almost any sizeable flea market at a hamfest or auction will reveal treasures like the SX-28A, SX-99, SX-100, SX-101, HQ-129X, HQ150, HQ145AX, HQ180, HQ170, RME 6900, Collins 75A3, 75A4, etc. The Hammarlund company produced a long series of military receivers based on the *Super-Pro* lines listed as HQ-models above (i.e., BC779, etc). There were also venerable military models such as the BC342/BC348 rigs, Navy TCS, and high-grade receivers such as the ARR-7 and R388.

Receivers, especially general-coverage models, manufactured prior to 1970 have fallen into disrepute amongst Amateurs because they lack the gloss and glitter of new technologies, may be a little troublesome when tuning single sideband (although this need not be true), or, they do not fit as nicely into the decoder of today's transceiver-oriented ham-radio station.

If, however, you are new to this game, i.e., a Novice or old timer "retread," seeking Morse code practice via W1AW (or its west-coast counterpart), or are on a tight budget that has little slack on a certain warm glow. Even for the more experienced or affluent Amateur, these oldies will serve admirably as a second, or "backup," receiver in a more luxurious station.

Unfortunately, there is often a bit of a problem . . . many of these old timers simply do not work. They may have been in working order when originally retired, but time and improper storage have taken their toll. In other cases, the receiver or transmitter went "belly up," and this was the former owners excuse to lay out some huge (so it seemed to the spouse) sum of money for the latest and fanciest (honest . . . I *need* it!).

Fortunately, receiver troubleshooting is not the terrible chore it is often thought to be, and CW or A-M transmitter servicing is even less of a chore. Servicing SSB transmitters is a bit more demanding, but even that onerous chore is easy to overcome, once you recognize it as *possible*. We are usually blessed by the fact that most problems are trivial in nature. This means that we can leave the really hard jobs to the pros with their bench full of multi-kilobuck instrumentation, and tackle the gravy jobs ourselves.

In this chapter we will discuss those repair problems that are peculiar to making old, discarded, equipment serviceable again.

Proverb: One persons trash is another's treasure.

Many pieces of equipment will sell for a sum considerably lower than the "fair market value," as judged from the used-equipment advertisements in the Amateur Radio magazines. If the item works perfectly, then expect to pay full price for it. But, if there is anything at all wrong with the equipment, then it is a better than fair bet that you can negotiate a lower price. The worse the problem seems, the more the owner will want to unload. So, while the owner thinks you are a chicken coming to get plucked, you are actually a fox with a feather coat!

The first job is to audition the rig, i.e., operate it as near normally as you can. Never pay top dollar for a rig that you cannot audition unless you have good reason to trust the seller! Turn the rig on and operate the controls. Determine exactly what is, or is not, working properly. Keep in mind the fact that most equipment that has been in storage for more than a few months will not be in top working order. If the set works on *any* band, then consider it a good candidate for rebuilding. A rig that does not work at all is not a total loss, merely a lesser candidate. Such equipment may require further evaluation, but the fact that it dead, or only goes "splutter-rrrr," does not eliminate it from consideration. Such merely serves to make the seller less confident about the asking price.

Proverb: The less appealing the rigs seems, the more you can negotiate.

Let us consider a case from my own past, when I resurrected a twenty-year-old Hammarlund HQ-145X. This receiver had been in poor storage for not less than fifteen years. The first step in any attempt at overhauling an ancient rig is to turn it on, and let it run for several hours a day, for a week or more. Note that you should never leave such a rig unattended, there might be a developing short circuit that could set the thing on fire in your absence.

The burn-in process is needed to drive out the inevitable accumulated moisture. It also serves to help the electrolytic filter capacitors in the power supply to reform, although you should be prepared to replace them as a matter of course. My own HQ-145X had a rather nasty audio hum coming from the loudspeaker, but this hum lessened to a tolerable (if not zero) level within a few days.

I noted the following problems in the receiver after the burn-in period was over:

1) The10-30 MHz band was dead. Only a scratching sound and some "birdies" were heard when the main-tuning dial was turned.
2) Occasional oscillation and a microphone condition. This condition seemed frequency dependent, and was worst on the modified 10 - 30 MHz band called *20 BS* on this model.
3) All potentiometers were noisy
4) The bandswitch was noisy
5) All other rotary switches noisy and/or intermittent.

The next step, after the one week (or so) burn-in is finished, is to remove the cabinet and begin cleaning. Please note that there are lethal voltages present inside of the cabinet! Safety dictates that you remove the AC-power cord from the wall receptacle, and then use an insulated alligator clip-lead (with one end grounded to chassis) to discharge the filter capacitors.

Word of wisdom. Turn the appropriate knobs to make sure all variable capacitors have their plates completely *meshed*. This prevents them from becoming damaged.

Corrollary to Murphy's Law. Any and all capacitor plates left unmeshed during repair and cleaning attempts will be accidentally struck . . . and permanently damaged.

Modified version of Murphy's Law of Selective Gravi-

tation. Any heavy tool or object accidentally dropped within ten feet of an unmeshed capacitor plate will strike the capacitor plate, thereby damaging it permanently.

The very first step in your cleaning efforts is to remove the layer of dust that seems to collect on all electronic-equipment chassis and printed-circuit boards. *Do not ever use steel wool or any similar product on electronic equipment!* The small particles that flake off the steel wool will cause all manner of short circuits! Remove the loose surface dust with a small (1.5 - 2 inch) paint brush. A small vacuum cleaner works wonders, once the dust has been loosened by the brush. Do not become too vigorous in cleaning nooks and crannies in the vicinity of the tuning and bandspread capacitors! Dust raised by cleaning can get in between the plates, causing a short circuit, or at least scratchy tuning later on.

The next job is not necessarily cleaning of the remaining caked-on gunk—that comes later, once we know that the receiver is worth saving. Use a spray cleaner to clean all potentiometers and rotary switches. Although there are many claims for one brand or the other of control/switch cleaner, it has been my experience that almost any brand will do nicely; even the cheapie stuff sold in the walk-in retail outlets of the national mail-order electronic giants. Immediately after spraying a control or switch, vigorously operate it through its range several times, to complete the cleaning action.

In cases where the cleaning fluid does not make a potentiometer work properly, then plan to replace the potentiometer. You probably cannot obtain an original part from the maker, but any electronic wholesaler that deals with the radio-tv-stereo service-shop trade can make a reasonable replacement if you can tell them the resistance, taper (audio, linear, reverse logarithmic, etc.) and the shaft style and diameter. These potentiometers are cheap enough to make them well worth replacing.

In the case where the cleaning fluid will not cut the black film on a rotary switch, however, a different tack may be required. Most of these switches are a real pain in the cathode to replace, so an alternative cleaning approach is needed.

Helpful hint. The best cleaner for "hopeless" rotary switches is an ordinary pencil eraser.

The use of a pencil eraser is time-honored, and is all but guaranteed to make the switch operate properly! But, this is not for free—you must be *very careful* when doing this job. These switches are delicate! In addition, in tuned circuits (bandswitches, especially) it is important not to move any wires attached to the switch. This

Fig. 30-1. Make sure the capacitor plates are fully meshed.

can cause a shift in alignment and turn a simple repair job into a real bear.

Cleaning procedure. Apply the eraser to the metal rotor surface of the switch, and gently rub the eraser back and forth, removing the dirt. Do this in all positions of the switch. In the case of bandswitches, and any other sensitive switch, hold the eraser still on the rotor and then operate the switch to effect cleaning.

The initial cleaning of my receiver completely cured problems 3 through 5 on the list above, and made a dent in problem no. 1; I could now hear stations on 20-meters, and 15 MHz WWV! The 20-BS oscillations, however, remained, and a similar oscillation had now appeared on the 10-30 MHz band.

Tunable oscillations can be a little hair raising! They are always cause for a little anxiety, provided that it does not make you unable to cope with the situation. Troubleshooting tunable oscillations is sometimes a lot like trying to nail Jello to the wall! I have looked for hours (and seen other professional servicers do the same) for tunable oscillations in receivers! But, before looking for the usual causes, such as open bypass capacitors, open suppressor grids, etc, you must first examine the rig for causes that are common to neglected receivers.

There are two main causes of oscillation and microphonic conditions in neglected receivers, and both are given the same name: D-I-R-T! In some cases, we find that dirt formed on the chassis allows regenerative-feedback paths between stages. These are almost impossible to troubleshoot in any logical way. The only easy way is to clean the stuff from the chassis. I have used Freon-propelled degreasers (aerosol cans) and the Lestoil/acetone concoction that is so popular. Be careful, however, not to slosh that concoction into trimmer capacitors, mesh capacitors, or I-F transformers, etc.

The second type of dirt is dried, dirty white grease (which is no longer white, but icky brown) in the variable-capacitor bearing-races around the shaft, and under the rotor-grounding springs on the main-tuning capacitor. Carefully burnish the capacitor frame and contact under these springs until bright, using a *small* screwdriver or relay-contact burnishing tool.

The main bearing race is located at the front-end plate of the capacitor, and is concentric with the tuning shaft. The race will be filled with tiny ball bearings. Clean this race out with a solvent such as Freon TF (*Miller-Stephenson* MS180). Never use any foam-type cleaners, or any radio-TV cleaner, that leaves a residue. This can be checked on a metallic surface prior to use. Also, do not allow the solvent to hit the tuning capacitor plates—it can be the very dickens to get out.

After the capacitor bearing race is cleaned, you will have to relubricate the bearings. Use a white grease such as *Lubriplate* (available at radio-TV parts wholesalers). This is not silicone grease or petroleum jelly, but a special kind of light, white lubricant often used in small mechanical and electronic-equipment applications.

Now it is time to worry about replacing components, not before. The receiver might actually work now, but there are some general guide lines that we call preventive maintenance.

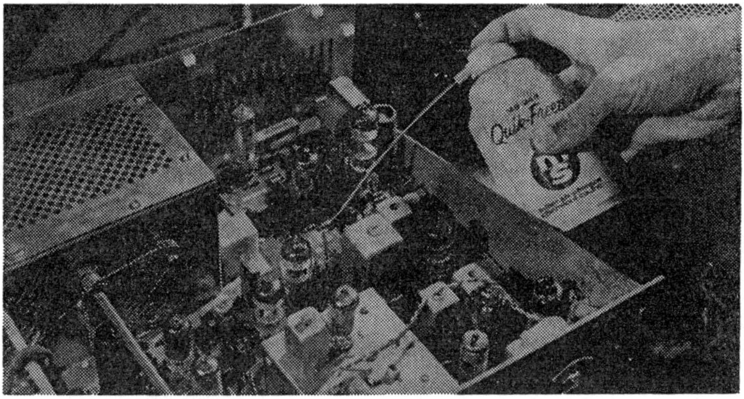

Fig. 30-2. Quick-freeze spray for thermally intermittent components.

For example, examine all electrolytic capacitors. These are found mostly in the power-supply section and the audio amplifier section of the receiver, or the modulator of a transmitter. Look for signs of swelling (an automatic indication for replacement) and the leakage of electrolyte around the seals. This is most frequently seen on the positive end (insulated from case) of the capacitor. It may take the form of a thick, viscous, fluid if the leakage is very recent, but is most likely a white, gray, or brownish paste or powder. Any leakage is a good reason to replace the electrolytic capacitor.

If there are no signs of problems, as in the previous paragraph, then look for any indication in the operation of the receiver that might indicate the unit is defective. In the case of power-supply capacitors, this sign might be 60 or 120-Hz hum on the output (too high a power-supply ripple component) or audio oscillations and motorboating. In the case of audio-coupling, or cathode-bypass, capacitors the sign might be low output.

You will hear some otherwise well-intentioned people advising you to not replace open multi-section filter capacitors. They will tell you that the old terminals make dandy soldering terminals, so you may bridge a good tubular capacitor across the defective section. This practice is especially appealing if only one section of a multi-section capacitor is bad. However, there is a hidden problem not foreseen by your poorer advisors: open sections often short out, far sooner than most people think possible! Do not fall for this advice, unless you want to kiss your filter choke and power transformer goodbye.

Rule of Wisdom. Bridging tubular capacitors across defective capacitors in-circuit is a *diagnostic tool*; it is not a repair procedure. Always replace the defective capacitor, even if it is a multi-section type!

Next, examine all the other capacitors in the set. On paper types (you know, those nasty looking little wax jobs), look for dried-out end-plugs, or missing end-plugs. If the end-plug is missing or in poor shape, then replace the capacitor. On those black-plastic-body capacitors (which are also paper), look for cracks in the body of the capacitor. Also examine these for any fluid on the case. The fluid may be either thick or watery. Replace it. Replace paper, and those black-plastic, capacitors with a quality dipped-mylar (Sprague Orange-drop or equivalent) or the same capacitance, and either the same, or higher, DC working-voltage ratings.

Ceramic and mica capacitors generally survive long storage well, at least a lot better than paper capacitors. But, they are not as

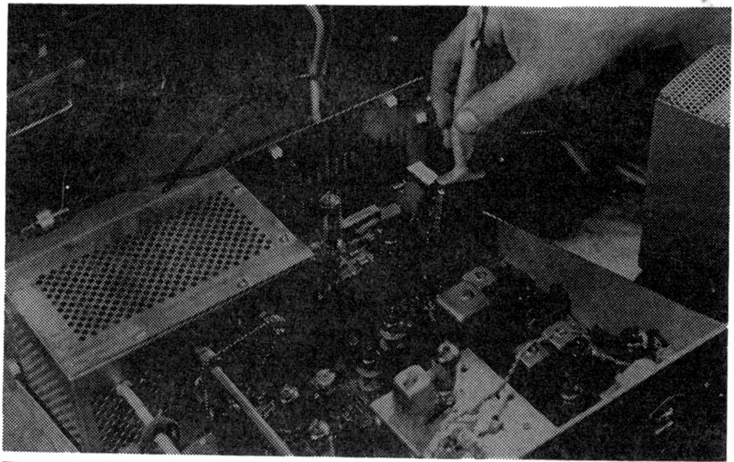

Fig. 30-3. Clean chassis of dust and grime.

easily replaced. It is necessary to replace mica or ceramic capacitors with identical units (i.e., same tolerance, temperature coefficient, etc.). The temperature coefficient is given in the form of color-code dots on mica capacitors, and a letter/number code (N750 or P330 or NPO) on ceramic types.

Examine all carbon composition resistors for any signs of overheating, cracking, etc. Being merely dirty is *not* cause for concern, but if charred or cracked, a resistor should be replaced.

Now let's talk about the quality of components that you use. Spend some money—support your local wholesaler. Do not defile your new wonder receiver (even if of yesteryear) with cast-off, surplus, or "hamfest-special" components. Do not use an old TV-receiver chassis as a source for small parts for the receiver. Such parts are ridiculous for any *really serious work.* Think about it for a minute; even if you replaced every capacitor and resistor in the old clunker, it would not cost you more than $15 - $25 to use all new components of quality manufacture. In case you are wondering, I do not own stock in, work for, or even especially like, component makers or wholesalers. This advice is given from sad, sad personal experience. Too often, hard work has been for nothing because of "bargain" components.

Once this work is all done, you may test the tubes if you want (of course, if operation has not been restored by now, this should have already been done). Do not replace tubes that are just slightly weak, i.e., within 15 percent of the proper transconductance. It has been found that green tubes (i.e., less than 90 days in operation) have a much higher failure rate than seasoned tubes. Once the burn-in period is over for these tubes, they can be expected to last out their life expectancy. Therefore, it is folly to replace mildly weak tubes. Replace only those that are seriously weak, or have defects such as shorted elements or microphonics.

Vacuum tubes are the only area where I relent in my advice to not use surplus components. If a tube is new, still in its blaaah-colored JAN box, then use it. There is some evidence that the "off-the-shelf" failure rates of these tubes is lower than most of the new tubes purchased in wholesale houses today. This may be due to the fact that the JAN tubes were burned-in at the factory, or that overall quality was better in those days. After all, some tube manufactuer of the decades of the 1940's, 50s and 60s are no longer in the tube business. Even a few of the biggies are actually reduced to marketing tubes bought from manufacturers overseas.

My Acquisition is a MOPA Transmitter! What Am I to Do?

Relax! The MOPA (master-oscillator, power-amplifier) transmitter is just about the easiest thing in ham radio to repair. After all, there are only three stages (oscillator, rf power-amplifier, and DC power supply) in a CW model, and four in an A-M rig (add the modulator).

Now that everybody wants to use a super-solid-state job with digital frequency readout, and Novices can run 250 watts, some of those old "Novice specials" can be bought for a song and good looks. I bought a Viking *Adventurer* MOPA transmitter for $4 at the famous Gaithersburg (MD) hamfest just a few weeks before writing this chapter. The rig didn't work, but the sole problem was lack of a 5U4GB rectifier tube. I bought this rig out of a delayed attack of puberty—the *Adventurer* was my first radio transmitter, bought new when I was an early teen!

You can look for many good buys on the hamfest and classified-advertisement markets. Models like the Johnson Viking *Adventurer*, Globe *Champion* and *Scout* models, Knight T-60, Eico 720, and Heath's wide line (DX-20, DX-35, DX-40 and DX-60 from more recent times, and the AT-1 if you remember the Korean war) run less than 90-watts DC input to the final amplifier (they were Novice rigs when the Novice was limited to 75 watts), and were crystal controlled. Most of them had a socket with 0.5 inch pin spacings to receive those ridiculous FT-243 crystals we all bought in those fine days. In many cases, these old MOPA transmitters will also come with an outboard variable-frequency oscillator, of which the Heath VF-1 and HG-10 are but common examples. These VFO models were connected to transmitters of all descriptions, a testimony to commonality of design, or lack of creativity, or something.

If you are a high-power freak (100 - 250 watts), then look for the Heath DX-100 and DX-100B, Heath TX-1 *Apache*, and Globe's *Globe King* models. If your tastes run to yesterday's caviar, then look for a Collins 32V2 transmitter . . . it was the Cadillac of 100-watt exciters!

The procedure for testing and repairing transmitters is pretty much the same as for receivers. Burn it in for a week (supervised . . do not leave the beast alone). Then clean it out. You will find this easier than on receivers. Check the tubes, if you can (power amplifier tubes are a bear to check on a drug store tube tester, and on most commercial testers).

Next, hook the monster up to a dummy load (don't be so coarse as to try loading it up on the air the first time . . . it's illegal and poor taste!). Plug in a crystal within the Amateur band (in case there is no operable VFO handy) and try operating the rig. Most of these rigs used 40-meter crystals on the 20- 15-, and 10-meter bands. The proper crystal for 20 is the operating frequency divided by two. For 15 meters, divide by three, and for 10 meters, divide by four.

If you are lucky, the rig will load up and operate on the first try. But, if not, then try some elementary troubleshooting.

Let your brain be the most frequently used piece of "test equipment" in your workshop. Find out what does or does not work, and proceed from there. For example, if there is not any rf output, yet the final amplifier grid current meter shows some current that responds to the *drive* or *oscillator* controls, the problem will be in the final amplifier or the power supply. Similarily, if the problem exists only on a single band, then look for something like a loose connection on the bandswitch (especially on kit-built transmitters), a bad solder connection, or no solder on a connection. Some of the latter work properly for years, until corrosion and vibration open the connection. If none are found, then look for a dirty bandswitch, or loose contact points on the bandswitch. Make sure all DC voltages are present. If you have rf, but the rig seems to produce a lousy-sounding signal ("you are T-zeros, old man"), then first replace the oscillator tube, and, if that doesn't work, replace the filter capacitors in the power supply. Neutralize the rig (instructions given in another chapter).

If the rig is an SSB transmitter transceiver, then you have a slightly harder problem, but not on that is unsolvable. In fact, it is even money that the thing will work properly after the initial work, as I described for the HQ145X receiver.

Caution

From time to time, I repeat warnings about electrical safety. There are lethal potentials inside of most transmitters, transceivers, and receivers. *Switch to safety* is the old ARRL slogan—and it is ever so true.

I once heard a medical doctor, who should have known better, claim that 115-volts AC from the wall outlet is harmless, *because it is the current that kills you.* The last part of that statement is true, but apparently Ohm's law is not taught in medical schools. In fact, 115-volt AC from the power company is the most common electrical source to cause death in the U.S.!

Similarly, the DC in the power supply of this equipment can be lethal. Always bleed off the charge of filter capacitors through an alligator clip lead, screwdriver, or what-have-you. I recall one fellow who claimed that the remnant charge after the power is turned off cannot kill you because it bleeds off when you touch it! This reasoning can get you killed. The remnant charge in power supply filters is potentially lethal. If you do not believe me, try asking your doctor what could happen if your heart received an electrical shock, particularly during the repolarization period (ECG T-wave feature). I bet he/she mumbles something about "V-fib."

Chapter 31
Scanners

VHF and UHF monitor receivers for many years were popular with a small army of police and fire buffs, volunteers, and so forth. Very few Amateur Radio operators became involved with these receivers. Most of those early models were blessed with the same problems that we saw in Amateur equipment of the same 1955-1970 era: drift, poor sensitivity, very difficult tuning, dial accuracy that was a joke, etc.

The introduction of solid-state, crystal-controlled models in the early 60s alleviated most of the drift and the dial-calibration error. In those, the owner would select the desired monitor frequency (usually a local police or fire emergency frequency), so the dial was replaced with a rotary switch marked "1, 2, 3. . ." or "A, B, C. : ." Popularity still suffered, however, because, being rockbound, the receivers could not listen to more than a few different frequencies. I recall that most commercial models were six- or twelve-channel jobs, while at least one used the same type of crystal deck then popular in presynthesized 23-channel CB sets. Many Amateurs who bought these receivers converted "few-channel" jobs into many-channel jobs with the addition of another channel-selector switch. One reason for the limitation of popularity in Amateur ranks was that 2- and 6-meter activity, outside of a few local nets, was scattered, and, for the most part, not channelized.

Newer monitor receivers, probably starting with the old *Bearcat I*, combine crystal control with the ability to scan several chan-

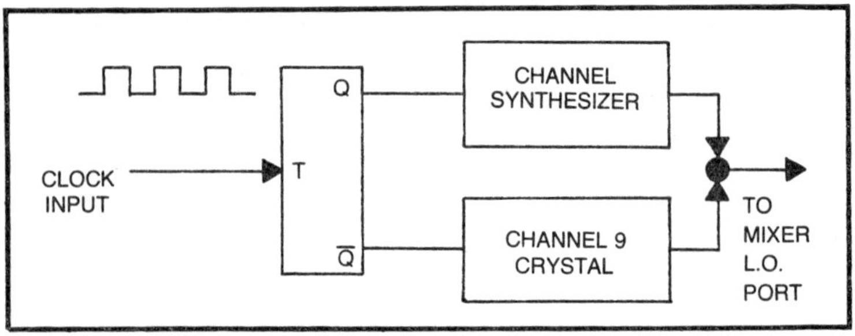

Fig. 31-1. CB receiver uses electronic switch to scan Channel 9 and main channel.

nels. This makes them more popular with Amateurs, and broadens the "police/fire buff" market. A scanner receiver sequentially looks through all of the channels, and stops on those that are active at the time. If no reception is sensed, then the unit continues to scan.

Older, and current low-price models, use individual crystals for each channel. The user specifies at the time of purchase, or subsequently, the frequencies to which he wants to listen. The "rocks" for those frequencies are then installed, and the receiver is customized to the customer's preferences. Of course, if those preferences change, or if the customer moves to another city where different frequencies are in use, then all eight crystals must be changed (at a cost of almost $50!).

Modern scanners use a phase-locked-loop local oscillator to set the frequency received. The "N-code" for up to eight or ten channels can be stored in registers. Digital circuitry then scans the registers, picking out first one frequency and then another, to perform the "scan" function.

If the user of one of these "programmable" scanners wishes to change the frequency of one channel, all that is necessary is to press the *program* button, and a channel button, and then enter the new frequency on the digital keyboard on the receiver front panel. The new frequency is then programmed into the machine and will be used in the next scan.

A SIMPLE SCANNER

Dual-receive CB equipment is neither VHF-FM nor Amateur, but is sufficiently simple as to warrant our consideration for instruction purposes.

CB scanners were developed so that a single receiver could be used to simultaneously monitor a regular working channel and the national emergency channel (9). The digital logic oscillator back and forth between the regular receive channel and channel 9. An increase in the automatic gain control (agc) voltage, indicating that a station is being received on channel 9, causes the scanner to stop switching and latch onto the channel-9 signal. Basic operation of the circuit is shown in Fig. 31-1. A digital-logic element, called a JK flip-flop, selects which local oscillator controls the reception at any given instant in time. An *inhibit* signal causes the circuit to latch when the agc voltage on channel 9 is above a preset level.

VHF Scanners

Most scanners used by Amateur Radio people are operated in the 50, 144, 220, or 430 MHz bands. The use of 144-MHz scanners is popular in areas where there are more than a couple of popular repeaters, so as to not miss a call (or a friend) on any one channel.

Most VHF/UHF scanners are double-conversion superheterodyne receivers, such as shown in Fig. 31-2. A local oscillator, operating from crystals selected by the scanner's digital-logic circuits, control the receive frequency at any given instant in time.

The local oscillator signal is applied to one input of the mixer, where it heterdynes down to a high I-F in the 9 - 13 MHz range, with most being at 10.7 MHz. A second mixer heterodynes the high I-F to the low I-F frequency, usually 455 kHz. An 11.155-MHz crystal

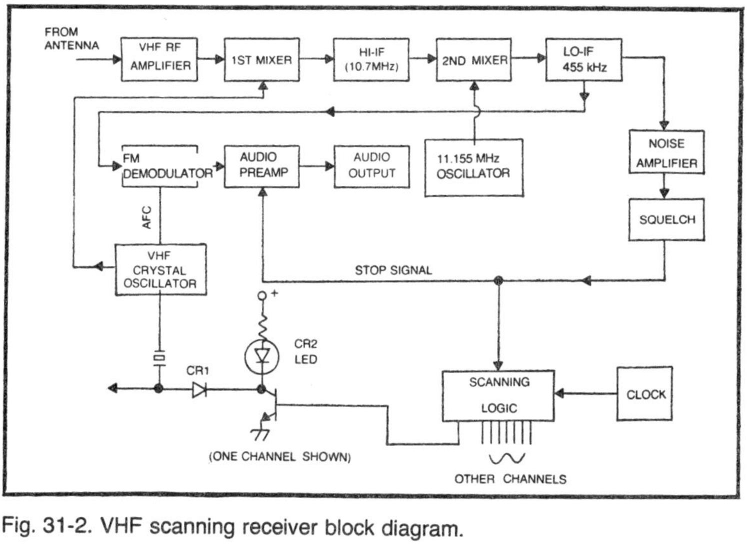

Fig. 31-2. VHF scanning receiver block diagram.

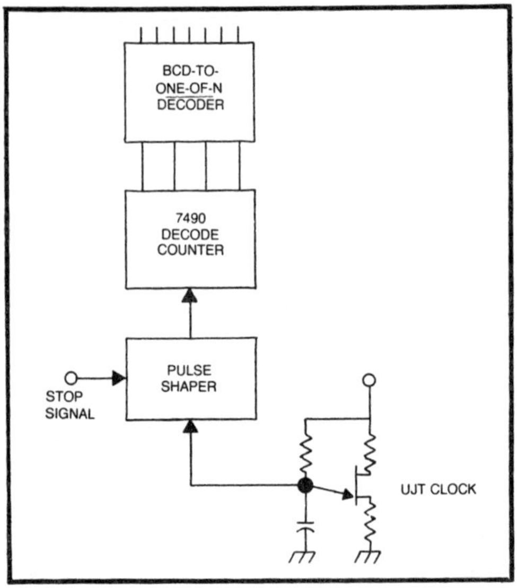

Fig. 31-3. One type of scanner circuit.

is used in this example. The low I-F is then processed in the receiver circuits to produce the audio-output, and an agc voltage that is proportional to received signal strength.

The squelch circuit in this type of monitor receiver does more than just keep the output quiet between transmissions: it provides the *stop* command to the scanner logic circuits. Without the ability to stop, scanning would be a useless feature.

SCANNING LOGIC CIRCUITS

Figure 31-3 shows the partial schematic of a typical scanning circuit. A unijunction transistor, or 555 IC timing circuit, operates as a low-frequency clock. This circuit supplies pulses to the control circuit. In that section, the pulses are changed so that the counter circuits see the nice fast rise and fall times that they require for proper operation. The control circuit will, on command, interrupt the flow of pulses to the type 7490 counter IC.

The binary coded decimal (BCD) outputs from the counter are fed to a decoder that selects one of several output lines each time a pulse arrives. Examples can be found where the decoder is a suitable connection of NAND gates, or an IC octal- or decimal-decoder IC (7446, 7447, etc.).

The simple four-channel scanner in Fig. 31-4A uses two J-K flip-flops (which are usually housed on the same IC; it is a dual J-K flip-flop). This circuit sequentially selects from a bank of four crystals.

Waveforms shown in Fig. 31-4B explain the operation of this circuit (in fact, always try to make a timing-waveform drawing when you are studying a digital circuit). The NAND gates are wired to the flip-flop in such a way that they produce a grounded output (logic state LOW) only when both inputs are HIGH. Notice the waveforms from FF1 and FF2 underneath clock pulse no. 1. At this time, only not-Q (Q) of FF1 and not-Q (Q) of FF2 are HIGH, so they are used to drive gate no. 1. When clock pulse no. 2 arrives, the Q output of FF1, and the Q of FF2 are HIGH, and all others are LOW. They are used to drive gate no. 2. This scheme continues for gates 3 and 4 on pulses 3 and 4.

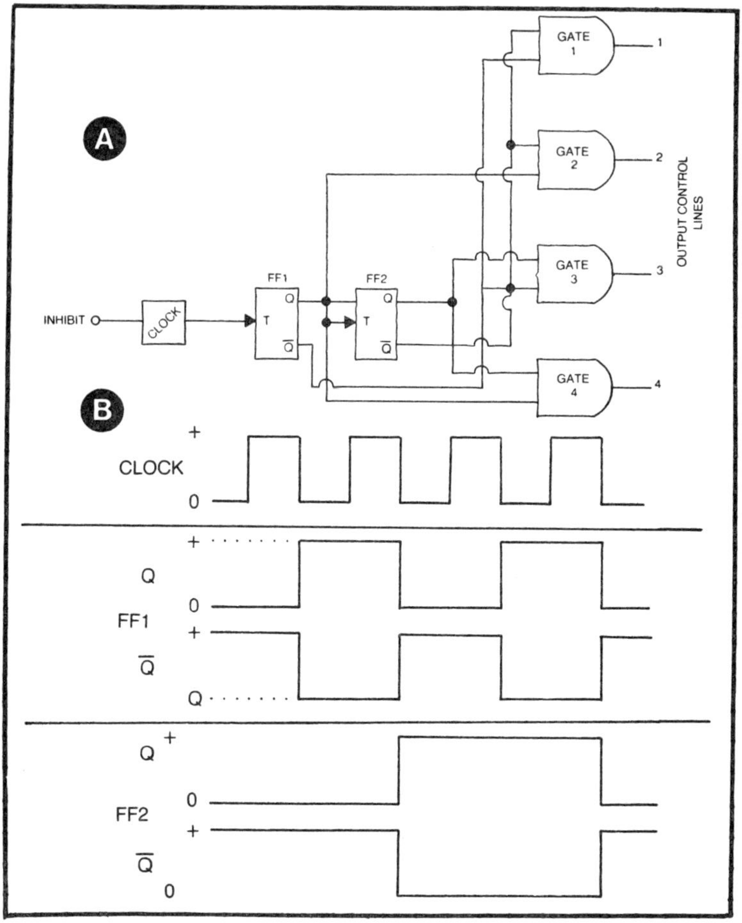

Fig. 31-4. (A), scanner logic; (B), waveforms.

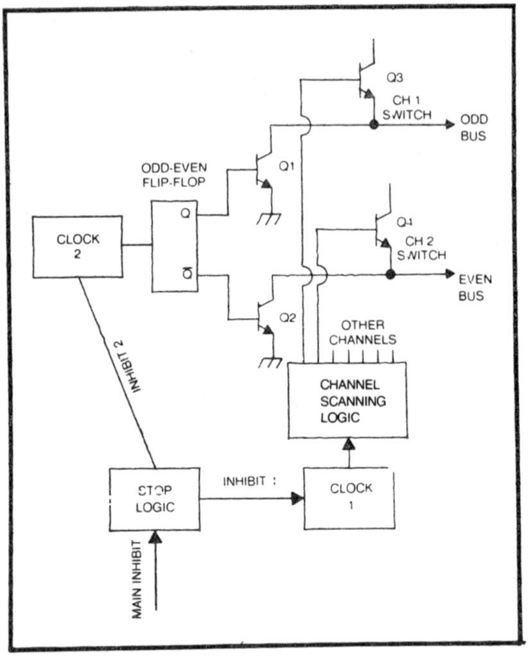

Fig. 31-5. Odd-even channel selection.

Most scanner receivers offer more than four channels (although a few pocket scanners are four channel jobs). In fact, the standard seems to be eight channels. Since the binary-number system is based on powers of two, it might be imagined that a mere doubling of the circuit in Fig. 31-4 would suffice. Actually, the gating of an eight-channel receiver is more complex.

A few receivers simultaneously scan two four-channel banks of crystals designated *odd* and *even*. One additional flip-flop sequentially selects from these two banks. An example of an odd-even circuit is shown in Fig. 31-5.

CRYSTAL SWITCHING

Transistor Q1 in Fig. 31-6 is the regular VHF overtone crystal oscillator used to heterodyne with the incoming rf signal in the first mixer. Although the circuit for only one channel is shown here, assume that each channel will have a similar arrangement. The cold end of the crystal, Y1, is grounded through transistor Q2 when Q2 is turned on by a command from the logic circuit.

When the logic circuit selects the channel, a positive voltage is applied to the base of the appropriate crystal-switching transistor. This saturates the transistor, causing a collector-emitter resistance of only a few ohms. Under this condition, diode CR1 is forward biased (allowing the crystal to be grounded), and the light-emitting-diode (CR2) finds a current path to ground. Depending upon design, you will sometimes find lock-in or lock-out switches, which will either manually select a channel or prevent it from being energized. Most scanners incorporate a small trimmer capacitor to "net" the crystals to their respective frequencies.

Crystal Selection

As many of us had to discover the hard way, crystals are not necessarily calibrated in absolute terms, despite case markings to three decimal places! Nor will crystals always remain on the calibrated frequency once placed in service. The exact frequency of operation depends upon both the operating temperature and the circuit parameters. It is, therefore, necessary to state precisely the requirements of your particular receiver or transmitter when ordering crystals. Scanner crystals are usually bought from either universal crystal suppliers, or from the scanner manufacturer directly. In the latter case, of course, the scanner maker is reasonably expected to supply the correct crystal, but, when ordering from a crystal supplier, you may or may not have to specify the electrical parameters. In many cases, the crystal supplier knows the specifications for your model, so all you need give them is the manufacturer and the model number (and sometimes the serial number if the manufacturer changed the design sometime during the life of the model).

High on the list of information which you must give to the crystal manufacturer, or use yourself, is the frequency to be received, and the i-f frequency of the receiver (the high I-F in dual-

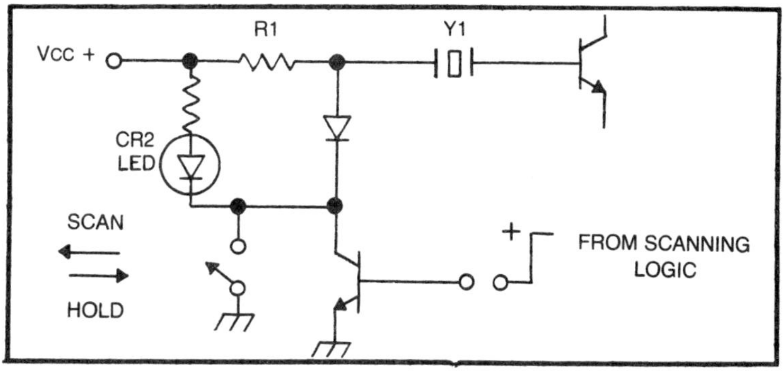

Fig. 31-6 Channel-hold switch.

conversion models). You must also determine whether the crystal will be used in the *fundamental* or *overtone* modes.

VHF-FM scanner oscillators typically (but not always) operate in the fundamental mode on low VHF band (30 - 54 MHz), and the third-overtone mode in the high VHF band (144 - 174 MHz). UHF models may use either the fifth or ninth overtone. Assuming these rules are true, use one of the following formulas to calculate the needed crystal frequency.

Low Band (30 - 54 MHz)

$$\text{Crystal Frequency} = \text{Channel Frequency} + \text{I-F}$$

(In some cases, the LO frequency may be below the channel frequency, in which case, *subtract* the i-f from the channel frequency).

High Band (144 - 174 MHz)

$$\text{Crystal Frequency} = \frac{\text{Channel Frequency} + \text{I-F}}{3}$$

UHF

$$\text{Crystal Frequency} = \frac{\text{Channel Frequency} + \text{I-F}}{9^{*}}$$

*Use 5 instead of 9 if the fifth overtone is used

In those receivers where the local oscillator frequency is below the rf channel frequency, subtract the I-F from the above channel frequencies, then use the same formulae.

Prepare a simple chart for the crystal manufacturer, listing the following data:

1. Make and model of receiver
2. Crystal frequency desired (use formulas)
3. Crystal holder style required (consult catalogue)
4. Operating mode (i.e., fundamental, and, if overtone, *which* overtone is required).
5. Circuit capacitance
6. Drive level of the oscillator, in milliwatts (mW)
7. Maximum allowable series resistance
8. Temperature of operation (if an oven is used)

Of course, not all of these data are available to all Amateurs. Most reputable scanner manufacturers will either publish this data in the service manuals for their product, or will have supplied it to the principal crystal manufacturers on request.

If a service manual is not readily available for your model, then you may be interested in the Howard W. Sams & Sons, Inc. series of books *Scanner-Monitor Service Data*. Sams has been supplying service data to the radio-TV service industry for several decades. They publish the famous *Photofact®* series of schematics. Their series on scanners and VHF/UHF monitors will probably contain the data for your model. You may have to ask the distributor to look up the specific model in the index, so the correct volume can be identified. Very little Amateur gear is in the Sams books, but scanners and monitors will be found. Incidentally, these books are available from electronic wholesalers who normally do business with the radio-TV repair trade.

It is worth noting that the cost of recrystalling a scanner can approach the cost of the scanner itself, and exceed it if bought used! It pays to shop around for "best buys." It has been my experience that the "bargain" crystals, such as those bought by "crystal certificates," will perform just as well as the high-cost custom types from an industrial supplier.

In years gone by, I would offer the advice that you should consider purchasing the scanner already "crystalled-up" from the dealer or manufacturer. In rigs that used separate crystals for each channel, this was the cheapest way to fully populate the rig. But, today, that advice is not quite so valid, because most fixed and mobile scanners being sold are programmable. Of course, these are not the total market. Some amateurs aren't rich as all get out, so will

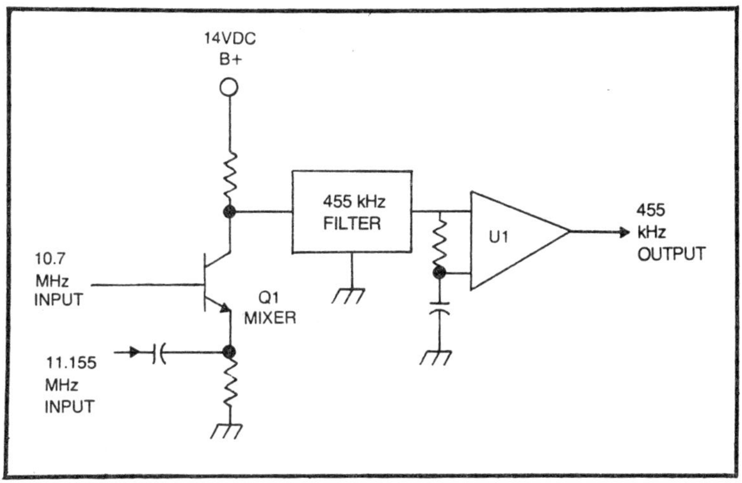

Fig. 31-7. Mixer stage.

buy either used equipment, or those four-channel portables that seem so cheap.

Other Scanner Circuits

For the most part, much of the remaining scanner receiver circuits are the same as those in any VHF/UHF radio receiver, including most 2-meter FM rigs. In fact, the typical scanner is little more than a VHF or UHF FM communications receiver in which the channels are scanned automatically instead of by hand.

The second mixer of one popular scanner receiver is shown in Fig. 31-7. Input signals to the mixer are coupled through a tank circuit and a ceramic-crystal bandpass filter. The output of the mixer is tuned to the lower I-F by another, similar, crystal filter.

Since most Amateur and commercial FM transmitters operating in the VHF/UHF region uses narrowband FM (± 5 kHz deviation), the required receiver bandwidth is 12 - 15 kHz. This allows the use of low-cost ceramic filters, such as the Murata line imported from Japan. These same filters are used in the receiver sections of most imported 2-meter FM transceivers, and in many FM broadcast receivers (but in a 200-kHz bandwidth.).

I-F amplification is almost universally supplied by a one- or two-stage IC amplifier. The detector might be an ordinary diode

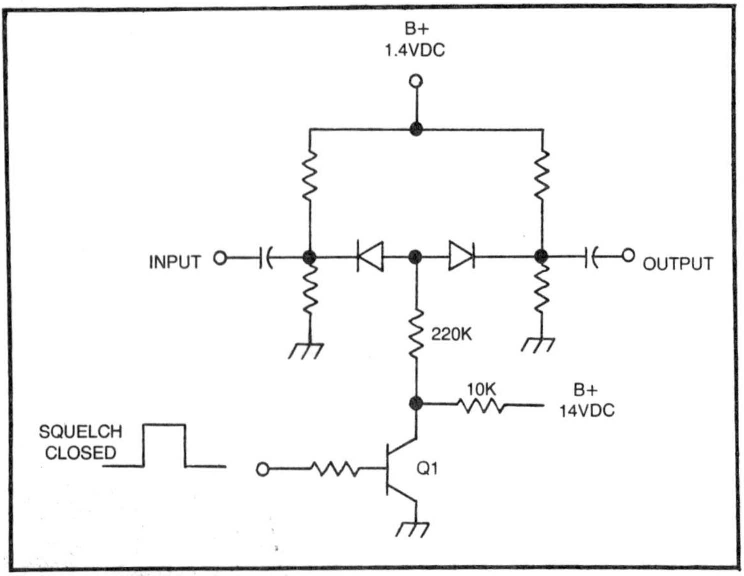

Fig. 31-8. Diode squelch circuit.

type (ratio detector or discriminator), or it might be an IC type. Some IC detectors use the two diodes inside of an IC amplifier, but require an external ratio-detector or discriminator transformer. An example of this class of detector/gain package is the RCA CA3043 device. Other equipment uses the IC quadrature detector, such as the ULN2111 or the Motorola MC1357P.

An unusual squelch circuit is shown in Fig. 31-8, which uses switching diodes to generate squelch action. This circuit produces little audio distortion because low level AC (audio) signals can ride on top of the DC which forward biases the diode. This circuit operates by causing the transistor to saturate. When that occurs, the B+ to the diodes is shunted to ground, causing the diodes to be reverse biased. This cuts off the audio path. When the transistor is inoperative again, the B+ biases the diode to permit audio to pass.

SERVICING SCANNERS

Other than the actual scan circuits, the scanner is exactly like any other VHF or UHF communications receiver. The techniques used to service them, therefore, are identical to ordinary receiver-servicing techniques.

Chapter 32
Transistor Substitution

There are literally thousands of different types of transitors on the market. When trying to locate a specific transistor, especially an older Germanium type, it is often impossible to find. Even with more recent devices, the supplier might not like to consider a one-piece order from an amateur. In fact, I know one semiconductor wholesaler who does 99.9 percent of the business by telephone and outside sales persons. You can only charge items (no way to accept and deal with cash), they have a $50 minimum order, and will only allow "walk-in" customers who have called their order in ahead of time (and even then may pick up only from 11 am to noon, or from 4 to 5 pm). This situation does not bode well for the individual looking to buy one tiny transistor to get the rig back on the air.

In this section, I will look at some alternatives. I'll tell you how to judge other transistors, either generic numbers or the seemingly universal (but not really) replacement lines offered by the major manufacturers (such as Motorola HEP, Sylvania ECG, General Electric GE-series, RCA SK-series). Transistor substitution is not easy as many will want you to believe—especially those with a vested interest in ripping you off—but neither is it an arcane art known only to nether-world engineers.

EXACT REPLACEMENTS

The easist, and by far the best, way to obtain a replacement transistor is from the original manufacturer of the equipment. Even if the transistor appears to be a standard type number, there might be something special in the way the OEM selected the actual transistors in the set. Transistor parameters are often not absolute, but follow a bell-shaped standard distribution curve. There are sometimes reasons why engineers find it cheaper, or better, to hand select transistors that meet certain specifications from a larger lot of so-called "standard transistors." When you want to replace the unit, it may be that the particular "2N-whatsis" that you obtain has a value that is someplace inside of the standard curve, yet outside of the range that the designer selected.

Unfortunately, the OEM may be unwilling, or unable, to supply transistors to you. Although Amateur Radio manufacturers are particularly good in this respect, not all will try to accommodate you in this matter. All too often, obtaining a replacement transistor from the OEM may take a long time. In some cases, you can count on being off the air for weeks, or even months, while awaiting your parts order to be filled. If the marker of your particular rig is super-good in the matter of parts, then sing hallelujah.

INDUSTRY STANDARD NUMBERS

If the defective transistor has a standard "2N . . . " number, then you will probably be safe in obtaining the same 2N numbered device from almost any source. In some critical circuits, the brand of 2N numbered transistor might be critical, but in most cases it is not. Fortunately, most OEMs who use standard 2N numbered transistor leave the designation printed on the device, or will give it in the service manual parts list or schematic diagram. A few unspeakable OEMs will code the devices with their own house numbers that are meaningless to anyone other than themselves. Although this is often done to denote specially selected transistors, it also has the effect of ripping you off for more nickels when you buy a replacement—and ensures that only the OEM can be a supplier.

CROSSOVER GUIDES

Now here's a real can of worms! Crossover guides are issued by companies who market lines of replacement-semiconductor devices. They purport to tell you the correct number transistor in their line that will replace a transistor removed from a set (guaranteed not to rust, bust, or gather dust for six months!). You will hear some people downing the crossover lines as being "cheap" replacements, while other people swear by them so much that all the transistors they ever use come from a crossover guide!

Let's put this mess into proper perspective. First, a reputable replacement-line manufacturer or marketer does not necessarily sell junk! Motorola HEP transistors, for example, are the same transistors as sold by Motorola Semiconductor distributors—but you pay 2 to 5 times as much for them (because of the cost of marketing onesey-twosey devices in blister packs, and publication costs on the crossover guide that tells you which to buy.)

The reason we can often often get away with using a replacement transistor that is not exactly identical is that transistors have a wide latitude in normal specifications. Even devices with the same type number will often have widely varying specifications. Crossover guides would seem to be a nearly perfect source of replacement transistors. They should be used wherever necessary, and especially for certain common, garden-variety, transistors. However, many gremlins can pop up unexpectedly when using crossover-guide replacement transistors. Theoretically, the crossmatching has all been done in advance, probably through the use of an "infallible" computer. When we follow those recommendations, however, we sometimes find some suggested replacements that have insufficient power or voltage ratings, too narrow a gain-bandwidth product, a different physical shape causing space or mounting problems, incorrect dimensions, or some other reason requiring modification of the circuit or the physical equipment. Many of these discrepancies occur because the crossovers are compiled from printed lists. It's an open secret that the recommended substitutions seldom are actually tried in any real circuits or equipment before the recommendation is made.

My policy is to return (or try to return), along with a note of explanation, any crossover transistors that require major reworking of the chassis or wiring of the circuit. If *everyone* did this, recalcitrant manufacturers might take the hint.

Another problem has nothing to do with electrical specifications, but rather with identification. Most manufacturers of equipment will occasionally use their own house numbers on parts. Usually this practice causes no great hardship, especially if the manufacturer of the equipment also publishes the type number.

But, what about the possiblity that each of two manufacturers accidentally might assign the same designation to two completely different, dissimilar, devices? It's not likely a crossover guide would solve this problem (it does, however, happen sometimes. In that case, the crossover guide lists the type number *twice* with each manufacturer's name in parenthesis).

I remember one case in which I needed a Delco type DS-25 PNP Germanium rf transistor. The DS-25 had been used for a dozen years as an rf amplifier, converter, and I-F amplifier transistor in most General Motors AM car radios. Unfortunately, a small hi-fi manufacturer also used the "DS-25" designation for a medium power PNP Germanium audio transistor in a TO-3 package. The crossover guide did not list *both* types, and did not indicate the maker of the "DS-25" listed.

Another example of poor crossovers involves the replacements often listed for Japanese transistors. There is, for example, a type 2SB492 that is used in the audio power-amplifier sections of low-power stereo equipment, two-way radios (mostly CB), etc. Many of the crossover guides list replacements that will safely dissapate only a few hundred milliwatts of audio power, although in many circuits the 2SB492 is called upon to deliver watts of power! Such underrated replacements will burn out after only a few minutes of operation . .

Incidentally, the loss of a 2SB492 often results in the loss of the interstage transformer as well. Better check it before replacing the transistor.

Some of the listed replacements for the 2SB492 (and other transistors; this is only one example) are unusable because they are physically different. In most cases, the original 2SB492 is mounted on a special flanged heat sink that accepts only TO-5 case styles. I strongly advise you to hold out for either the original, or at least a replacement that fits electrically *and* mechanically!

Special Problems In Mobile Rigs

Most crossover guides select transistors that will replace the indicated device in the normal temperature range of 0 °C to +70 °C; the normal "commercial" temperature range for semiconductor devices. These devices will often perform well in base-station equipment, but fail in mobile installations. The reason: environmental temperature ranges in mobile gear are much greater than base-station temperature ranges.

There are actually two problems here: too cold and too hot. We often hear complaints, from amateurs and CBers alike, that the rig will not work in the morning during the winter months. But, after they have been driving for twenty minutes or so, the rig will work perfectly! Some transistors will fail to operate when it is too cold, so, when the operator came out in the cold morning air, the transistor was too cold to work. After the engine warms up, and the car's heater begins to work, then the transistor has reached its normal operating temperature. These transistors can be spotted on the work bench with freeze (Freon) spray.

The inverse problem is also seen. The rig works perfectly on summer mornings, but when the user tries it on the return trip home in the afternoon, it refuses to work. Either that, or the operator notes that repeated failures of the same device always occur on summer afternoons! There is no occult reason why this happens; The cause is simple: too much heat.

One automobile manufacturer became concerned about the excessive number of transistor failures in their first all solid-state car radios (back in 1962!). They decided to investigate the under-dashboard heat as the cause, because all of the complaints occurred in the warm months. They asked plant employees to leave their car door unlocked for one day. During that day of 90 °F weather, engineers from the radio plant measured the temperatures inside of the cars after they had been sitting in the sunlight for four hours. The *average* temperature was 160 °F in the air above the driver's seat, and a few peaked at 180 °F. The underdash temperatures tended to be 10 -15 °F higher than the front-seat air temperatures. It's no wonder that mobile rigs can have troubles. In fact, given the limited temperature range of transistors, it is possible a wise act to not use the two-way radio until the air conditioner has been working for a few minutes.

DERATING TRANSISTOR SPECIFICATIONS

The collector dissipation (power) ratings of transistors are usually specified at room temperature, i.e., 25 °C or 77 °F. If transistors are to be used in environments of substantially higher temperature, then the maximum wattage must be reduced to prevent extra failures which could occur even when all of the other electronic specifications are fulfilled.

A curve typical of those for derating transistors is shown in Fig. 32-1. Notice that a transistor having a collector-power dissipation rating of 600 milliwatts at 25 degrees centigrade can safely deliver only 300 milliwatts at 85 degrees centigrade. That is 185 °F—the possible behind-dashboard temperature in a mobile rig!

This explains why a transistor that is operating slightly below its supposed collector dissapation rating can be destroyed by operating the circuit inside of a hot car. Watch for these hazards in all crossover transistors, but especially in the bargain types that sell

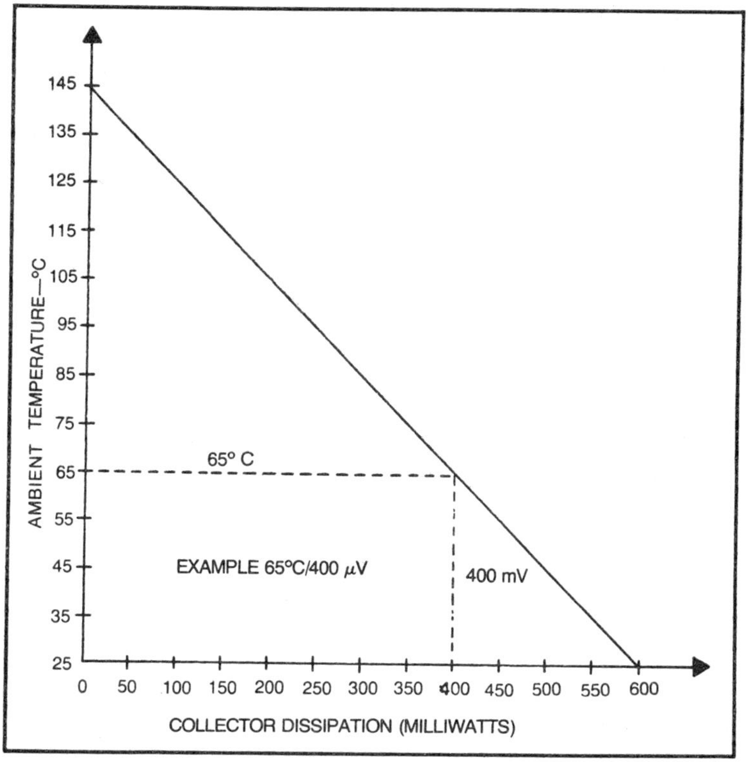

Fig. 32-1. Transistor collector dissipation derating curve.

for "several for a buck". Where transistor devices are concerned, a bargain that is "too good to be true . . ., " probably isn't a bargain at all.

BEYOND CROSSMATCHING

Often, the numbers on a bad transistor seem meaningless. There is no way to crossmatch the device in any of the crossover guides, and you cannot locate an OEM replacement.

The next step is to locate a "universal" replacement from one of the crossover/replacement lines, or another OEM, or "2N . . ." device. You must, it seems, become an electronic detective and find out these things about the defective transistor:

- Is it a Silicon or Germanium device?
- Is it PNP or NPN?
- What frequencies must it amplify?
- What are the power dissipation requirements?

- Are there any special requirements for mechanical mounting (such as case style)?
- Are there any electrical parameters of special importance (such as an unusually high collector-emitter voltage rating, or an exceptionally low noise-figure, as in UHF/VHF devices)

After you have answered those questions, you are in a position to make a reasonably accurate selection from most brand-name lines of replacement devices.

SILICON OR GERMANIUM

Silicon transistor PN junctions generally measure higher resistances than do Germanium junctions. In fact, Silicon transistors usually read open on all measurements except base-emitter and base-collector forward polarity. If even one junction remains intact in the failed device, you can often tell which material is used.

Forward bias voltages for stages other than oscillators, agc amplifiers, squelch circuits, and pulse amplifiers should be 0.2 to 0.3 volts DC for Germanium devices, and 0.6 to 0.7 volts DC for Silicon devices. Check the schematic, or other nearby stages, to see if accurate base-emitter voltages can be derived.

PNP OR NPN?

When the collector voltage is more positive than the emitter, the transistor is a NPN type; and if the collector is more negative than the emitter, then it is a PNP type. Most schematics give these voltages. On the other hand, you can measure collector-emitter voltages, accurate enough for this purpose, right in the circuit in most cases.

If even one junction of the defective transistor is intact, you can determine the NPN or PNP polarity by using a simple ohmmeter (assuming the ohmmeter is not a low-voltage digital type). If you obtain the normal diode-type reading with the positive ohmmeter lead on the base and the negative lead on the emitter it is an NPN type. But, if you must reverse the leads to obtain a diode reading, then the transistor is a PNP type.

FREQUENCY RESPONSE

Suppose you have installed a universal replacement transistor and it doesn't amplify. According to the DC voltage measurements, it is drawing current, and it doesn't heat excessively. It simply fails to operate as expected.

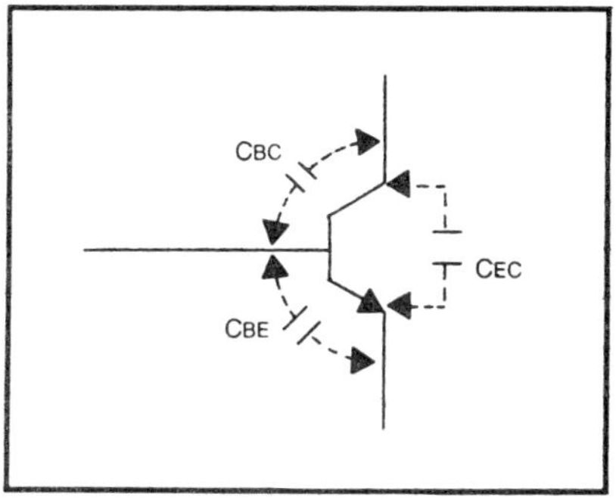

Fig. 32-2. Junction capacitance in transistors.

Chances are the transistor has a bandwidth that is too narrow, and so the amplification is insufficient.

As you know, there is no easy way of measuring the frequency response of transistors. Worse yet, the manufacturer's spec sheets do not always agree on the correct method for denoting frequency response. In fact, I have seen examples of three different methods used in the same crossover guide!

Probably that's the reason you can install a transistor rated as simply "50 MHz" and yet find that it won't operate properly in a 10.7 MHz I-F amplifier. Of course, the manufacturer didn't lie, even though some might claim that you were misled by using a rating system that did not fit the circuit at hand.

As shown in Fig. 32-2, one factor is the capacitances between junctions. Another, not shown, is the thickness of the base region, and the time it takes for the majority carriers to cross the base region. If the capacitances of the replacement transistor are too far out of tolerance, the gain will be reduced at high frequencies.

In addition, there is the *Miller effect* in which the effective capacitance is multiplied by the gain of the stage. A small difference in internal capacitance, in a high-gain amplifier, can create a large change in effective input capacitance because of this effect.

ALPHA AND BETA

Transistor alpha is the ratio of the collector current to the emitter current, and can never exceed unity. Beta is the ratio of collector current to base current.

The frequency-cut-off point is greatly different for these two kinds of ratings, as shown in Fig. 32-3. A manufacturer who rates transistors by the common-base method might correctly list them as having a far wider response than is possible by the common emitter method.

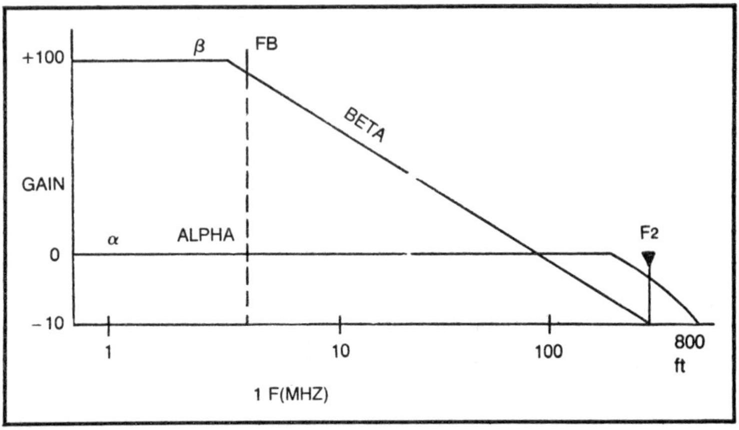

Fig. 32-3. Frequency response.

Gain-bandwidth Product

Another misleading (although not intenionally so) method of rating the frequency response of transistors is the gain-band-width product method. It is defined as the frequency at which the common emitter gain drops to unity. For example, let's assume a transistor with a low-frequency (usually taken at 1000 Hz) beta of 50, and a gain-bandwidth of 50 MHz. The gain-bandwidth product equals the beta times the common emitter cut-off frequency. So the common emitter cut off frequency is found to be only 1 MHz. That's why you can sometimes act on the data given in spec sheets and still obtain a dud that won't amplify.

Analyzing Maximum Ratings

Care must be used in analyzing the manufacturer's maximum voltage and current ratings. Just because a transistor is listed for certain maximum collector voltages and currents, doesn't mean that you may safely operate those devices at the stated levels in all cases. The graph of Fig. 32-4 shows that the transistor can safely withstand either the highest voltage, or the highest current, but not both simultaneously. Maximum wattage (collector power dissipation), the product of both current and voltage, must not be exceeded.

MECHANICAL PROBLEMS

Problems of physical size, connecting leads, and methods of mounting some substitute transistors are equally as vexing as those of finding suitable electrical characteristics.

Fig. 32-4. Operating curves.

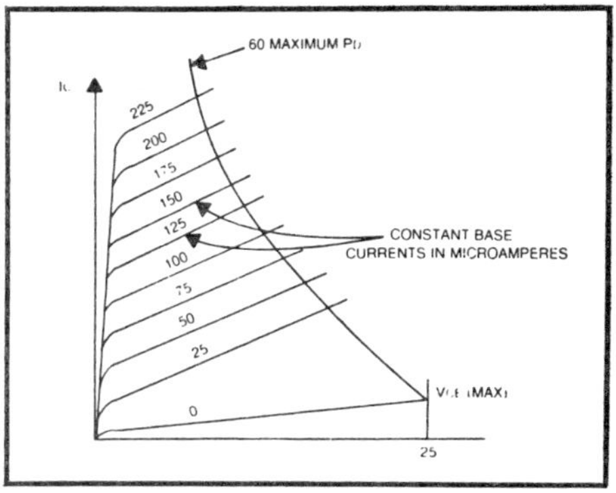

Figure 32-5 pictures four types of plastic power transistor cases. There are transistors listed as replacements for each other that come in all four case styles and for both TO-3 and TO-66 case styles! In some equipment, however, the premachined heat sink, or other mounting problems, will determine which is to be selected—regardless of the recommendation of the crossover guide.

Incidentally, on these plastic transistor cases you are admonished not to bend the leads from side to side during mounting. Also, do not bend the leads directly at the case. Either of these actions will possibly cause the internal connections to break, ruining the transistor.

Make sure that the mounting screws are tight, and that a generous film of silicone heat-transfer grease is used between the transistor case and the mounting surface. This is especially impor-

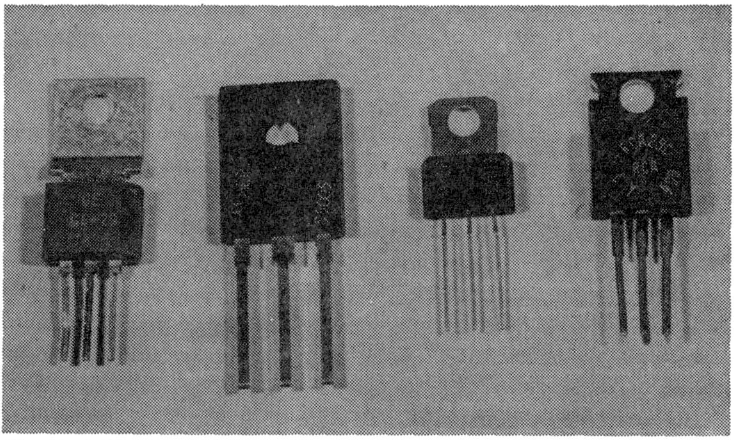

Fig. 32-5. Transistor case styles can pose a problem when they are listed as "substitutes."

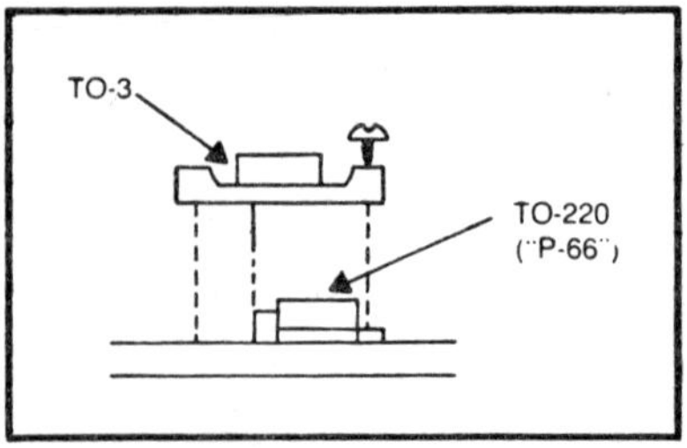

Fig. 32-6. Using a TO-220 (P-66) plastic-case transistor to replace a TO-3 case transistor.

tant if the transistor uses an anodized suface, instead of bare-metal surface.

TO-66 Replacement

In many instances, we find that we can replace TO-66 (and sometimes TO-3) transistors with plastic models using the tactic shown in Fig. 32-6. Use a heatsink, and mount the plastic transistor by one screw. Bend the leads and insert them into the socket.

Chapter 33 Troubleshooting Single-Sideband Gear

Single-sideband communications has almost completely eliminated ordinary A3 amplitude modulation. In the early 1950s, only a few intrepid experimenters used SSB transmission. By the late 50s, the bands heard more and more SSBers, but these were usually the more technically oriented amateurs. During the 60s, SSB really blossomed, to the point where an A-M transmitter on the bands today is an unwanted intruder.

Interestingly enough, the SSB transceiver is often considered "off limits," even to the many amateurs who are willing to attempt to repair other complex gear. I have never been able to figure out just why a fellow once told me that he would have no trouble troubleshooting a ham-band receiver like the Collins 75A3, yet he would not want to tackle the later model "because it is an SSB receiver." The only significant difference between the two receivers was that one had a product detector to demodulate the SSB signal, while the other had an envelope detector for A-M and a crude semi-product detector for CW reception. SSB gear is not so technically advanced, in most cases, that any reasonably intelligent ham cannot troubleshoot most problems. It is unfortunate that to many, the suggestion that they try troubleshooting is enough to snap their mind clean out of its socket.

WHAT IS SSB?

Single sideband is nothing more than a form of amplitude modulation. Fig. 33-1 shows the voltage vectors for AM and two forms of SSB transmitter. In Fig. 33-1A, we see an AM signal centered around 1000 kHz (I always seem to use 1000 kHz because it makes the arithmetic easier)! The voltage vector for each sideband is ½ that of the carrier vector, when a sine wave is used to 100-percent modulate the carrier. Since the power is proportional to the square of the voltage, we find that only $(½)^2$, or ¼ of the total power is in the sidebands. The rest of the signal is going along for a free ride, and is essentially doing little or no work.

One form of sideband operation is called *double sideband suppressed carrier* (DSSC), or DSB for short. In this form of transmitter, the carrier is suppressed to almost zero, and only the sidebands are transmitted. This results in an increase in the "effective power" of the transmitter, because only an information-containing signal is sent out. However, the bandwidth is the same as the A-M signal of Fig. 33-1A. This signal is modulated by a 1000 Hz (1 kHz) sinewave, producing sidebands 1 kHz either side of the carrier. The bandwidth of the signal is 2000 Hz, whether the signal is AM or DSB.

In the single-suppressed-carrier (SSSC, or SSB for short) signals of Fig. 33-1B and 33-1C, the carrier and one of the two sidebands are suppressed. Only one sideband is transmitted. We can get away with this because all of the modulation information is present in *both* sidebands (only one need to be transmitted for total intelligibility). In Fig. 33-1B, the upper sideband (USB) is transmitted, while in Fig. 33-1C, the lower sideband (LSB) is transmitted.

But enough of background, we have covered this material in an earlier chapter anyway. Let's look at the details of the SSB rig.

There are two basic methods for generating the SSB signal: phasing and filtering. Of these, both were popular several years ago (when good filters were terribly costly),but, today most Amateur SSB equipment uses the filter technique. Since this is not a discourse on Amateur transmitters, we will use the common filter method as our example, and then deal with troubleshooting aspects.

Figure 33-2 shows the basic block diagram for a filtering type of SSB generator. Note that the circuit uses a heterodyne method. It is much easier, and cheaper, to generate the SSB signal at some single frequency, and then heterodyne the output of the generator to the various Amateur bands. In most Amateur-equipment designs, the intermediate frequency (yes I-F is used in transmitters too!) for the SSB generator will be 9MHz, 455 kHz, or 3385 kHz (used in Heathkit equipment for years).

The heart of any SSB generator is the balanced-modulator stage. The purpose of this circuit is to suppress the carrier signal,

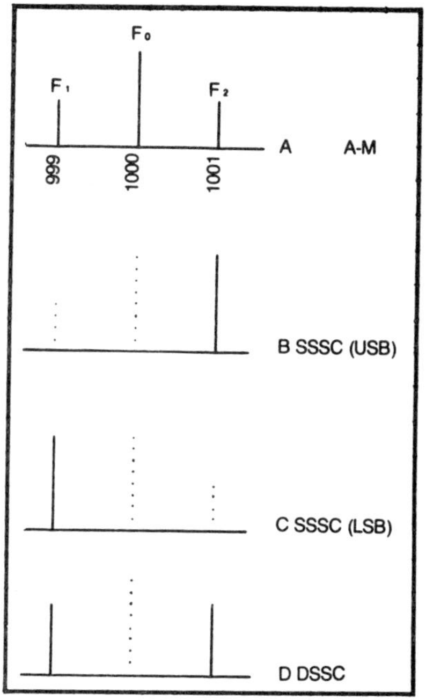

Fig. 33-1. Relationships between the carrier (F_0), and the sidebands, F_1 and F_2.

and leave only the sidebands. The filter following the balanced modulator is used to remove the unwanted sideband.

There are two philosophies to the problem of sideband selection. Some military SSB equipment uses only one carrier frequency, and then two filters are used to select the USB or LSB. Referring to Fig. 33-1A, one filter would have a bandpass of F_0 - F_1, while the other would have a bandpass of F_2 - F_0. This is expensive. Most Amateur gear that I am familiar with uses one (expensive) filter, and two carrier oscillators. To select the USB, we would produce a carrier oscillator signal that is about 1.5 kHz below the center of the filter passband. Similarly, the LSB is selected by generating a carrier oscillator signal 1.5 kHz above the center of the filter passband (Fig. 33-2B).

Once the SSB signal is created, at an I-F frequency, it can then be heterodyned up/down to the appropriate Amateur band for transmission. We cannot use frequency multiplication, as in FM VHF transmitters, because the modulated SSB signal must be processed with as little distortion as possible. Therefore, all stages after the balanced modulator must be linear amplifiers.

Power output on SSB transmitters has been a bone of contention for years. But it is safe to say that the peak envelope power

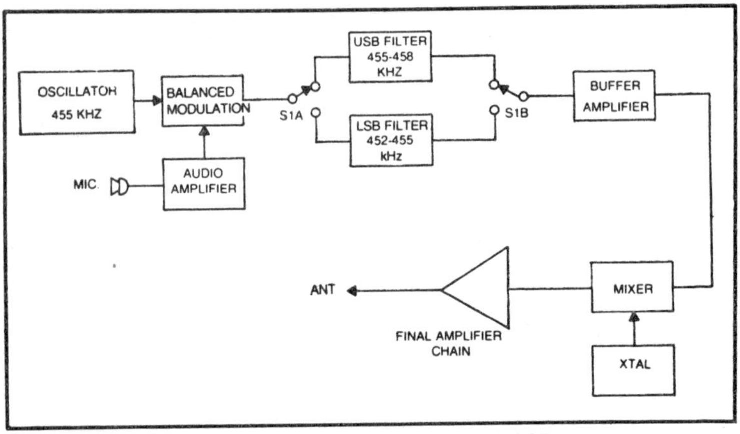

Fig. 33-2A. Block diagram of a basic filtering type of SSB transmitter.

(PEP) of an SSB signal is approximately twice the DC input power to the final amplifier, if the transmitter is being modulated with ordinary speech (assumed to have a 50-percent duty cycle).

BALANCED MODULATORS

There have been almost any number of balanced-modulator circuits over the past two or three decades. In all of them, however, the one criteria is that there must be zero rf-output voltage when there is no modulating-signal voltage present. In many designs, this is done by allowing a balanced rf-carrier signal source (such as two signals 180 degrees out of phase) cancel each other unless an audio signal is present to unbalance the network. In other cases, perhaps the most common, there are electronic switches provided that will not pass rf signal to the output unless turned on by the modulating signal.

Figure 33-3 shows three common types of switching balanced-modulator circuit. Fig. 33-3A is the shunt-switch type of

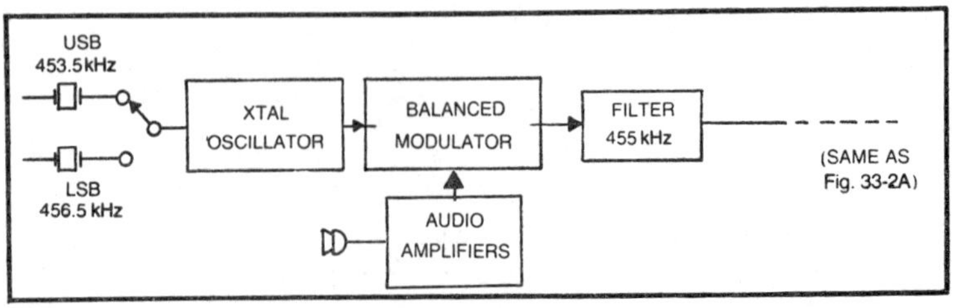

Fig. 33-2B. Switching the crystals to place the carrier on either side of the filter "window."

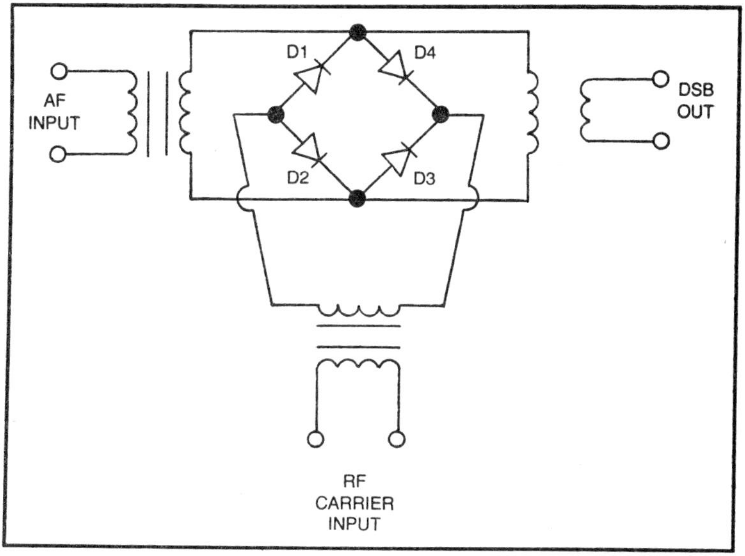

Fig. 33-3A. Basic shunt-switch type of balanced modulator.

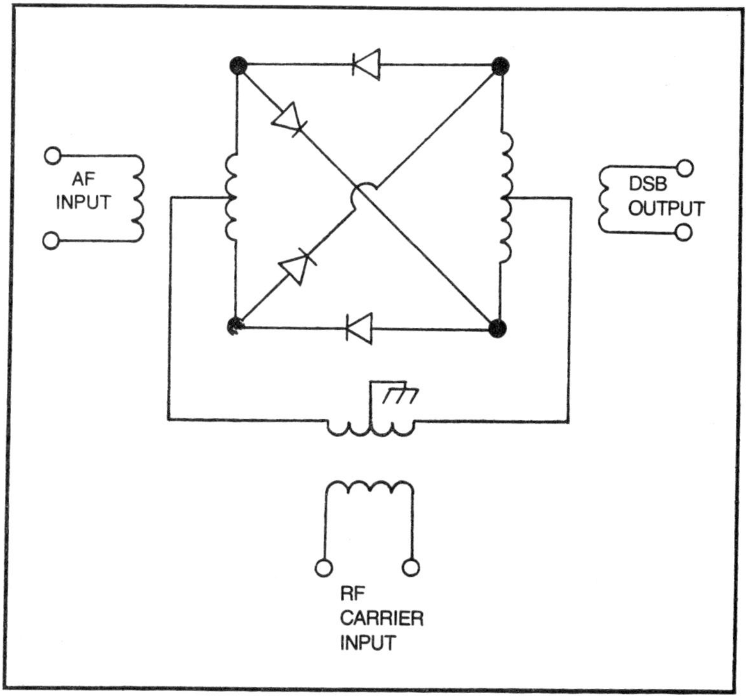

Fig. 33-3B. A Ring modulator type of DSB generator.

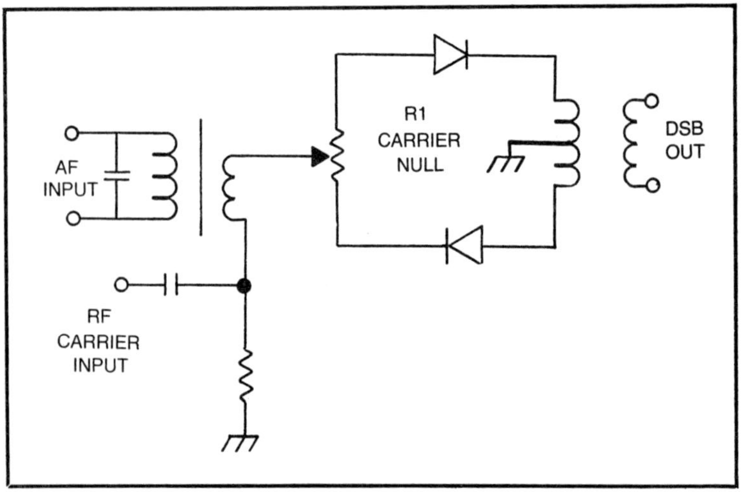

Fig. 33-3C. The half-ring, or semi-ring, DSB generator.

circuit which the secondary of the oscillator signal transformer is alternately shorted out and then opened up, but signal does not pass to the DSB transformer. The audio signal, however, alternately turns on D1/D3 and D2/D4 to provide the modulating action.

A ring modulator is shown in Fig. 33-3B, while a semi-ring, or half-ring, is shown in Fig. 33-3C. Both of these circuits operate on principles similar to Fig. 33-3A.

Figure 33-4 shows two vacuum-tube balanced modulators. The circuit in Fig. 33-4A is a twin-triode circuit, ordinarily using a tube such as the 12AU7, 12AT7, 12AX7, etc. The audio signal drives the grids of the tubes in push-pull, and the rf carrier signal drives the cathodes in parallel, in a grounded-grid (for rf) configuration. When there is no audio signal present, the contributions of V1 and V2 produce equal but opposite currents in the primary of transformer T2. This causes the two rf currents to cancel each other when no audio is present, thereby suppressing the carrier signal. But when an audio signal is present, it drives grids of V1/V2 out of phase (they are push-pull connected for audio), and this produces an unbalanced current in T2. The rf output, then, is no longer zero and the DSB signal is taken from the secondary of T2.

A special beam-switching tube (the 7360) is used in the circuit of Fig. 33-4B. This type of tube uses two anodes that share a common cathode. Deflection plates in front of each anode can allow the cathode beam of electrons to be switched back and forth between the two anodes.

When the beam switching tube is used as a balanced modulator, the rf-carrier oscillator signal is applied to the control grid, and the audio signal is applied to the deflection plates in push pull. When the audio signal is zero, both deflection plates are at the same potential, so each anode gets approximately one half of the cathode current. Since these anodes are connected in push-pull at transformer T2 (as in the previous case), equal but opposite current flow, cancelling each other out. But, when an audio signal is present, one deflection plate will go negative and the other will go positive with respect to ground. This unbalances the circuit, allowing the audio signal to switch the electron beam back and forth between the two anodes.

SSB RECEIVERS

An SSB receiver is more complicated than an AM broadcast receiver, for sure, but it is no great mystery-instrument. The SSB receiver will be a little better built than many other types, however,

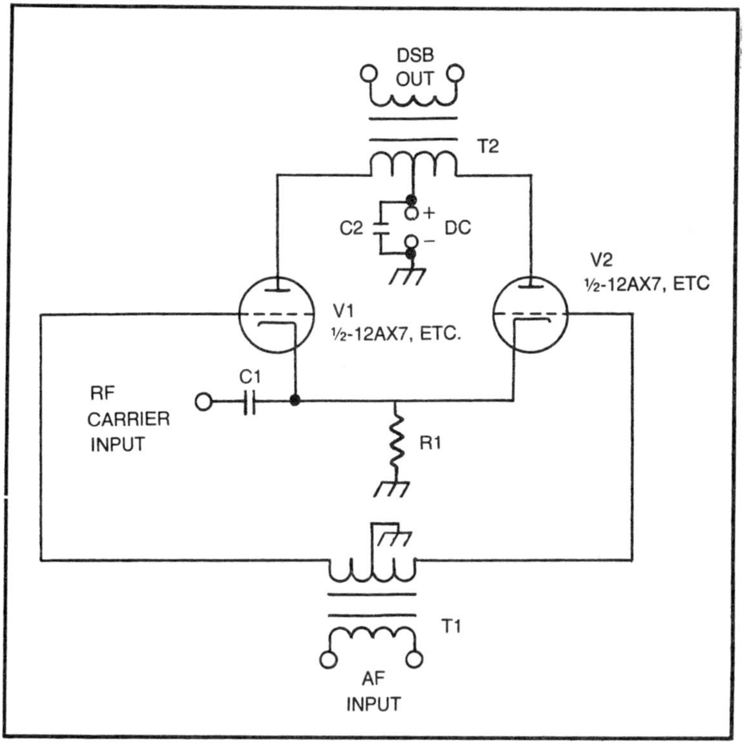

Fig. 33-4A. A vacuum-tube balanced modulator.

because SSB reception tolerates little drift in either the local oscillator or the beat-frequency oscillator used in the product detector. Even that detector, incidentally, is not all together unique to the SSB receiver. CW receivers, after all, require a kind a product detector. In fact, many SSB receivers do nothing additional to receive CW, than that which is done to receive SSB. True, the beat frequency oscillator (BFO) used on older non-SSB receivers tended to be VFO-like (that is, continuously variable), and, almost all SSB receivers use a selector switch to choose two different BFO frequencies for USB and LSB.

The only real point of difference that the so-called SSB receiver has over the non-SSB receiver is the I-F filter. The SSB receiver will have an *LC* or crystal I-F filter that is 1.5- to 3-kHz wide, while the non-SSB receiver may use only ordinary I-F transformers.

Troubleshooting SSB receivers is not different from troubleshooting other receivers, and the reader is therefore referred to Chapters 2, 3, 10, 11, 12, 13, and 14.

SSB TRANSMITTERS

There is nothing in the SSB transmitter that warrants any special fear on the part of the new troubleshooter. All is knowable! With that upbeat note, let's consider just what is needed.

The first thing to do is take stock of the *symptoms*; what does it do, or not do? Does it do what it is supposed to do? Does it do something that it is not supposed to do? Or does it do nothing at all? In Chapter 2, we discussed general troubleshooting, and used a receiver for the example. Recall Fig. 2-4, which gave a troubleshooting flow chart that could be used as a general guide to troubleshooting the receiver. For the most part, this same method is usable in the SSB transmitter. Fig. 2-4A was a list of preliminaries, which, if adapted for the SSB transmitter would look like:

1. Turn set on, does it light up? If not, then suspect something in the primary side of the power supply.
2. Instead of noting if any sound comes from the speaker, examine the rf output power. Do not use an antenna for this. Besides being illegal, and annoying to fellow hams who want to use the band, it is also impractical. An antenna defect, producing a high SWR can affect ALC circuits and SWR-detector circuits, causing the rig to operate improperly.

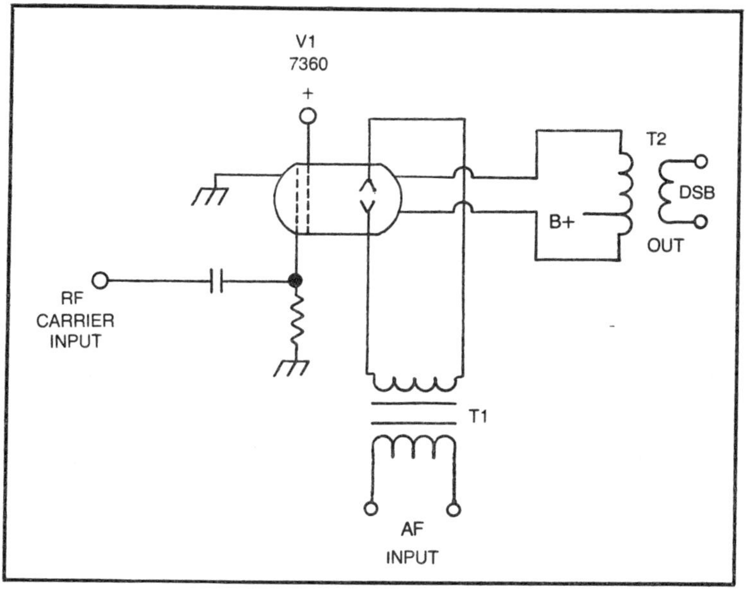

Fig. 33-4B. A balanced modulator using a special gated-beam tube.

3. Note the readings on the plate-current, or rf-output meter. Are they normal? If not, how are they abnormal?

Let's assume that the problem is no transmitter-output power. Note whether the problem exists on CW and SSB, or just one mode of operation. If the transmitter functions on CW, the problem could be in the audio amplifiers or balanced modulator. But, if the problem is common to both SSB and CW, then the problem is in the carrier oscillator, or one of the stages between the output of the balanced modulator and the antenna. Note that a little thinking, and desk troubleshooting with the schematic or block diagram, goes a long way in this business.

If the rig is equipped with a milliammeter for both plate and grid of the final amplifier, then use it to determine if there is drive to the final amplifier. No plate current, with good grid current, usually indicates that there is drive to the final amplifiers, but that the finals are not developing any output power. In that case, suspect the output amplifier tubes, or transistors, first. This is especially true if the rig uses television sweep tubes in the final amplifier, instead of a good hearty rf power-amplifier tube. Most of those sweep-tube final amplifiers go through tubes like they are a dime -a-dozen!

If the tube(s) is not the problem, then look for the DC voltages on the plate and screen grids. There is not much in a typical power amplifier, so this should be easy. One word of caution, however: *Be careful*! The potentials in the final amplifier can kill you! This is not said in order to paralyze the newcomer with fear and trembling—it is intended only to make sure newcomers live to become experienced pros!

If there is no grid current, on the other hand, then it is a safe bet that one of the stages prior to the final amplifier is bad. Loss of grid current indicates loss of the driving signal. It can mean that the final amplifier tubes were destroyed too! Many amateur transmitters use self-bias, in which the negative DC bias to the final amplifier tubes is developed from the rf driving signal. Loss of drive, then, places the tubes in the final at zero bias, allowing a possibly destructive plate current to flow. Don't be surprised, then, if restoring drive by repairing a fault fails to "fix" the rig! Those finals may have "gone west" when drive failed. Again, this is very common in sweep-tube designs, although I have lost a few tubes in my "pair of 6146B" transmitter when drive was lost.

In receiver servicing, we often have the option of using either signal tracing, or signal injection, techniques to locate the bad stage. But in transmitters, the signal source, (the oscillator/balanced modulator), is already producing a signal. Here we ordinarily choose signal tracing. Three methods present themselves (remember these are rf stages): rf voltmeter, oscilloscope, and DC voltmeter.

If we have an rf voltmeter, or a DC voltmeter equipped with a demodulator or detector probe, then we can look for the existence of an rf signal at each transistor base or tube grid back from the final. The last stages (towards the final), in which signal is found, is probably the defective stage. A voltage check, resistance check, or tube/transistor substitution will probably find the defective component. If not, then suspect capacitors, resistors, etc.

We can use the oscilloscope as a signal tracer, provided that it has a vertical bandwidth sufficient to view the rf signal! Keep in mind that the vertical bandwidth need not be as high as the Amateur frequency being tested. . . provided the signal is much higher amplitude than the scope's vertical sensitivity. I have often used a 15-MHz scope to view 27-MHz CB transmitter signals. The Tektronix scope I used had a −3 dB point of 15 MHz, but it still had enough gain at 27 MHz to see, if not accurately measure, signal voltages in a CB transmitter.

The DC voltmeter, not equipped with an rf probe, can often be used as a signal tracer. The signal will usually create a slight negative voltage on the grid of each tube, or a slight voltage drop across the emitter resistor in transistor stages. These voltages can be used as indications of the presence or absence of drive. Of course, if the transistor stage is biased by a resistor network, then you may have to look for a *change* in the emitter-conduction voltage rather than its mere existence. That usually requires some previous experience, or the service manual. I never hesitate to call the manufacturer's service technician for information. Most of them are hams too, and are (like most ham-radio people) very friendly and eager to help. Of course, common decency requires that you not make a pest of yourself. . . and that will mean that you and others will be able to get help in the future. Surprisingly, most companies that sell ham gear would rather see you happy, even if that means giving away free advice, than to have you unhappy. Ham radio is a limited market, in which a few unhappy owners can do a lot of damage to sales (both their own future purchases and those of hams they talk to. . . I have often selected one rig over another because of the comments overheard on the bands!). That's leverage; use it if you have to!

Power Supply

It has been generally true, in my twenty years of servicing electronic equipment, that the more "flaky" a symptom or problem seemed, the more likely the fault to be in the power supply. I once kept the service records of a repair shop. It was surprising how often the power supply was indicted in a troubleshooting effort! For this reason, I always tell the servicer (and that means you, too!) to examine the power supply *first*. After all, no rf output may be merely the *symptom* of "no B+!" Look at all of the DC voltages (under load) with a good DC voltmeter. If you are using almost any type of DC voltmeter other than an old fashioned VOM, then the strong rf field may cause improper readings. In those cases, an rf choke in series with the voltmeter probe (in a shielded box, please) often cleans up the reading enough to make your measurement.

It is also wise to look for the amount of *ripple* from the power supply present on the DC level. Use an oscilloscope for this purpose (set the sweep to show 60 or 120 Hz). Excessive ripple is an indicator of a problem. Keep in mind that the ripple right at the high-voltage rectifier will be much greater than the ripple measured through several sections of *RC* or *LC* smoothing filter. Don't be alarmed at several volts of ripple on a high voltage (300 - 1000 volt) power supply.

If the supply is regulated, as will usually be the case in solid-state transmitters, then the ripple should be almost zero! It should be difficult to see on most oscilloscopes. If you see 25 - 100 mV of ripple, where I mV is more likely, then suspect a bad regulator. Also suspect a bad regulator if the DC voltage is incorrect.

Once power-supply problems (and that includes bad filters and decoupler capacitors) are cleaned up, then, and only then, should you go to the other stages.

Remember the following points: 1) an audio amplifier in an SSB transmitter is the same as any similar stage in a receiver, 2) a low-level rf, or I-F, amplifier in a transmitter is the same as an I-F/RF amplifier stage in a receiver, and 3) most of the same techniques can be used on both receivers and transmitters.

TROUBLESHOOTING THE SSB TRANSCEIVER

The transceiver is a complex electronic-communication device that contains both a full transmitter and a full receiver. Therefore, it should be harder to troubleshoot than either a receiver or transmitter alone. Right? WRONG! In most cases, it is at least no more difficult, and in many it is easier to locate the problem in a transceiver! Does this surprise you? Consider the fact that most transceivers take economic advantage of the fact that certain stages are common to both (audio amplifiers, I-F amplifiers, crystal filters, BFO, etc.). In general, we can make the following statement with regards to SSB (or any other) transceiver troubleshooting:

1. Problems common only to the transmitter are found in those stages that are used only in the transmit mode.
2. Problems common only to the receiver are found in those stages common to the receive mode.
3. Problems common to both modes are usually found in those stages used for dual roles.

In regard to number 3, above, however, let it be known that some stages perform differently on transmit and receive. The symptom, therefore, may be different on different modes! This may lead you to think that you have two separate problems instead of one (sigh. . . another night shot, and all that juicy DX). Also, when it seems that two or more problems exist, then this is also a good reason to suspect the power supply (we keep coming back to that power supply!).

The transceiver is one area where you are advised to study the block diagram in the service manual closely before beginning to troubleshoot. A little insight as to the stages used on the two modes can be worth an hour or so of bench troubleshooting. I could philosophize all day about those who simply jump into the job, voltmeter and oscilloscope probe waving in the wind, without ever bothering to think about the problem. A little desk troubleshooting can make bench troubleshooting a whole lot easier!

Alignment

The alignment of an SSB transceiver or transmitter is not any more difficult than the alignment of any other piece of radio equipment. However, I advise you not to attempt it unless you are suitably equipped with proper test equipment. A little error goes a long way in SSB. Where the A-M equipment would be a little more tolerant of sloppy equipment or procedures, the SSB device will not be at all tolerant.

It is all too commonly believed that alignment is something to be done often. . . one authority quoted "yearly," and many hams believe it. I have found very few rigs over the years that were helped by alignment. It seems that the deterioration of the alignment on modern equipment is so slight from year to year that I have to suggest "several years" as the criteria for periodic realignment (if at all!). For this, it is usually less costly to send it to a servicer, or the factory, than it is to properly equip your own shop.

And to repeat my earlier advice: *don't ever use alignment as a troubleshooting tactic*! I can only wince in pain at the number of times that I have seen, or heard about, hams who diddle with alignment adjustments when troubleshooting. I suppose that the idea is to twiggle a capacitor screw, or coil slug, and observe what happens. What *usually* happens is that the adjustment so fouls things up that no amount of troubleshooting will find the fault until a complete (professional) realignment has been done. If you are afflicted with a disease I call "nervous diddle-stick," then you have my sympathy. But, if you diddle alignment adjustments just for the fun of it, then I'm sure you'll get exactly what you deserve (Ye reap what ye sow!).

Index